JONES, D W
INTRODUCTION TO THE SPEC
HCL   QP801.P64.J75

Introduction to
the Spectroscopy of
Biological Polymers

Introduction to the Spectroscopy of Biological Polymers

Edited by

D. W. JONES
*School of Chemistry,
University of Bradford,
Bradford, England*

1976

Academic Press
London · New York · San Francisco
A Subsidiary of Harcourt Brace Jovanovich, Publishers

ACADEMIC PRESS INC. (LONDON) LTD.
24/28 Oval Road,
London NW1

United States Edition published by
ACADEMIC PRESS INC.
111 Fifth Avenue
New York, New York 10003

Copyright © 1976 by
ACADEMIC PRESS INC. (LONDON) LTD.

All Rights Reserved
No part of this book may be reproduced in any form by photostat, microfilm, or any other means, without written permission from the publishers

Library of Congress Catalog Card Number: 76–1089
ISBN: 0–12–389250–3

Printed in Great Britain by
Willmer Brothers Limited, Birkenhead

List of Contributors

S. Ainsworth   Department of Biochemistry, The University, Sheffield S10 2TN, England
D. G. Dalgleish   The Hannah Research Institute, Ayr KA6 5HL, Scotland
C. E. Johnson   Department of Physics, University of Liverpool, Liverpool L69 3BX, England
D. W. Jones   School of Chemistry, University of Bradford, Bradford BD7 1DP, England
J. H. Keighley   Department of Textile Industries, University of Leeds, Leeds LS2 9JT, England
J. L. Koenig   Division of Macromolecular Science, Case Western Reserve University, Cleveland, Ohio 44106, U.S.A.
J. S. Leigh, Jnr.   School of Medicine, Johnson Research Foundation, University of Pennsylvania, Philadelphia, Pennsylvania 19104, U.S.A.
T. R. Manley   Department of Materials Science, Newcastle upon Tyne Polytechnic, Newcastle upon Tyne, NB1 8ST, England

Preface

The last four decades have seen the emergence of the several main branches of molecular spectroscopy. There has been a gradual expansion in the scope of application of each spectroscopic technique to molecular structure from problems in physics and chemical physics, via physical and structural chemistry, to embrace problems in biophysical chemistry and biology. Even for physiological problems, a wide range of advanced physical methods is now reinforcing classical physical procedures. Sometimes, the application of spectroscopy to chemistry and biology has led to greater refinement of the experimental technique than occurred in the original application in physics. Indeed, the association of the specificity in many biological processes with small structural details often demands spectroscopy of the highest sensitivity. The development of homogeneous high-field magnets and Fourier-transform spectrometers, for instance, has been stimulated by the requirements and hopes for n.m.r. spectroscopy in enzyme chemistry and other biochemical applications. Spectroscopic methods, with their ability often to allow the molecular system to be examined in solution and sometimes *in vivo*, are playing a major part in the progression in the study of biological processes from the mere description to the molecular interpretation of function.

The aim of this volume is only partly to enable the senior undergraduate or beginning research worker to apply spectroscopic methods directly. Rather, it is to present the principles, limitations and scope of the spectroscopic techniques that are being focused on biological molecules to undergraduates or more experienced scientists whose background is primarily in chemistry or biology. It is hoped that this may facilitate future collaboration between chemists, physicists and life scientists in appreciating and solving structural and other problems of biological polymers, whether in biophysics, biochemistry or medicinal chemistry.

Following the introductory survey of spectroscopic methods and some of their common features, each of the main chapters, written by an authority experienced in the field, outlines the principles and emphasizes the value of one technique; it does not purport to be a review of biological applications. After the contents of a chapter on one topic have been mentally digested, the authors hope that the reader will at least be able to discuss with a specialist the relevance and application of that technique to a specific problem involving a biological polymer. Chapter 2 covers symmetry, principles of vibrational spectroscopy and experimental aspects of infrared spectroscopy in some detail. Two chapters on the Raman and far-infrared branches are followed by two on spectroscopies observed in the ultraviolet-visible regions. Chapters 7 and 8 present the powerful magnetic resonance spectroscopies, which utilize interactions with the magnetic vector. Chapter 9 introduces Mössbauer spectroscopy, one of the later branches of spectroscopy to be exploited for biopolymers and more restricted in its field of applicability.

The last chapter illustrates the merits of combining two or more of the above approaches in studies of particular biological systems. Finally, emerging applications to biological macromolecules of the main spectroscopies and certain other spectroscopic techniques are outlined.

Multiple authorship almost inevitably introduces delays. The editor thanks all the authors for their contributions and for their acceptance of some changes designed to improve the unity of the volume. He is especially grateful for the patience shown by contributors for whom publication of completed chapters was delayed until later manuscripts were available. The editor and authors also express appreciation to many colleagues for helpful discussions and for permission to adapt tables and diagrams. While thanks are due especially to Professor G. Allen, Dr. B. M. Chadwick, Dr. V. Fawcett and Dr. J. Feeney for constructive comments on drafts of individual chapters, the responsibility for remaining errors, of course, remains that of the authors and editor. Finally, the editor and authors thank their families for their tolerance and help.

Harden, 1976   *D. W. Jones*

Contents

List of Contributors v
Preface vii

1 Introduction to Molecular Spectroscopy D. W. JONES

1.1 Introduction 1
1.2 The spectroscopic process. 2
    1.2.1 Definitions; 1.2.2 The electromagnetic spectrum; 1.2.3 Thermal populations and relaxation processes; 1.2.4 Units
1.3 Regions of the electromagnetic spectrum 8
    1.3.1 γ-rays; 1.3.2 Visible and ultraviolet; 1.3.3 Infrared; 1.3.4 Raman spectroscopy; 1.3.5 Resonance spectroscopy at microwaves: e.s.r.; 1.3.6 Radiofrequency spectroscopy: n.m.r.
1.4 Characteristics of a spectral line 12
    1.4.1 Intensity; 1.4.2 Width; 1.4.3 Position; 1.4.4 Structure
1.5 Detectability of a spectrum 14
    1.5.1 Resolution; 1.5.2 Improvement of the signal-to-noise ratio; 1.5.3 Fourier-transform spectroscopy

2 Infra-red spectroscopy J. H. KEIGHLEY

2.1 Theoretical aspects 17
    2.1.1 Introduction; 2.1.2 Gross selection rules; 2.1.3 The origins of rotational spectra; 2.1.4 Selection rules; 2.1.5 The origins of vibrational spectra; 2.1.6 The anharmonic oscillator; 2.1.7 Polyatomic molecules; 2.1.8 Molecular symmetry and group theory; 2.1.9 Point groups; 2.1.10 Group theory; 2.1.11 Application to point groups; 2.1.12 The total number of vibrations; 2.1.13 Infra-red and Raman activity; 2.1.14 Overtone and combination bands
2.2 Experimental aspects 32
    2.2.1 Polymer spectra; 2.2.2 Experimental techniques
2.3 General factors relevant to the investigation of molecular structure by i.r. spectroscopy 50
    2.3.1 Deuterium exchange; 2.3.2 Hydrogen bonding; 2.3.3 Dichroism and crystallinity measurements
2.4 Application to polypeptides and proteins 54
    2.4.1 Amide bands; 2.4.2 Vibrational interaction; 2.4.3 Overtone bands; 2.4.4 Derivative techniques; 2.4.5 Water absorption by proteins; 2.4.6 Chain conformation in proteins; 2.4.7 The chemical bonding of metal ions to proteins
2.5 Polysaccharides 74
    2.5.1 α-D-Glucose; 2.5.2 Cellulose; 2.5.3 Chitin
2.6 Conclusion 78
References 78

3 Raman spectroscopy J. L. KOENIG

3.1 Introduction 81
    3.1.1 What is Raman spectroscopy? 3.1.2 Advantages of Raman spectroscopy over i.r. spectroscopy; 3.1.3 Experimental Raman spectroscopy
3.2 Raman spectral identification of amino acids 86
3.3 Raman spectra of chain conformations 88
    3.3.1 Vibrational spectroscopy of the α-helical conformation; 3.3.2 Raman spectra of α-helical polypeptides (amide modes); 3.3.3 Vibrational spectroscopy of the β-conformation; 3.3.4 Vibrational spectroscopy of other helical forms of polypeptides; 3.3.5 Vibrational spectroscopy of the random-coil configuration; 3.3.6 Raman spectra of aqueous solutions of polypeptides
3.4 Raman spectra of proteins 101
3.5 Raman spectra of polynucleotides and nucleic acids 104
3.6 Conclusion 117
References 117

4 Far infra-red spectroscopy T. R. MANLEY

4.1 Introduction 119
    4.1.1 General; 4.1.2 Historical
4.2 Equipment for far infra-red spectroscopy 120
    4.2.1 Materials; 4.2.2 Filters; 4.2.3 Instruments
4.3 Applications of far infra-red spectroscopy 131
    4.3.1 Investigation of skeletal vibrations; 4.3.2 Peptides; 4.3.3 Polyalanines; 4.3.4 Copolymers of D,L-alanines; 4.3.5 Polarized far infra-red measurements; 4.3.6 Calculation of Young's modulus for helical polymers
4.4 Conclusions 142
References 143

5 Electronic absorption and emission spectroscopy S. AINSWORTH

5.1 The nature of light and its interactions with molecules 145
5.2 Energy states of molecules 146
    5.2.1 Electronic energy levels
5.3 Processes of absorption and emission 148
    5.3.1 Absorption; 5.3.2 Paths of molecular excitation loss; 5.3.3 Lifetimes of states and extinction coefficients; 5.3.4 Fluorescence excitation spectra and fluorescence assay
5.4 Polarization of fluorescence 158
5.5 Environmental effects 160
References 163

6 Optical rotatory dispersion and circular dichroism D. G. DALGLEISH

6.1 Introduction 165
6.2 Measurements 166
6.3 Sources of optical activity 169
6.4 Polypeptides and the secondary structures of proteins 170

6.5	Nucleic acids.	176
6.6	Extrinsic Cotton effects of proteins	179
6.7	Extrinsic Cotton effects caused by ligands	183
	6.7.1 Proteins; 6.7.2 Nucleic acids	
6.8	Summary	186
References		187
Additional References		188

7 Nuclear magnetic resonance J. S. LEIGH

7.1	Introduction.	189
	7.1.1 What is n.m.r.?; 7.1.2 A short history of n.m.r.; 7.1.3 Electromagnetic spectrum; 7.1.4 Spins in a magnetic field	
7.2	Principles of n.m.r. spectra	193
	7.2.1 General features; 7.2.2 Chemical shifts; 7.2.3 Spin-spin splitting; 7.2.4 Relaxation times	
7.3	Experimental methods	202
	7.3.1 The continuous-wave high-resolution n.m.r. spectrometer; 7.3.2 Computer averaging; 7.3.3 Fourier-Transform n.m.r.	
7.4	Paramagnetic interactions	208
	7.4.1 Unpaired electrons; 7.4.2 Paramagnetic relaxation effects; 7.4.3 Paramagnetic shifts	
7.5	Chemical exchange	212
7.6	Multiple resonance experiments	214
	7.6.1 Double resonance; 7.6.2 Spin decoupling; 7.6.3 Nuclear Overhauser effect; 7.6.4 Saturation transfer	
Appendices		216
References		218

8 Electron spin resonance J. H. KEIGHLEY

8.1	Introduction	221
	8.1.1 The basic principles of electron spin resonance	
8.2	Interpretation of spectra	224
	8.2.1 Intensity; 8.2.2 g-values; 8.2.3 Electronic splitting; 8.2.4 Line width; 8.2.5 Stability; 8.2.6 Hyperfine splitting; 8.2.7 Anisotropic splitting	
8.3	Experimental methods	234
	8.3.1 The e.s.r. spectrometer; 8.3.2 The detection system; 8.3.3 The magnetic field; 8.3.4 Sample preparation	
8.4	Amino-acids, peptides and polypeptides	241
	8.4.1 Amino acids; 8.4.2 Peptides and polypeptides	
8.5	Proteins.	250
	8.5.1 Structure and radical formation in proteins; 8.5.2 Resonances from unirradiated materials; 8.5.3 Free radicals induced by ionizing radiation.	
8.6	Applications of e.s.r. to the action of blood	262
	8.6.1 Studies of haemoglobin; 8.6.2 Studies of fibrinogen	
8.7	Conclusion	269
References		269

9 Mössbauer spectroscopy C. E. JOHNSON

9.1 Introduction 271
9.2 Resonant absorption of radiation 272
9.3 The Mössbauer effect 274
9.4 Experimental apparatus. 277
9.5 The Mössbauer spectrum 278
    9.5.1 The chemical (or isomer) shift; 9.5.2 The quadrupole splitting; 9.5.3 Magnetic hyperfine structure
9.6 Applications to biology. 288
9.7 Measurements on biological molecules 289
    9.7.1 Haem proteins; 9.7.2 Fe-S proteins; 9.7.3 Haemerythrin; 9.7.4 Iron-storage proteins
9.8 Conclusion—future prospect 293
References 293

10 Combined applications and other techniques D. W. JONES

10.1 Spectroscopic methods and biological polymers 295
10.2 Combined spectroscopic applications to biological systems . . 297
    10.2.1 Conformation of polypeptides and proteins: helix/random-coil transitions; 10.2.2 Structure and action of proteins and enzymes: paramagnetic probes; 10.2.3 Binding of drugs and other small molecules to biomolecules; 10.2.4 Combined studies of paramagnetic biological macromolecules; 10.2.5 Study of membrane structure by magnetic resonance: spin labels; 10.2.6 Polysaccharides
10.3 Emergence of new or modified spectroscopic techniques . . . 309
    10.3.1 New spectroscopic developments; 10.3.2 Additional techniques
References. 315

Subject Index 319

1. Introduction to Molecular Spectroscopy

D. W. Jones

1.1. Introduction

As with some other branches of physics, the techniques of molecular spectroscopy have been increasingly applied to problems in chemistry and subsequently in biology. At first, variations with chemical properties have sometimes appeared to disturb the simplicity of a physical experiment. In nuclear magnetic resonance (n.m.r.), for instance, as with Mössbauer spectroscopy, the term "chemical shift" seems to have been introduced almost in a pejorative sense. Gradually the chemical utility of the n.m.r. technique has been exploited until chemical and then biological applications to large molecules have provided a major impetus in the development of high-field (220 and 300 MHz) and pulse spectrometers.

Advances in a branch of spectroscopy have frequently proceded concurrently at two levels: (a) in spectroscopists' spectroscopy, predominantly of small monomeric (archetypically diatomic) molecules, fine spectroscopic details are measured at high resolution and interpreted to reveal accurate molecular parameters; (b) in a more empirical approach, typified by the fingerprint (or psychiatric*) region (700–1400 cm^{-1}) in organic i.r.† spectroscopy, no attempt is made to associate an absorption line with a transition, but the profile of part of the spectrum is taken as characteristic of a large grouping or specific molecule; the spectrum of the unknown may be compared (directly or by computer) with that of a standard.

While this second, more empirical, approach has often perforce been followed, some applications of spectroscopy to biological polymers have benefited from rather subtle interpretations, for example of relaxation processes in n.m.r. Moreover, progress in even the approximate interpretation of spectra of biological polymers often rests on much more detailed complementary study of relevant monomers or low-molecular-weight oligomers. Thus Chapter 2 on i.r. vibrational spectroscopy begins with a consideration of diatomic molecules.

In the present chapter, some of the terms, units and parameters common to

---

* Dr. L. J. Bellamy. † i.r. = infrared; u.v. = ultraviolet.

several branches of spectroscopy are introduced. The relevant regions of the electromagnetic spectrum are then summarized, as a preliminary to the more detailed discussion in the succeeding chapters of the applications of each spectroscopic technique to biological molecules. Finally, characteristics of spectral lines and factors affecting detectability are discussed.

Any physical method literally measures a physical quantity; as it stands, this may or may not be of direct value to the chemist or biologist. In the simplest case, a monoparametric study of a spectroscopic parameter may allow the change in concentration of one molecular species to be studied unambiguously in a kinetic experiment, for example. In some other cases, for example high-resolution n.m.r. in contrast to u.v. spectroscopy, a single technique can, in principle, allow several different chemical groups or properties to be monitored simultaneously and thus permit much more detailed structural information to be deduced. With large molecules containing a variety of chemical groups, situations will always arise when one technique is insensitive to the parameter in question; therefore, the combination of several spectroscopic (and other) techniques can often facilitate the study of biological polymers. Chapter 10 gives examples of such combined attacks on biological problems, as well as mentioning some techniques not dealt with in detail in earlier chapters.

1.2. The spectroscopic process

1.2.1. DEFINITIONS

The essence of the spectroscopic process is an interaction with matter of either of the mutually perpendicular alternating electric (E) and magnetic (B) component vectors in electromagnetic radiation whereby energy is absorbed or emitted at characteristic frequencies, v. Experimentally, v or some related parameter is generally varied to realize a record of absorption versus frequency, with features as discussed in Section 1.4. Observation of a line at a particular frequency is associated with a *transition*, for which the energy difference $\Delta E = hv$ corresponds to the separation between quantized energy levels or so-called stationary states. One of the levels involved will often be the *ground state*, that of lowest energy for the molecule, while at least one of the others will be an *excited state* of higher energy.

Spectroscopic processes of inelastic scattering, which involve a change in frequency, are to be distinguished from processes such as diffraction by beams of neutrons, electrons of X-ray photons, or from Rayleigh scattering (Section 1.3.4), in which the radiation is elastically scattered with frequency unchanged. As a result of much effort, X-ray diffraction has enjoyed considerable success in the study of the three-dimensional structure of biological molecules, including proteins, in the solid state. However, it is not necessarily the case that the structure *in vivo* is the same as that in the crystalline state; moreover, X-ray diffraction is less well suited than spectroscopic techniques to the study of, for example, molecular dynamics and short-range interactions relevant to biological action, or membrane systems with appreciable motional freedom. Despite the relatively high concentrations which may be necessary, spectroscopic measurements on solutions, even if less detailed, can be more relevant to biology and pharmacology than crystallographic measurements; they are also less time-consuming.

1.2.2. THE ELECTROMAGNETIC SPECTRUM

Electromagnetic waves are known and studied throughout the vast range from audiofrequencies up to cosmic rays of 10^{24} Hz or more. Since their uses, methods of generation and detection, and other properties vary enormously, relatively narrow regions are somewhat arbitrarily designated by a variety of names, not all of which are relevant to spectroscopy. The components to be taken into account in assessing the total energy of a molecule or other moiety depend somewhat on its nature, but typically they may be represented:

$$E_{tot} = E_{trans} + E_{nucl\ orient} + E_{elec\ spin} + E_{rot} + E_{vib} + E_{elec}$$

Leaving aside translational energy, for which the close levels may be regarded as continuous, Table 1.1 summarizes the branches of spectroscopy corresponding to the main groups of internal energy levels (Section 1.3). The utility to chemistry and biology of allowed transitions in these several regions stems largely from the ways in which local interactions with the surroundings can perturb the energy levels and so influence the appearance of the spectra.

With practicable magnetic fields, the newer resonance spectroscopies (Chapters 7 and 8) n.m.r. and (electron spin resonance) e.s.r., which arise from interaction with the magnetic component of the radiation, are observed in the radio, T.V. and microwave regions, and they involve small energy separations. Nuclear quadrupole resonance (n.q.r.) requires no laboratory magnetic field and the splitting arises from interaction between the nuclear quadrupole moment and the electric-field gradient present in a solid sample; it occurs in the same part of the spectrum but has not been applied much to the study of polymers. In principle, the low-energy processes such as n.m.r. (and, for electric-vector spectroscopy, pure rotational) spectroscopy have a simplicity in that they should be uncomplicated by simultaneous transitions of still lower energies. While most spectroscopic techniques can be non-destructive, magnetic resonance, with its low-energy interaction with atomic nuclei or unpaired electrons sensing the local environment and causing the minimum of disturbance, is a useful probe of a chemical system, since it does not interfere with chemical processes and can be applied *in vivo* (the flow rate of human blood has been followed by n.m.r.). It is also possible to interpret magnetic resonance line shapes in terms of molecular motion. The restriction imposed

TABLE 1.1. Main regions of the electromagnetic spectrum

Spectral region	γ-rays	X-rays	u.v.	v_i v_b v_e	i.r.		Microwaves		Radiowaves	
Typical kinds of transition in molecular spectroscopy	Nuclear—Mössbauer	Inner electrons	Outer electrons		Vibrations	rotations	Electron spin (e.s.r.)		Nuclear spin (n.m.r.)	
Energy (kJ mol^{-1})	10^9 10^8 10^7	10^6 10^5	10^4 10^3	10^2 10^1	1	10^{-1}	10^{-2} 10^{-3}	10^{-4} 10^{-5} 10^{-6}		
Frequency (Hz)	10^{21} 10^{20} 10^{19}	10^{18} 10^{17}	10^{16} 10^{15}	10^{14} 10^{13}	10^{12}	10^{11}	10^{10} 10^9	10^8 10^7 10^6		
Wavenumber (cm^{-1})	10^{11} 10^{10} 10^9	10^8 10^7	10^6 10^5	10^4 10^3	10^2	10^1	1 10^{-1}	10^{-2} 10^{-3} 10^{-4}		
Wavelength (nm)	10^{-4} 10^{-3} 10^{-2}	10^{-1} 1	10 10^2	10^3 10^4	10^5	10^6	10^7 10^8	10^9 10^{10} 10^{11}		

by the need for relatively large concentrations (10^{-3} M) for observation, serious for a biological sample containing many different units, is diminishing in importance with improved instrumentation.

The optical spectroscopies (electronic and vibrational transitions; rotational spectra are of little relevance to biological polymers restricted to the condensed phases), covered in Chapters 2–6, have larger energy separations between ground and excited states. In contrast to n.m.r. and e.s.r., in which lines can be assigned to very specific parts of the molecule, they tend to sense larger sets of vibrations and other motions, or even transitions involving the whole molecule.

1.2.3. THERMAL POPULATIONS AND RELAXATION PROCESSES

Sample conditions apart, the intensity of a spectral line depends on two factors: the population of the initial energy state, and the probability of the system changing from one state to another. At equilibrium, the relative populations, N, of two or more energy levels at temperature T are determined by their energy separations. Thus, for two states i and j ($j > i$), $N_j/N_i = \exp\{-\Delta E/kT\}$, where ΔE is the energy separation and $k = 1\cdot38 \times 10^{-23}$ JK^{-1} is the Boltzmann constant. It follows that, at low frequencies and small energy differences as in the microwave and magnetic resonance regions, excited states are heavily populated, indeed very nearly as highly populated as lower states. If the *probabilities* (see below) for upward and downward transitions are the same, the *net* absorption observed (Section 7.1.3) may correspond to perhaps 10^{-5} of the sample, since excited states are being stimulated to emit radiation just as lower states are absorbing energy. With equal populations of adjacent states, appropriate to an infinite temperature, the system is *saturated*, with absorption balancing emission. At high frequencies, as in u.v. spectroscopy, excited states are negligibly occupied at normal temperatures, so that any strong absorption observed must almost certainly involve transitions starting from the ground state. Since at 300 K the energy kT of $4\cdot1 \times 10^{-21}$ J mol^{-1} corresponds to a frequency of $6\cdot2 \times 10^{12}$ Hz, the intermediate case with upper levels appreciably occupied is in the border region between microwave and far i.r. frequencies.

The probabilities for both *induced* (or *stimulated*) *emission* and absorption are proportional to the radiation energy at frequency v and to the population N of the initial state, as well as to the *probability coefficient*, B_{ji} or B_{ij}. The probability for *spontaneous* (non-radiative) *emission* depends only on the number of molecules N_j in the upper state and on the probability coefficient A_{ji}. Since it is related to B_{ji} ($= B_{ij}$) in the ratio $8\pi h v^3/c^3$, coefficient A_{ji} is negligible except at high frequencies. Conversely, despite the magnitude of B_{ji}, induced emission is significant only when N_j is large, i.e. when the energy-level separation is small, as in magnetic resonance. The probability coefficients, B, are proportional to the squares of *transition moments*, p, which may be derived quantum-mechanically and lead to *selection rules* governing the occurrence of spectral lines. If p is strictly zero, the transition is *forbidden* or *inactive* and the line intensity will be zero; if, as sometimes happens, more detailed calculation shows p to be near, but not quite, zero, the transition is only weakly forbidden. If p deviates much from zero, the transition is *allowed* or *active* and the intensity of absorption or emission is likely to be appreciable (although it will not necessarily be detectable experimentally).

Inequality of population states has been seen to be a pre-requisite for spectroscopy. For its continued observation, some *relaxation* mechanism is required, whereby entities in the excited state return to the lower states. In magnetic resonance, the need for spin-lattice relaxation (emission induced by interaction of spins with appropriate frequency components of fluctuating magnetic fields from the motion of neighbouring molecules in the sample) is especially pronounced. With slower motion of nuclei, as in a polymer solution, spin-lattice (T_1) and spin-spin (T_2) relaxation times shorten and the lines broaden.

At higher frequencies, radiationless processes predominate. Thus a molecule which has been excited electronically to a high vibrational state (lifetime 10^{-6}–10^{-9} s) may lose the excess vibrational energy by collision and then exhibit a *fluorescence* spectrum with emission at a lower frequency than that incident, v_{inc}, whereby the molecule reverts to the ground state (Fig. 1.1a).

Fluorescence (Section 5.3.2.3) can also occur via radiationless transfer from the excited electronic state to an intermediate excited electronic state;

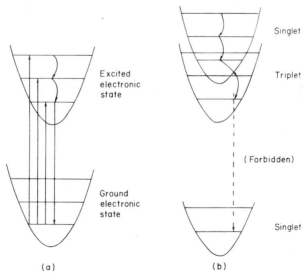

FIG. 1.1 (a) Electronic excitation from ground electronic and vibrational state to excited levels; internal conversion and direct fluorescence. (b) Phosphorescence (radiative fluorescence from lower excited state by forbidden transition). (See Figs 5.3 and 5.4.)

evidently the radiative fluorescence frequency will be appreciably lower than that of the exciting radiation, v_{inc}. In particular, *phosphorescence* occurs when the two excited states have close energies but different spins, as in singlet (total spin, $S = 0$) and triplet ($S = 1$) states (p. 147), so that spectroscopic transitions between them are forbidden. Radiationless transition lines near the intersection of the potential-energy curves and further loss of vibrational energy can bring the molecule to the lowest vibrational level of a triplet state. The low probability of the subsequent (formally forbidden) transition to the ground state (or long lifetime of excited state) ensures that the phosphorescent emission continues long after the original absorption (Fig. 1.1b; see also Fig. 5.2).

Fluorescence and phosphorescence, which involve absorption of visible or u.v. radiation to form an excited electronic state, are to be distinguished from *Raman spectroscopy*, which also involves high-frequency (visible u.v.) monochromatic incident radiation (but *not* the frequency appropriate to electronic excitation; for mention of resonance Raman spectroscopy, see Section 10.3.2.4) and provides information about molecular energy levels of much smaller separation (vibrational and, for small molecules, rotational). In Raman spectroscopy, no excited electronic state is involved. Inelastic collisions with incident photons cause the molecule either to gain energy (more likely) and go to a higher vibrational (or rotational) level so that the frequency of the scattered radiation is reduced (Stokes' lines) to $v_{\text{inc}} - v_{\text{vib}}$, or (less likely, since it involves a higher initial state) the molecule loses energy and radiation of frequency $v_{\text{inc}} + v_{\text{vib}}$ is scattered (anti-Stokes' lines). (Most of the light is scattered elastically at frequency v_{inc} (Rayleigh lines) without energy transfer.) *Brillouin* scattering is another inelastic process involving incident energies much higher than those about which information is obtained, for example about phase transitions in polymers; frequency shifts from the exciting line are smaller than in Raman spectroscopy.

1.2.4. UNITS

S.I. (Système Internationale) units are used predominantly (Table 1.1). Thus frequency v is expressed as Hz. At the low-energy end of the electromagnetic spectrum, in the radiofrequency region, typical frequencies are kHz and MHz for n.m.r., and GHz for e.s.r., while at the high-energy end of the spectrum Mössbauer frequencies may be as high as 10^{18} Hz. According to the part of the electromagnetic spectrum, the wavelength λ is expressed in cm or mm for the microwave region; μm (1 micron) for the i.r. region; and nm (10 Å) for the visible and u.v. region. For the i.r. region, however, it is convenient and customary to specify the reciprocal wavelength or wavenumber σ or $\tilde{v} = 1/\lambda = vc^{-1}$ in reciprocal centimetres (formerly called Kaysers), where the velocity of light, $c = 3 \times 10^8$ m s^{-1} (in S.I.) $= 3 \times 10^{10}$ cm s^{-1}. At a wavelength of 1 cm, $\tilde{v} = 1$ cm^{-1} and $v = 3 \times 10^{10}$ Hz (0·03 THz); the energy corresponding to this frequency is 12 J mol^{-1}, which is equivalent to 2.86×10^{-3} kcal mol^{-1} or 1·24 (1 THz = 10^{12} Hz) $\times 10^{-4}$ eV. Some authors refer loosely to wavenumbers as "frequencies".

1.3. Regions of the electromagnetic spectrum

1.3.1. γ-RAYS

The high frequencies of γ-rays are known with greater precision ($\delta v/v \approx 10^{-12}$ compared with 10^{-8} in a high-resolution n.m.r. spectrum) than in any other part of the electromagnetic spectrum. They are produced by a kind of spectroscopic process, namely radioactive disintegration of certain atomic nuclei. While γ-ray spectroscopy for trace elements, following radioactivation of the sample in a nuclear reactor, has some analytical applications, γ-rays have proved more useful for biological materials in Mössbauer (MB) spectroscopy (Chapter 9). The effect can be detected in over one hundred species of atomic nuclei, of which the most readily studied are ^{57}Fe and ^{119}Sn. While the bulk of the chemical applications of MB spectroscopy might therefore be expected to be in inorganic chemistry, it happens that the most favoured nucleus, ^{57}Fe, also plays an important role in biology, e.g. in haemoglobin (see Section 9.7.1).

As radiation source, the MB technique uses γ-rays (or hard X-rays of frequency 10^{18} Hz upwards) emitted by nuclei in a short-lived excited state (life-time 1.5×10^{-7} s, for example, for ^{57}Fe), itself derived from a parent radioactive source of conveniently long lifetime (220 days for ^{57}Co), so that the intensity is effectively constant during an experiment. The principle is then to observe the reabsorption of γ-rays by nuclei of the same species (here at a frequency of 3.5×10^{18} Hz corresponding to an energy of 14 keV or 1.4×10^9 J mol^{-1}) contained in a stable ground energy level of the sample in the solid state (to increase the apparent mass and so the probability of some emission effectively without recoil); the need for a solid sample is a major limitation for physiological studies. The ^{57}Fe nucleus becomes a probe of its surroundings, since the latter influence the exact frequency and line structure of resonant absorption. To enable the structure of the Mössbauer line to be examined, frequencies are varied by Doppler shifts resulting from relative motion between source (or emitter) and sample (or absorber). Occasionally, it may be advantageous for the unknown sample to be in the emitter rather than the absorber. Brief reference is made in Section 10.3.2.4 to softer X-ray energies of displaced electrons.

1.3.2. VISIBLE AND ULTRAVIOLET

The visible region extends from 750 to 400 nm (10^{-9} m); the near u.v. (for which glass prisms and light paths in air are satisfactory) covers 400–200 nm and the far (or vacuum) u.v. (little used except for photo-electron spectroscopy (Section 10.3.2.4); gratings are required and air absorption must be eliminated by evacuation) involves wavelengths below 200 nm and extending into the X-ray region. Lamps provide intense radiation sources and detection utilizes the photo-electric effect, often with a double-beam arrangement whereby an unknown sample is compared with a reference sample.

In electronic spectroscopy (which occasionally extends below the visible into the i.r.), electric-dipole interactions with the electric vector of the radiation enable changes to be detected between molecular orbitals in ground and excited states (Chapter 5). With widely separated electronic energy levels (typically 10^6 J mol^{-1} or 2.4×10^2 kcal mol^{-1}), the many vibrational possibilities can yield broad absorption bands (and also rotational fine structure for simple molecules). Transitions between tightly bound σ electrons can occur below 200 nm, whereas the more useful transitions involving π and non-bonding electrons occur at lower, more accessible, frequencies. Selection rules prohibit transitions between ground and excited states of different J multiplicity, where $J = |2S + 1|$ and S is the sum of the spins (each $\pm\frac{1}{2}$) in the molecular orbital; thus singlet-to-singlet ($J = 1$) and triplet-to-triplet ($J = 3$) transitions are allowed (p. 147). The symmetry selection rule predicts low intensities for transitions between non-overlapping orbitals. Many polymers have few chromophores (chemical groups, typically unsaturated and covalent, associated with parts of the u.v. spectrum); for a single chromophore, information is more specific than when several absorptions overlap.

Another branch of electronic absorption spectroscopy is applicable to dissymmetric molecules, i.e. optically active molecules, lacking both a symmetry plane and a symmetry centre (Chapter 6). The extent of rotation of plane-polarized light as a function of frequency is the optical rotatory dispersion (ORD) curve, while the corresponding difference absorption (between left- and right-circulatory-polarized beams), a measure of the ellipticity of polarization, is the circular dichroism (CD). Such measurements around the frequencies of electronic absorption bands, made with gratings and photo-multipliers, can yield stereochemical information for biological polymers; the changes in light polarization (Cotton effects) brought about by the chiral sample are influenced by the geometrical distribution of chemical groups with respect to the chromophore.

1.3.3. INFRARED

The overall infrared spectroscopic region extends in wavelength from the long end of the visible region (0.8 μm or μ) up to about 10^3 μ. Within this range, the far infrared region from about 30 μ (corresponding to a frequency of 10^{13} Hz) down into the microwaves was formerly confined to the study of rotations of quite small molecules with permanent dipole moments (Section 2.1.1.3); Chapter 4 describes how the advent of Fourier-transform spectroscopy should facilitate the study of biological low-frequency vibrations (for example, from heavy groups or weak bonds) in polymers. The main i.r.

region (Chapter 2) has frequencies typically of 3×10^{14}–3×10^{13} Hz, more often given in wavenumbers, 10^4–10^3 cm^{-1}. The corresponding energies, around $10^5 - 10^4$ J mol^{-1}, are those of molecular vibrations involving dipole-moment changes. Fundamental bands ($\tilde{v}$ generally 3300 cm^{-1} or less) originate from normal modes (vibrations at a single frequency), while generally less intense bands arise from multiples (overtones, Section 2.1.6) of them, or from combinations (sums and differences). The gross selection rule for observation of i.r. spectra is that the dipole moment must change during the vibration. Instrumentation (Section 2.2.2.1) in the nearest i.r. region (wavelength 0·8–2·5 μm, wavenumbers $1·25 \times 10^4 - 0·4 \times 10^4$ cm^{-1}) is closely analogous to that for visible-u.v. spectroscopy, using quartz cells. With a Nernst filament or Globar source, i.r. detection is generally by a thermocouple, a bolometer (resistance thermometer), or gas expansion (pneumatic) detector (Section 4.2.2.1).

1.3.4. RAMAN SPECTROSCOPY

In Raman spectroscopy (Section 3.1.1), optical frequencies from a powerful monochromatic source are utilized to study molecular vibrations and rotations, i.e. the same kind of motions as studied by i.r. spectroscopy, but with the condition that, during the motion, there is a change in polarizability rather than in dipole moment of the molecule. Raman (occasionally called combination-diffusion) spectroscopy is a form of inelastic light scattering, in contrast to elastic Rayleigh scattering at the same frequency as the exciting line (which process can, incidentally, be used for the study of polymers in solution). The availability of laser sources has led to remarkable expansion of Raman spectroscopy, which is often complementary to i.r. vibrational spectroscopy; however, laser beams can cause some degradation. Furthermore, while the Raman scattering is always much weaker than the Rayleigh scattering, it is proportionately even smaller for large biological molecules. Raman spectroscopy covers an overall wavenumber shift range 10–4000 cm^{-1} and has a major advantage over i.r. of negligible interference from aqueous solvents. Raman scattering is proportional to the number of molecules and can be used to measure concentration. Resonance Raman is increasing in usefulness (Section 12.3.2.4).

1.3.5. RESONANCE SPECTROSCOPY AT MICROWAVES: E.S.R.

Electron spin resonance (e.s.r.) spectroscopy (Chapter 8) or electron paramagnetic resonance (e.p.r.) involves the absorption of electromagnetic radiation at microwave frequencies as a result of reorientation of magnetic moments of unpaired electrons. It requires a laboratory magnet for observation, like n.m.r. (Section 1.3.6), and it involves interaction with the magnetic (rather than the electric) component of electromagnetic radiation. While the operating frequencies (9·5 GHz or about 3 cm wavelength for X-band in a field of about 0·3 T, 35 GHz for Q-band) depend on the magnitude of the magnetic field, they are typically higher than the radiofrequencies used for n.m.r. In e.s.r., peak area is proportional to the number of unpaired spins, in n.m.r. to the number of resonating nuclei (see Section 1.4.1).

In biology, the requirement that samples for e.s.r. should possess unpaired electrons leads to the following main kinds of application: (a) free radicals, for example, identification of intermediates in enzyme reactions or following formation and decay of radicals induced in proteins by high-energy radiation (Section 8.5.1); and (b) paramagnetic metal atoms (Fe, Cu, Co, Mn) present in haem proteins (Section 8.6.1), metalloenzymes and the cytochromes. The comparative rarity of paramagnetism in biology can, however, be mitigated by deliberate introduction of a paramagnetic reporter group or label which acts as a probe of the biomolecule (Section 10.2.5). It must be accepted that, in principle, binding any kind of reporter group (for detection by e.s.r., fluorescence or any other technique) to the macromolecule must perturb the system in some way. While typical e.s.r. free-radical spin labels are relatively bulky and require somewhat higher concentrations than some other labelling techniques (notably radioisotopic labelling), they have been very widely used in biological applications and are being developed with improved specificity as to binding location.

1.3.6. RADIOFREQUENCY SPECTROSCOPY: N.M.R.

Nuclear magnetic resonance (n.m.r.) spectroscopy (Chapter 7) has close analogies with e.s.r. and is confined to such atomic nuclei as possess a non-zero spin quantum number I. It arises from interaction of the magnetic component of electromagnetic radiation with the very small magnetic moments (about 10^{-3} of the electron moment) possessed by a number of isotopes. As with electron orientation levels in e.s.r., nuclear orientation levels have an energy separation, and hence a resonance frequency, proportional to the magnetic field employed. According to the size of this field, (commonly within 0·5–7 T) and the nuclear species at resonance (^{1}H with a large moment, high abundance and $I = 1/2$ is the most favourable), frequencies are usually in the radiofrequency range 1–300 MHz, i.e. the wavelength range 300–1 m. The small size of the quanta (typically 10^{-3} cal, or 10^{-5} kT) helps to minimize disturbance of the system by a nuclear probe and to eliminate the complication of any lower energy interactions. On the other hand, magnetic resonance spectroscopy, especially n.m.r., has an inherent insensitivity in that the small separation between energy levels results, by the Boltzmann distribution (Section 1.2.3), in only a small population excess in the lower levels. Since lack of sensitivity and of chemical-shift (Section 7.2.2) resolution has been felt most acutely in the n.m.r. study of biological polymers, it is from potential applications in such systems that the major impetus has come for development of high magnetic fields, time-averaging and Fourier-transform spectrometers (Sections 1.5.3 and 10.3.1.1). Arising from the larger energy differences between ground and excited states, the reciprocal frequencies in optical spectroscopies are briefer than random correlation times (10^{-10}–10^{-12} s) in liquids, whereas in the resonance spectroscopies (e.s.r. and n.m.r.) reciprocal frequencies are longer, so that weighted average spectra of interconverting species will often be obtained. High-resolution n.m.r. can be a rich source of structural information, especially perhaps for medium-sized molecules. The chemical shift, derived from the precise frequency at which a nucleus resonates, characterizes its nature and surroundings. In addition, coupling constants between nuclear spins (Section 7.2.3) can provide indications about the relation of one group to another within a molecule, while relaxation-time measurements (Section 7.2.4) often facilitate the study of motions of ions and small molecules attached to polymers.

1.4. Characteristics of a spectral line

In general, spectroscopic transitions may be described by four kinds of experimental parameters: intensity, width, position and structure; polarization (Section 2.2.5) may also be characterized in some branches of electric-vector spectroscopy.

1.4.1. INTENSITY

In addition to the transition probability and the population of the energy state (Section 1.2.3), the amount of material in a sample influences the intensity of a line. For spectroscopy based on the fixed magnetic moments of the electron (e.s.r.) and nucleus (n.m.r.), the intensity is proportional to the numbers of unpaired electrons or nuclei resonating; peak areas have provided a reliable estimate of the free-radical concentration, or relative numbers of chemically different nuclei, respectively. In n.m.r., the ^{1}H spectrum of a solid polypeptide may be several kHz wide compared with 1–2 Hz for a similar material in solution. Consequently, the strength of the signal at a given value of field from such a broad line may be very little above that of the background noise. Low sensitivity can be a serious problem with n.m.r. since biomolecules and cell structures may not be able to supply the required concentration of about 0·001 M; even where practicable, the use of high enough concentrations to facilitate n.m.r. measurements may mean the utilization of a system relatively remote from *in vivo* reality.

For optical absorption spectra involving partly induced electronic transition moments, the ratio of transmitted intensity, I, to incident intensity, I_0, at a given frequency, is related experimentally to the path length l, in cm, and concentration c, in mol l^{-1} of the sample by the Beer-Lambert law

$$I/I_0 = \exp(-Kcl)$$

where K is a constant.

If logarithms to the base 10 are taken, to give the optical density D (Section 2.2.2), or absorbance (Section 5.3.3), $A = \log_{10} I_0/I$, then the new constant,

$$K/2\cdot 303 = \varepsilon = A/cl$$

is the molar extinction coefficient, with units $l\ \text{mol}^{-1}\ \text{cm}^{-1}$. In the u.v., visible and i.r. regions, extinction coefficients measured from peak areas (or often heights) can be as low as 1 for weak bands or, for strong bands as much as 100, 1000 or higher. Their magnitudes are useful both for identification and for quantitative measurements.

1.4.2. WIDTH

The natural width of a spectral line (generally regarded as the breadth at half maximum height or sometimes between points of maximum gradient), independent of instrumental imperfections and limitations, in a molecular system, is associated with the spread of the relevant energy levels. Ultimately, the narrowness of a line is limited, via the Heisenberg Uncertainty Principle, by the life-time of the excited state. Since the several contributions to broadening of a line can vary with the physical state, with the environment, and with motion, line width can be used to investigate motion, kinetics and interactions. Often what matters is the frequency width *relative* to the frequency of the transition: a spread of 10^7 Hz would correspond to a typically broad e.s.r. line at 10^9 Hz but would be a minute fraction of a u.v. line frequency. Since high-resolution n.m.r. lines can be as narrow as $0\cdot 2$ Hz in 10^8 Hz, more homogeneous magnetic fields are required than for e.s.r.

1.4.3. POSITION

The position of a spectral line (or, in magnetic resonance, the position relative to that of a standard, so as to allow for the size of laboratory field) is its most immediately useful characteristic. It is measured as frequency, wavelength, or wavenumber (Section 1.2.4), or some quantity proportional to frequency, as often in magnetic resonance spectroscopy. In e.s.r., the departure of the g-factor (extracted from the ratio of magnetic field to frequency at the line centre, see Section 8.2.2) from its free-electron value arises from the orbital magnetic moment contribution (Section 8.2.2) and is greater most for transition-metal compounds than for free radicals. For large molecules, fairly simple consideration of i.r. and n.m.r. frequencies can yield valuable structural parameters—not so much by providing a so-called fingerprint, but rather by confirming the existence of suspected groups.

1.4.4. STRUCTURE

Some spectra, even of biological molecules, consist of only a single line. However, it is from the multiplicity and fine structure of lines, at least for molecules of only moderate complexity, that spectroscopy has yielded most detailed information about structure. In n.m.r., chemical groups can be assigned from chemically-shifted peaks, while spin-spin multiplets enable the relative numbers and positions of distinct groups of magnetic nuclei to be deduced; in e.s.r., nuclear fine structure enables free radicals to be identified. At the other end of the electromagnetic spectrum, magnetic hyperfine interactions can be utilized in the Mössbauer spectroscopy of biological molecules (Section 9.5.3).

1.5. Detectability of a spectrum

The existence of a known energy-level difference in conformity with selection rules does not ensure that the corresponding spectroscopic transition will be observed experimentally. Ability of a spectrometer to detect a signal, an essential prerequisite to assignment and interpretation, depends partly on the breadth and structure of the spectral line—a narrow line gives a bigger signal—and partly on the instrumental parameters which define the resolution and sensitivity.

1.5.1. RESOLUTION

The resolving power of an instrument, a measure of its ability to distinguish two close lines, is usually taken to be the ratio, $v/\Delta v$, between the mean frequency and the minimum frequency separation that can be discerned. In many branches of spectroscopy, resolution, Δv, is controlled by the slit width, the minimum of which is limited in turn by the need to enable adequate radiation to reach the detector. In magnetic resonance, the resolution is usually restricted by the variation in magnetic field over the sample. A 100 MHz n.m.r. spectrometer might have a resolution of 0·3 Hz and a resolving power of $10^8/0·3$, so that effectively the magnetic field must stay constant to about three parts in 10^9 over the sample during a run.

1.5.2. IMPROVEMENT OF THE SIGNAL-TO-NOISE RATIO

All spectra have to be observed against a background of random fluctuations or "noise", whether generated by the source-spectrometer-amplifier assembly (and, as such, dependent on design) or the more fundamental thermal or Johnson electronic noise from the detector. When genuine signals are weak and further increase in sample size is not feasible, increasing the power of the source (while retaining monochromaticity where necessary) may be advantageous. However, in the lower frequency regions of the electromagnetic spectrum, excessive power levels can saturate the system, or tend to equalize the populations of energy levels (Section 1.2.3), so that the signal strength diminishes, even to zero.

In the double-beam technique, employed in several branches of spectroscopy, the incident beam is split into two, one passing through the test sample cell and the other through a reference cell. Comparison of the two beams eliminates spurious signals generated, for example, from variations in the source and has the intrinsic advantage of a null method. Infra-red double-beam spectrometers are discussed in Section 2.2.2.1.

The effect of noise at the detection stage may be minimized by lock-in amplification or phase-sensitive detection. In this, the equal amplification of signal and noise (which would give no net advantage) is avoided by converting the signal into an alternating voltage at a frequency well away from the dominant frequency components (near zero) constituting the noise, and then using an amplifier tuned to the new signal frequency. In i.r. spectroscopy, the source beam is chopped by a rotating sector, at a frequency of a few Hz or a few hundred Hz, according to the response time of the detector, and the square wave is then smoothed. In magnetic resonance spectroscopy, modulation of the transmitter frequency (or, by means of subsidiary magnet coils, of the magnetic field) is effected at a few hundred Hz for n.m.r., often 100 kHz for e.s.r. With a modulation amplitude smaller than the line width, the carrier is modulated in proportion to the difference in absorption over the modulation range, i.e. to the rate of change of absorption with frequency (or field) (Fig. 8.18). Random signals will tend to be averaged out. Such differential curves appear in the usual presentation of most e.s.r. spectra and of broad-line n.m.r. spectra.

Lock-in detection and frequency-sensitive amplification imply relatively slow scanning across the spectrum; usually a choice of time constants is available. Indeed, the method of minimizing the effect of random noise fluctuations by averaging over a long time has been applied in nearly all branches of spectroscopy; this evidently presupposes sufficient long-term stability of the spectrometer. Increasing the scanning time allows the use of a longer time constant and enables bandwidth, and thus the range of noise frequencies transmitted, to be reduced. In this way, the improvement in signal-to-noise is proportional to the square root of the scan time. However, in some cases, e.g. pulsed n.m.r., the bandwidth cannot be reduced.

A variant of the principle, particularly valuable when (as in magnetic resonance) very slow sweep might induce saturation, or when constancy of spectrometer conditions over many hours cannot be guaranteed in advance, is to make a large number, n, of sweeps through the spectrum, and add the results. Whereas the total signal has an amplitude n times that of an individual run, the noise occurs in different parts of the spectrum in successive runs and increases in total only in proportion to $n^{1/2}$. In a computer of average transients (CAT), a digital storage device, the spectrum (total signal and total noise) may be monitored at perhaps 10^3 discrete points of frequency (or channels) and the improvement in signal-to-noise is proportional to $n^{1/2}$ (p. 204).

Incidentally, the same small computer as is used for signal averaging can often also be utilized for other computations on the digitized spectral data. Thus, a first-derivative e.s.r. spectrum (Section 8.3.2) may be integrated once to the absorption curve or twice to give areas (proportional to concentrations); or the spectrum of one component may be subtracted from a composite spectrum.

1.5.3. FOURIER-TRANSFORM SPECTROSCOPY

In the far i.r. region, development of spectroscopy (Section 4.1.2) was slow because of detection problems with weak radiation sources. A major instrumental advance was the utilization, through the ready availability of computers (Section 4.2.3.5), of fast Fourier-transform spectroscopy from the distance domain (Section 4.2.3.3) whereby, in effect, a wide range of energies (and frequencies) is sampled simultaneously. The same principle of frequency differentiation by means of a two-beam interferometer, but with rapid scan and fast Fourier transformation by computer, is being applied to i.r. absorption spectra and to the study of emission spectra from weak sources in the mid i.r., e.g. from human skin *in vivo*.

Signal-to-noise problems in magnetic resonance spectroscopy (e.g. for n.m.r. of less abundant nuclei such as ^{13}C) are being overcome by Fourier-transform (FT) spectroscopy from the time domain (Section 7.3.3). Slow consecutive sweeping through frequencies with time in continuous-wave n.m.r. is replaced by short bursts or pulses of high radiofrequency power which excite an extensive band of frequencies simultaneously. A series of similar intense pulses is applied, the transient responses of the system are added in a time-averaging computer, and the stored total envelope is Fourier-transformed to yield a conventional absorption spectrum of high signal-to-noise ratio. In addition to the greatly reduced time (or improved signal-to-noise, if the time is utilized for more scans), FT is much more effective than continuous-wave n.m.r. for the measurement of relaxation rates.

2. Infra-red Spectroscopy
J. H. Keighley

2.1. Theoretical aspects

2.1.1. INTRODUCTION

The examination of biological materials by i.r. spectroscopy has concerned chemists, physicists and biologists for many years. Most biological materials are structurally complex; consequent difficulties in the analysis of the results meant that, in many cases, the investigator has had to await the development of new or more sophisticated experimental methods for a more complete understanding of underlying theory. Thus, application of Raman spectroscopy (Section 1.2.3 and Chapter 3) to biological specimens has followed the development of the laser (a powerful illuminating source) and the acceptance and understanding of the nature of vibrational modes which participate in both i.r. absorption and Raman scattering.

The theoretical bases of i.r. and Raman spectroscopy are closely related and will be developed concurrently in this chapter and illustrated first by application to small molecules; the applications of Raman spectroscopy to biological materials are developed more fully in Chapter 3. Fig. 2.1 shows the experimental arrangement of a typical i.r. spectrometer.

2.1.2. GROSS SELECTION RULES

The total energy of a molecule arises from rotational, translational, vibrational and electronic motions (Section 1.2.2). Electronic motions, or transitions, are associated with the absorption and emission of radiation in the u.v. and visible regions of the electromagnetic spectrum (Section 1.3.2), while molecular rotations and vibrations give rise to absorptions and emissions of radiation at characteristic frequencies in the far and near regions of the i.r. (Section 1.3.3).

As indicated in Chapter 1, the energy, E, of a quantum of radiation is given by

$$E = h\nu$$

where h is Planck's constant, ν is the frequency of the absorbed radiation, and the frequency, v, of the absorbed radiation is related to the wavelength, λ, by the equation

$$v = c/\lambda$$

where c is the velocity of light:
In i.r. and Raman spectroscopy, the absolute frequency, v, is generally replaced by wavenumber, $\tilde{v}$, which has units of reciprocal cm and which is defined by

$$\tilde{v} = \frac{1}{\lambda}\,\mathrm{cm}^{-1}$$

where λ is measured in cm, so that

$$E = hc\tilde{v}$$

In order that an atomic group can absorb i.r. radiation, it must possess an electric dipole moment which changes during the motion. Similarly, the change in polarizability with motion is a necessary criterion for the detection of Raman lines.

Thus molecules such as H_2 and Cl_2 do not absorb i.r. radiation, since there is no changing dipole moment, whereas HCl molecules, which are formed from two atoms with differing electro negativities, will give rise to absorption owing to the inherent polarity of the system. If two charges q^+ and q^- are separated by a distance l, then the dipole moment, M, is given by

$$M = ql$$

Thus, as the distance l varies during a vibration, so does the dipole moment; the dipole moment can also vary for some direction during rotational motion to give rise to further absorption of radiation.

2.1.3. THE ORIGINS OF ROTATIONAL SPECTRA

The simplest form of molecular rotation occurs for the rigid diatomic molecule, which may be likened to a dumbell with unequal masses m_1 and m_2 joined by a bar of length r.

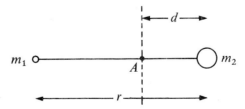

If the centre of mass occurs at A, then by the lever principle,

$$m_1(r-d) = m_2 d$$
$$\therefore d = m_1 r/(m_1 + m_2)$$

If the molecule rotates about an axis through the centre of mass, the moment of inertia, I, is given by

$$I = m_1(r-d)^2 + m_2 d^2$$

$$\text{or } I = m_1 m_2 r^2/(m_1 + m_2)$$

Thus the unequal dumbell can be replaced by a so-called reduced mass, μ, rotating in a circle of radius r,

$$\text{i.e. } \mu = m_1 m_2/(m_1 + m_2)$$

Hence

$$I = \mu r^2$$

The energy of this rotation is given by $E = \tfrac{1}{2} I \omega^2$, where ω is the rotational angular velocity.

According to classical mechanics, the system has rotational energy which depends on the value of ω. Quantum-mechanical principles, however, show that only discrete energy levels can occur. By solution of the Schrödinger wave equation,

$$\frac{\partial^2 \Psi}{\partial x^2} + \frac{\partial^2 \Psi}{\partial y^2} + \frac{\partial^2 \Psi}{\partial z^2} + \frac{8\pi^2 \mu E \Psi}{n^2} = 0$$

with the restriction that the values of Ψ must be single-valued, finite and continuous, it can be shown that the energy levels are given by

$$E = \frac{h^2}{8\pi^2 \mu r^2} \cdot [J(J+1)]$$

where the rotational quantum number, J, has integral values of 0, 1, 2, 3 . . .

Thus the energy levels of the rotator are not equally spaced. A transition from a low-energy to a high-energy level is effected by the absorption of a quantum of radiation of frequency v, as given by

$$v = (E_2 - E_1)/h$$

where E_2 is the higher energy level. Substitution for E in the above equation gives the relation

$$v = \frac{h}{8\pi^2 \mu r^2 c} [J_2(J_2+1)] - \frac{h}{8\pi^2 \mu r^2 c} [J_1(J_1+1)],$$

or

$$\tilde{v} = B[J_2(J_2+1) - J_1(J_1+1)]$$

where B is called the rotational constant.

For $\tilde{v}$ to be calculable, the nature of the transition $J_2 \to J_1$ must be established.

2.1.4. SELECTION RULES

When radiation with electric vector E interacts with a dipole M, the resultant interaction is given by EM. With this energy inserted in the Schrödinger wave equation, the probability of a transition from an energy state m to a state n is proportional to the vectors $\mathbf{R}^{mn}$, called transition moments (Section 1.2.3) or matrix elements of the electric-dipole-moment operator. Provided such a transition moment is not zero, then the transition between m and n is possible. For i.r. rotational transitions, $\mathbf{R}^{J_1 J_2}$ is zero (for all values) except when $J_2 = J_1 \pm 1$, i.e. when $\Delta J = \pm 1$. This is a selection rule for i.r. rotational transitions. (The corresponding vibrational-spectrum rule is $\Delta v = \pm 1$, where v is the vibrational quantum number.) Hence,

$$\tilde{v} = B(J_1+1)(J_1+2) - B(J_1)(J_1+1)$$

$$\therefore \tilde{v} = 2B(J_1+1)$$

On this basis, the rotational i.r. spectrum of a heteronuclear diatomic molecule would consist of equally spaced (by $2B$) lines which correspond to $J = 0, 1, 2$, etc. In practice, a bond joining the two atoms is not totally rigid but is deformed during rotation by centrifugal-force effects, which cause a change in the moment of inertia. Hence, at high values of J, B decreases and the line separation consequently decreases.

2.1.5. THE ORIGINS OF VIBRATIONAL SPECTRA

A suitable model for the molecule for this purpose is two masses m_1 and m_2 joined by an elastic spring.

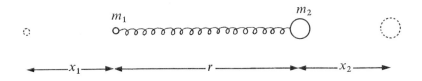

Just as the rotator could be considered as a particle of mass μ rotating at a distance r from the centre of mass, so the vibrator may be considered as a particle of mass μ vibrating about a point, distance r from the centre of mass, with a maximum amplitude given by

$$x_0 = x_1 + x_2$$

If the vibrational motion is assumed to be simple harmonic, the displacement x from the mean, at a time t, is given by $x = x_0 \sin 2\pi vt$.

The restoring force F is given by

$$F = -kx = \mu \delta^2 x / \delta t^2$$

where k, the force constant of the bond, is a measure of the bond strength. The solution to the above equation is given by the preceding equation, from which the vibrational frequency can be determined by

$$v = \frac{1}{2\pi} \sqrt{(k/\mu)}$$

The potential energy, P, of the system is given by

$$P = \tfrac{1}{2} k x^2 = 2\pi^2 \mu v x^2$$

so that the potential-energy curve is a parabola with its minimum at the position of equilibrium, r_e, of the vibrating system. If the expression for P is inserted into the Schrödinger equation, the eigenvalues are given by

$$E = \frac{h}{2\pi} \sqrt{\frac{k}{\mu}} \, (v + \tfrac{1}{2})$$

Hence

$$E = hv(v + \tfrac{1}{2})$$

so that the frequency of the radiation absorbed during a transition is given by

$$\tilde{v} = \frac{hv}{hc}(v_2 + \tfrac{1}{2}) - \frac{hv}{hc}(v_1 + \tfrac{1}{2})$$

If the selection rule $\Delta v = \pm 1$ applies,

$$\tilde{v} = \frac{v}{c}(v_1 + \tfrac{3}{2}) - \frac{v}{c}(v_1 + \tfrac{1}{2}) = \frac{v}{c}$$

Hence the spectrum of a simple harmonic oscillator consists of a single band at a wavenumber which corresponds to the vibrational frequency of the molecule.

2.1.6. THE ANHARMONIC OSCILLATOR

For a diatomic molecule, the curve of energy against displacement is not, as suggested above, a parabola, but has the Morse-potential form; the two curves are similar near the equilibrium position but deviate at higher energies. A closer approach to the experimental curve is obtained from the equation

$$P = fx^2 - gx^3 \ldots$$

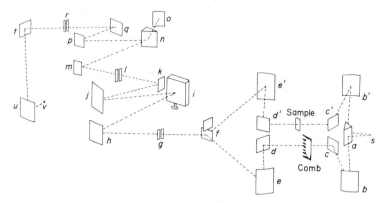

FIG. 2.1. Optical diagram of an i.r. spectrometer system (see p. 34).

This introduces a maximum at higher energies and is interpreted as the superimposition of fundamental and overtone vibrations in a Fourier series. The energy levels are now given by the relation

$$E = hc[\tilde{v}(v+\tfrac{1}{2}) + \tilde{v}_x(v+\tfrac{1}{2})^2]$$

where $\tilde{v}_x$ is the anharmonicity constant. Evidently, the inclusion of this additional squared term in v causes E to decrease as v increases. For example, for the transition $v = 1$ to $v = 2$, the energy difference is given by

$$\Delta E = hc\tilde{v} - 4hc\tilde{v}_x$$

and for $v = 2$ to $v = 3$

$$\Delta E = hc\tilde{v} - 6hc\tilde{v}_x$$

Thus two bands will appear with a separation of $2\tilde{v}_x$ cm^{-1}. In addition, the rigid selection rule $\Delta v = \pm 1$ breaks down and transitions which correspond to $\Delta v = \pm 1, 2, 3, \ldots$ can occur. This gives rise to overtone bands which have approximately $2, 3, \ldots$ etc. times the frequency of the fundamental; in practice, as before, the system is perturbed by anharmonicity effects which cause small departures from integral values.

2.1.7. POLYATOMIC MOLECULES

For a diatomic vibrator, the frequency of the absorption may be calculated accurately from a knowledge of the molecular parameters. Polyatomic molecules, however, present a more complex problem since the relative motion of two atoms is no longer so simple owing to the perturbation of neighbouring atoms, which superimposes wave motions.

In order to describe motions of a molecule of N atoms, a total of $3N$ coordinates is required (e.g. the x, y and z coordinates for each atom). During molecular motion, changes in these are used to specify the motion which occurs. Three coordinates, however, are necessary to specify each of rotational and translational modes for a non-linear molecule, while for a linear molecule only two are required to specify rotation.

Thus $3N-6$ ($3N-5$ for linear molecules) coordinates describe the molecular vibrations. In order to enable the nature of these vibrations to be determined, the mathematical theory of groups may be applied.

2.1.8. MOLECULAR SYMMETRY AND GROUP THEORY

Polyatomic molecules may exhibit a certain regularity or symmetry. In the ethane molecule, C_2H_6,

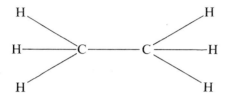

each CH_3 group is the mirror image of the other if as, a preliminary illustration, we consider the symmetry properties of a flat two-dimensional drawing of this molecule.

If a molecule shows regularity, it is said to possess a *symmetry element*. Thus for the above figure, one symmetry element is a plane of reflection perpendicular to the C—C axis. The *symmetry operation* corresponding to this element is reflection in this plane. The drawing has other symmetry elements such as a plane along the C—C axis and perpendicular to the page.

For spectroscopic purposes, there are five major types of symmetry element.

1. The identity I. The symmetry operation is such as to leave the molecule unchanged. This trivial element, possessed by all molecules, is introduced for the purpose of group theory.
2. A centre of symmetry i. The symmetry operation is inversion of the molecule through the centre, so that any atom with the coordinates x, y and z measured from the centre will have an equivalent atom at $-x, -y$ and $-z$. Thus the ethane molecule in this conformation has a centre of symmetry i.
3. A p-fold rotation axis C_p. The corresponding symmetry operation is rotation of the molecule through $360°/p$ to give an identical configuration. The figure above indicates that the molecule as shown has a C_2 axis passing through i perpendicular to the C—C bond. Another example is the axis bisecting the H—O—H angle of an isolated water molecule.

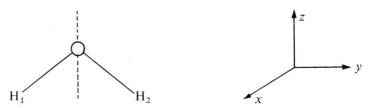

Rotation of the molecule about the z-axis through $\frac{360°}{2}$ gives an equivalent configuration if hydrogen atoms H_1 and H_2 are identical. If they are not, as in HOD, then rotation through 360° is necessary to produce the identical molecule. The rotational symmetry element for HOD is thus only C_1, which is possessed by all molecules and it is the same as the identity I (HOD also possesses a plane of symmetry; see (4)). Ammonia has a C_3 axis, since rotation through $\frac{360°}{3} = 120°$ produces an equivalent configuration.

4. A plane of symmetry σ. The symmetry operation is a reflection of the molecule through a plane to produce an identical configuration, i.e. the plane bisects the molecule into two mirror images. The water molecule has a plane of symmetry xz and also a second plane of symmetry yz. The ammonia molecule has three symmetry planes which are included at 120° to each other.

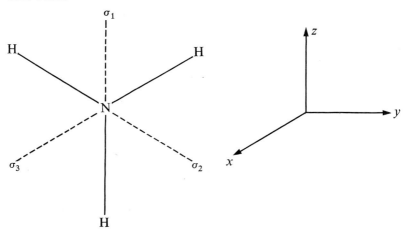

It should be noted that reflection through a plane is not the same as rotation by $\frac{360°}{p}$. Consider the water molecule,

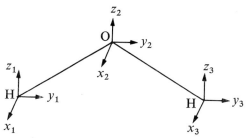

The molecule is fixed in space with the coordinates $x_1, y_1, z_1; x_2, y_2, z_2; x_3, y_3, z_3$. Rotation through 180° gives a molecule with coordinates $-x_3, -y_3, +z_3; -x_2, -y_2, +z_2; -x_1, -y_1, +z_1$, while reflection through the xz plane gives a molecule with coordinates $x_3, -y_3, z_3; x_2, -y_2, z_2; x_1, -y_1, z_1$.

In both water and ammonia molecules, the vertical symmetry planes contain the rotation axis and are designated σ_v. Consider the benzene molecule; this has a centre of inversion i and a six-fold rotational axis through i perpendicular to the molecule.

There are six vertical symmetry planes and, in addition, one plane of symmetry designated σ_h perpendicular to the rotational axis (the plane of the molecule).

5. *p*-fold rotation-reflection axis, S_p. This symmetry operation is rotation of the molecule through $360°/p$, followed by reflection through a plane perpendicular to this axis. For the methane molecule:

If the four hydrogens are equivalent, the original configuration is obtained and the molecule possesses an S_4 symmetry element along each H—C—H bisector.

2.1.9. POINT GROUPS

The presence of certain combinations of symmetry elements can require the presence of other elements. Thus, a molecule with a C_2 axis and a C_3 axis coincident must also possess a C_6 axis (as in benzene). Also the presence of certain symmetry elements forbids the presence of others. An allowed combination of symmetry elements about a point is called a *point group* and all molecules in isolation belong to one of them. Among the most important point groups for polymeric molecules are those designated C_{pv}, for which the symmetry elements are a C_p rotational axis and p vertical planes containing the rotation axis. Thus water belongs to the C_{2v} point group and ammonia to the C_{3v} point group. In biological work, molecules belonging to the C_v or C_s point group, which has just one plane of symmetry, are of importance; HOD and water molecules singly-bonded to organic solvents are normally classified in

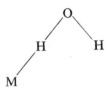

this group. Linear molecules often belong to point group $C_{\infty v}$ or, if centrosymmetric to $D_{\infty h}$; since an infinite number of rotations leave the configuration unchanged, these molecules possess an infinite number of symmetry planes. The *factor group* (Section 3.3.3) of the space group of the unit cell of a crystal, or polymer, is closely related to (and is said to be isomorphous with) the corresponding point group.

2.1.10. GROUP THEORY

A set of elements, I, A, B, C, is said to form a group if it satisfies four conditions.
1. There is an identity I such that

$$AI = IA = A$$

2. The product of any two elements in a group is also an element of the group.
3. The elements obey the law of association

i.e. $A(BC) = (AB)C$

4. Every element has an *inverse* such that the product of the element and its inverse is the identity I.

Consider the four elements, $1, -1, i, -i$, where $i^2 = -1$. To check that these elements form a group, it is convenient to use a multiplication table.

	1	−1	i	−i
1	1	−1	i	−i
−1	−1	1	−i	i
i	i	−i	−1	1
−i	−i	i	1	−1

If the identity I is set to 1 and A to -1, then $AI = IA = A$ gives $(-1) \times 1 = 1 \times (-1) = -1$; and if $B = i$, then $i \times 1 = 1 \times i = i$. In addition, the product of two elements is a third element,

i.e. $(-1) \times i = -i$; $A \times B = C$

Also $A(BC) = (AB)C$ is equivalent to $-1(i \times -i) = (-1 \times i) \times (-i) = -1$
Thus the inverse of 1 is 1; of $-1, -1$; and of $i, -i$, so that $A \times A^{-1} = I = A^{-1} \times A$ gives

$$(-1) \times (-1) = 1 = (-1) \times (-1) \text{ and } i \times (-i) = 1 = (-i) \times i$$

The four elements $1, -1, i, -i$ thus form a group.

2.1.11. APPLICATION TO POINT GROUPS

The elements of the C_{2v} point group are the identity I; rotation through 180° (or C_2); reflection through the xz plane (σ_v'), and reflection through the yz plane (σ_v''). The multiplication table for these elements is

	I	C_2	σ_v'	σ_v''
I	I	C_2	σ_v'	σ_v''
C_2	C_2	I	σ_v''	σ_v'
σ_v'	σ_v'	σ_v''	I	C_2
σ_v''	σ_v''	σ_v'	C_2	I

The first condition is satisfied by the presence of the identity I. The second condition is also satisfied since, for example, rotation through 180° followed by reflection across one plane of symmetry is equivalent to reflection across the other plane. This is shown for a C_2 rotation followed by reflection in the xz plane:

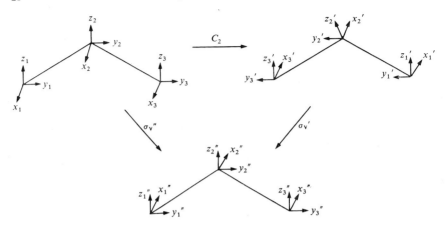

Application of the multiplication table gives

$$(\sigma_v'.\sigma_v'') = C_2 \text{ and } (C_2.\sigma_v') = \sigma_v''.$$

Thus $C_2.C_2 = \sigma_v''.\sigma_v''$ and, from the table, $C_2.C_2 = I = \sigma_v''.\sigma_v''$. Thus C_2 is the inverse of C_2, and σ_v'' is the inverse of σ_v'' (condition 4 of Section 2.1.10).

For each symmetry operation, it is possible to construct a matrix describing the changes in the coordinates of each atom during the symmetry operation. Thus, for the C_2 rotation of the water molecule above,

$$\begin{array}{lll} X_1' = -X_3 & X_2' = -X_2 & X_3' = -X_1 \\ Y_1' = -Y_3 & Y_2' = -Y_2 & Y_3' = -Y_1 \\ Z_1' = Z_3 & Z_2' = Z_2 & Z_3' = Z_1 \end{array}$$

These equations represent a 9×9 matrix

$$\begin{vmatrix} X_1' \\ Y_1' \\ Z_1' \\ X_2' \\ Y_2' \\ Z_2' \\ X_3' \\ Y_3' \\ Z_3' \end{vmatrix} = \begin{vmatrix} 0 & 0 & 0 & 0 & 0 & 0 & -1 & 0 & 0 \\ 0 & 0 & 0 & 0 & 0 & 0 & 0 & -1 & 0 \\ 0 & 0 & 0 & 0 & 0 & 0 & 0 & 0 & 1 \\ 0 & 0 & 0 & -1 & 0 & 0 & 0 & 0 & 0 \\ 0 & 0 & 0 & 0 & -1 & 0 & 0 & 0 & 0 \\ 0 & 0 & 0 & 0 & 0 & 1 & 0 & 0 & 0 \\ -1 & 0 & 0 & 0 & 0 & 0 & 0 & 0 & 0 \\ 0 & -1 & 0 & 0 & 0 & 0 & 0 & 0 & 0 \\ 0 & 0 & 1 & 0 & 0 & 0 & 0 & 0 & 0 \end{vmatrix} \times \begin{vmatrix} X_1 \\ Y_1 \\ Z_1 \\ X_2 \\ Y_2 \\ Z_2 \\ X_3 \\ Y_3 \\ Z_3 \end{vmatrix}$$

The *character* of the matrix χ is the sum of the diagonal elements; for the C_2 operation $\chi = -1$. The matrices for all the symmetry operations in the group form the *representation* of the group. In the above matrix, it can be seen that not all of the non-zero elements lie on the diagonal. It may be possible, however, to find a coordinate transformation which puts all the elements on the diagonal. If the matrix can be so transformed, then the representations is said to be *reducible*, but if the matrix cannot be further reduced, it is known as the *irreducible representation*. (More generally, diagonal elements may be blocks.)

Now there are as many irreducible representations as there are classes of symmetry operations and there are as many different vibrational types or species as there are irreducible representations. Hence, for the C_{2v} point group, there are four vibrational types which are signified by A_1, A_2, B_1 and B_2. A and B designate whether the vibration is symmetrical or not, respectively, with respect to the C_2 axis, and the subscripts 1 and 2 designate symmetry or asymmetry with respect to the σ_{xz} symmetry plane.

The *character table* for the C_{2v} point group is shown below:

	I	C_2	σ_v'	σ_v''
A_1	1	1	1	1
A_2	1	1	−1	−1
B_1	1	−1	1	−1
B_2	1	−1	−1	1

The figures 1 and −1 indicate whether the vibration is symmetrical or antisymmetrical, respectively, with respect to each symmetry operation. Thus vibrations of the type A_1 are symmetrical to all symmetry operations.

Further consideration of matrix characters allows the determination of the total number of vibrations in each species and also allows a determination of whether or not the vibration is infrared-active (p. 31).

2.1.12. THE TOTAL NUMBER OF VIBRATIONS

The total number of possible vibrations N_v for a molecule is given by

$$N_v = 3N - 6$$

However, this gives no information about the distribution such vibrations between the various species.

The number of vibrations N_i, of the type i, is given by

$$N_i = (1/N_g)\Sigma n_\varepsilon \beta(R)\chi_i(R)$$

where N_g is the total number of symmetry elements in each group and n_ε is the number of symmetry elements in each class. For the C_{3v} point group there are two types of C_3 rotation and three symmetry planes, so that

$$n_\varepsilon \text{ (rotation)} = 2$$

and

$$n_\varepsilon \text{ (reflection)} = 3$$

$\chi_i(R)$ is the character of the vibration, of type i, for the symmetry operation and is obtained from the group tables.
$\beta(R)$ is given by

$$\beta(R) = (u_R - 2)(1 + 2\cos\phi)$$

where u_R is the number of atoms left unchanged by the symmetry operation R and ϕ represents the angle of rotation during the symmetry operations.

The above equation in $\beta(R)$ relates to *proper rotations*, i.e. rotations around symmetry axes; *improper rotations* are rotations followed by reflections through a plane perpendicular to the rotational axis. Reflections such as σ_v are improper rotations through $0°$ and for these $\beta(R)$ is given by

$$\beta(R) = u_R(-1 + 2\cos\phi)$$

For the isolated water molecule, H_2O, the following table may be constructed.

R	I	C_2	σ_v'	σ_v''
u_R	3	1	1	3
ϕ	0	180	0	0
$\cos\phi$	1	-1	1	1
$2\cos\phi$	2	-2	2	2
$\beta(R)$	3	1	1	3

For the four C_{2v} vibrational types

$$N_{A_1} = \tfrac{1}{4}[1\times3\times1 + 1\times1\times1 + 1\times1\times1 + 1\times3\times1] = 2$$
$$N_{A_2} = \tfrac{1}{4}[1\times3\times1 + 1\times1\times1 + 1\times1\times(-1) + 1\times3\times(-1)] = 0$$
$$N_{B_1} = \tfrac{1}{4}[1\times3\times1 + 1\times1\times(-1) + 1\times1\times1 + 1\times3\times(-1)] = 0$$
$$N_{B_2} = \tfrac{1}{4}[1\times3\times1 + 1\times1\times(-1) + 1\times1\times(-1) + 1\times3\times1] = 1$$

Hence the isolated water molecule has three vibrations: two of type A_1 and one of type B_2. The A_1 vibrations are symmetrical with respect to the C_2 axis and the xz and yz planes. The only vibrations which can give this are an in-phase OH stretching (symmetrical v_1 mode), and the deformation mode v_2.

The B_2 vibration is symmetrical with respect to the σ_{xz} but not σ_{yz} planes.

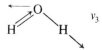

This is the anti-symmetrical (or v_3 stretching) mode.

INFRA-RED SPECTROSCOPY

The HOD molecule, point group C_s, has two vibrational species, A_1 and A_2. The above equation in N_i, can be used to show that there are three vibrations of type A_1 and none of type A_2; these are the OD stretch v_1, the OH stretch v_3 and the deformation mode v_2.

2.1.13. INFRA-RED AND RAMAN ACTIVITY

The i.r. activity of each vibration type is given by

$$N_i(M) = (1/N_g)\Sigma n_c \chi_m(R)\chi_i(R)$$

where $\chi_m(R)$ is the character of the dipole moment for the operation R and is given by

$\chi_m(R) = 1 + 2\cos\phi$ for proper rotations
and $\chi_m(R) = -1 + 2\cos\phi$ for improper rotations

For the C_{2v} point group,

$N_{A_1}(M) = \frac{1}{4}[1 \times 3 \times 1 + 1 \times (-1) \times 1 + 1 \times 1 \times 1 + 1 \times 1 \times 1] = 1$

$N_{A_2}(M) = \frac{1}{4}[1 \times 3 \times 1 + 1 \times (-1) \times 1 + 1 \times 1 \times (-1) + 1 \times 1 \times (-1)] = 0$

$N_{B_1}(M) = \frac{1}{4}[1 \times 3 \times 1 + 1 \times (-1) \times (-1) + 1 \times 1 \times 1 + 1 \times 1 \times (-1)] = 1$

$N_{B_2}(M) = \frac{1}{4}[1 \times 3 \times 1 + 1 \times (-1) \times (-1) + 1 \times 1 \times (-1) + 1 \times 1 \times 1] = 1$

Thus the vibration types A_1, B_1 and B_2 are all i.r. active while A_2 types are inactive.

Raman activity is determined in a similar manner with $\chi_m(R)$ replaced by $\chi_\alpha(R)$, the character of the polarizability for the operation R. Full tables calculated for all the point groups may be found in Woodward (1972).

2.1.14. OVERTONE AND COMBINATION BONDS

The selection rules for the overtone and combination bands may be deduced from the direct product of the characters of the fundamentals. If the character product of the overtone or combination is the same as the character of one of the allowed fundamentals, then the overtone is also allowed. For example, for the A_2 vibration of the C_{2v} point group, the fundamental vibration is forbidden in the i.r. and is characterized by $+1, -1, -1$ for the symmetry operations C_2, σ_v' and σ_v''.

	I	C_2	σ_v'	σ_v''
χ_{A_2}	1	1	-1	-1
χ_{A_2}	1	1	-1	-1
$\chi_{A_2}^2$	1	1	1	1

Thus the character of the first overtone is the same as the character of the fundamental A_1 which is an allowed vibration; the first overtone of the A_2 vibration is therefore also allowed, as is the first overtone of the A_1 vibration. However, the second overtone of A_2 is forbidden, since further multiplication by $1, -1, -1$, gives a character identical with the forbidden fundamental. Thus only odd-numbered overtones of A_2 vibrations are owed.

Combination bands are found in a similar manner. For a combination of A_1 and A_2 vibrations, it follows that:

	I	C_2	σ_v'	σ_v''
χ_{A_1}	1	1	1	1
χ_{A_2}	1	1	-1	-1
$\chi_{A_1 A_2}$	1	1	-1	-1

This has the same character as the type A_2 fundamental and is forbidden.

Of importance in the water spectrum are combinations of A_1 and B_2

	I	C_2	σ_v'	σ_v''
χ_{A_1}	1	1	1	1
χ_{B_2}	1	-1	-1	1
$\chi_{A_1 B_2}$	1	-1	-1	1

The combination $A_1 \times B_2$ has the same character as the allowed B_2 fundamental and is therefore allowed.

For the C_{2v} point group, all combinations and overtones are allowed except $A_1 \times A_2$, $B_1 \times B_2$ and A_2^n, where n is odd. For water, all overtones are possible since there are no vibrations associated with A_2 or B_1 vibration types. Such conclusions are equally relevant to i.r. and Raman activity. For fuller treatment of vibrational spectroscopy, see Woodward (1972).

2.2 Experimental aspects

2.2.1. POLYMER SPECTRA

Since a molecule containing N atoms has $3N-6$ modes of vibration (Section 2.1.7), one might expect that a high-molecular-weight polymer with N very large would yield a spectrum which is too complex for interpretation. To some extent, the reverse is the case in that, e.g. in $(glycyl)_n$ glycine, fewer absorptions are detected as molecular weight increases. The explanation requires that the molecule XAY must be considered in terms of (a) its end groups (X and Y) and (b) its non-end-group chemical structure (A). Now the two-unit polymer of this molecule, with the structure $X-A-A-Y$, would have its total absorption made up from end groups X and Y together with two A groups; and similarly the trimer $X-A-A-A-Y$, has the end groups together with three A groups. For a high-molecular-weight homogeneous polymer, $X-A-(A)_n-A-Y$, the effective absorption of the end groups becomes insignificantly small. The total absorption exhibited by group A is influenced by the end groups X and Y in the molecule $X-A-Y$, and in the molecule $X-A-A-Y$ by adjoining A as well as by X or Y. At high molecular weights, the great majority of the A groups, influenced only by their two neighbouring A groups, will absorb in identical ways. Consequently, the absorption spectrum may be the relatively simple one of the unit A.

The frequency of an absorption is governed by the bond strength and length and by the masses of the atoms present in the absorbing group. In addition, the absorption coefficient depends on the dipole moment, $q.l$, where l is the distance between centres of positive and negative charge and q is the magnitude of that charge. While these factors are usually considered to be independent, they are in reality inter-related. As is shown in Section 2.4, if bond strength decreases as a consequence of molecular interactions, bond length will increase and absorption frequency decreases. At the same time, however, an increase in dipole moment is likely due to the change in l. The decreased frequency can therefore be accompanied by an increase in the optical density (pp. 31 and 35) of the band.

While band assignments can be made from mathematical calculations based on atomic masses, force constants, etc., the majority of peaks in large molecules are assigned from experimental observations. Certain classes of compounds exhibit absorptions for specific groups that occur within a narrow frequency range (and, as such, are in many cases specific). However, the range can be so modified by molecular interactions, for example H bonding, and the effects of crystallinity, that it may be difficult in practice to differentiate between H-bonded hydroxyl groups and H-bonded NH groups as a consequence of differences in H-bond strength. While the majority of characteristic vibrational modes arise from side groups or end groups, absorptions in the 1400–1000 cm^{-1} region arise from skeletal modes which appear sufficiently unique to be termed "fingerprint spectra".

2.2.2 EXPERIMENTAL METHODS

2.2.2.1. *The spectrometer*

A spectrometer is an instrument for recording variations in intensity of radiation when some other parameter, such as frequency or magnetic field, is varied. An i.r. spectrometer, therefore, plots a graph of light intensity against frequency in cm^{-1} or, in some older instruments, against wavelength in μm. Most commercial i.r. spectrometers of whatever complexity operate within the range 20 000–400 cm^{-1}; the optical arrangement of a typical instrument is shown in Fig. 2.1.

Any spectrometer requires (a) a source of radiation, (b) a radiation detector and (c) a device to select the conditions (e.g. frequency) at which the detector gives a measure of the intensity of the radiation incident on it. In the i.r. region, the radiation source is usually a Nernst filament,* made from a mixture of rare-earth oxides which, when heated above 800°C by passing an electric current through it, emits a range of frequencies in the i.r. and visible regions. While the detector is usually a thermopile, some instruments employ a Golay detector or a bolometer; this is incorporated into a monochromator unit which, with the use of a prism or grating, diffracts the radiation so that the detector measures the intensity of the radiation incident on it within a narrow frequency range. Interferometers (Section 4.2.2.3) are now available for the i.r.

The optical diagram (Fig. 2.1) can be considered in two sections: the radiation unit (s to f); and the spectrometer (g to v). In the radiation unit, i.r. light from the source s is split into two equivalent beams, of equal optical path length, by reflection from mirrors a, b, c, d, e and f and a, b', c', d', e' and f. These two beams are then combined by the reciprocating mirrors f which reflect each beam alternately through the monochromator entrance slit g. "i" is the diffraction grating, n is a KBr prism which separates other orders of diffraction, while the widths of the slits g, l and r determine in part the performance of the instrument. Mirrors h, j, k, m, o, p, q, t and u reflect the light to the detector v and the whole unit is enclosed by a cover, the inner surface of which is blackened to reduce scattered stray light to a minimum. Now the reciprocating mirrors f accept light alternately from path $a, b \ldots f$ and $a\ b' \ldots f$ so that, if no light is absorbed or if light is absorbed equally in both beams, then the detector output will be constant in magnitude. However, if light is preferentially absorbed in beam $a\ b' \ldots f$, then the output of the detector will have an alternating component; the frequency will depend on the frequency of reciprocation of mirrors f, and the intensity will depend on the amount of light absorbed. Hence, if an absorbing sample is placed in beam $a\ b' \ldots f$, the detector will give an alternating output. This is amplified, rectified and used to control a servo motor which drives a comb into beam $a\ b \ldots f$ until the intensity balance is restored, when the detector output has no a.c. component. When the absorption by the sample is decreased, an alternating component is again produced, but of opposite phase, so that the servo motor is energized to drive the comb out of the beam until balance is again restored. As the percentage of light transmitted by the comb is a linear function of displacement, its position at balance provides a direct measure of the absorption of the sample (Section 2.2.2.2). The beam in which the sample is placed is termed the sample beam and that attenuated by the comb is the reference beam.

* Alternatively a SiC rod or a nichrome helix embedded in a ceramic sheath is used.

The servo motor is also directly coupled to a wire-wound potentiometer; from this, a voltage proportional to the sample transmission is fed to a potentiometric chart recorder. The frequency at which the detector operates is varied by rotation of the diffraction grating i to give a constant rate of frequency change, i.e. $\dfrac{d\tilde{v}}{dt}$ = constant. Thus, a graph of servo-potentiometer output against time by the chart recorder is equivalent to a graph of sample absorption plotted against wavelength since, if the chart speed is $\dfrac{dz}{dt}$ m s^{-1}, then $\dfrac{d\tilde{v}}{dt} \propto \dfrac{dz}{dt}$.

The effective operation of a double-beam spectrometer involves the comparison of the sample and reference beams. While this is not essential in a spectroscopic system (Section 1.5.2), single-beam i.r. instruments, which record the absorption of the sample only, are now uncommon because water vapour and carbon dioxide (present within the case of the instrument) unfortunately absorb strongly in the frequency regions of greatest interest. It is impracticable to reduce the concentration of the water vapour and CO_2 to a low level but, in an optically balanced double-beam system, the absorption due to these entities is the same in both beams. However, it must be recognized that, in spectral regions where absorptions in each beam are balanced out, the sensitivity of the spectrometer is diminished due to the lower intensity of the radiation incident on the detector. In practice, however, the resolving power of an instrument (Section 1.5.1) will depend on its ability to make the range of frequencies incident on the detector as narrow as possible.

2.2.2.2. Optical density

The absorption of light by a material may be defined (Section 1.4.1) by the Beer-Lambert relation

$$D = \log_{10}(I_0/I) = \varepsilon c l$$

where, *at a given frequency*,

D is the *optical density* or *absorbance* (Section 5.3.3);
I_0 is the intensity of light entering the sample;
I is the intensity of light leaving the sample;
ε is the molar extinction coefficient;
c is the concentration of absorbing material in the sample;
l is the path length through the sample.

The absorption of a sample plotted against the frequency can be expressed in terms of optial density, D, as above or in terms of *transmission*, T (Fig. 2.2), or *absorption*, A,

$$\text{i.e. } T = I/I_0 \quad \text{and} \quad A = \frac{I_0 - I}{I_0} = 1 - T$$

Hence $D = \log_{10} 1/T$.

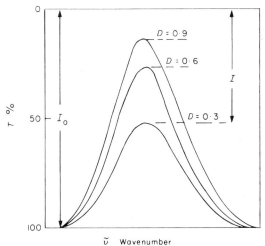

Fig. 2.2. Transmission and optical density.

As an illustration, the $D = 0.3$ curve in Fig. 2.2 shows that 50% of the incident light has been transmitted and 50% has been absorbed by the sample. If a second identical sample is also placed in the sample beam, it will absorb 50% of the remaining light so that the net percentage transmission will now be 25% and percentage absorption 75%. Similarly, if a third sample is added, T and A will be $12\frac{1}{2}$% and $87\frac{1}{2}$%, respectively. It is clear, therefore, that the percentage of the light transmitted decreases non-linearly with increase in the amount of absorbing material in the sample. However, in terms of absorbance D, the corresponding values are 0.30, 0.60 and 0.90, respectively, i.e. there is a linear relation between sample thickness or quantity and optical density. More generally, when several absorbing species are present, the absorbance due to each species at a given frequency can simply be summed arithmetically. Since most instruments are precalibrated in terms of percentage transmission, values of D must be calculated from measurements of the spectra.

Before a sample is inserted, the spectrometer should be adjusted to give accurate values of 0% and 100% transmission. Few samples are completely free from reflection and light scatter at any wavelength so that the absorption curve is superimposed on a continuous background absorption. Several methods have been devised to overcome this problem. First with liquid samples, one can use matched pairs of cells (Section 2.2.2.6), one of which is placed in the reference beam to ensure that the same background absorption is present in both beams, so that only the absorption of the solute is recorded by the spectrometer.

Second, when quantitative measurements are made (Fig. 2.3), the point method of intensity measurement is often used if the background absorption is unchanging throughout a series of measurements. In this, the percentage transmission or absorbance is measured directly from the spectrum at the absorption maximum and includes the background absorption. However, when a calibration graph of peak height D against concentration c is plotted for a series of samples of known but differing concentrations, the constant background absorption will lead to a graph of the form $D = (\varepsilon l)c + b$, where b is the background absorbance at zero concentration. Since the path length l is known, the extinction coefficient ε can be determined from the gradient, εl, of this graph.

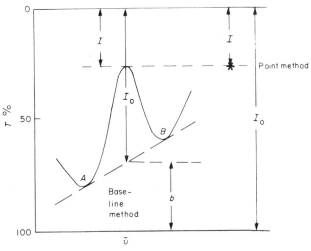

FIG. 2.3. Methods of peak intensity measurement.

The third method, often used when the background absorption is likely to vary or when solid samples are being studied, is the baseline method. This (Fig. 2.3) assumes that the background on which the absorption is superimposed approximates to a straight line. Measurements of I_0 and I are made from this line; any error due to deviations from this assumption will be systematic and hence will be accounted for in any calibration curve based on this method. A more likely source of error occurs when the positions of A and B vary markedly because of a rapidly changing background; this error may be random due to inadequate spectrometer performance.

2.2.2.3. *Polarized light*
Whilst the direction of change of the dipole moment of a bond is unimportant when molecules are randomly aligned, when molecules are in regular orientation, the direction of change may be along a preferred direction. Since the interaction between the changing dipole and the electric vector E of the incident radiation will be greatest when these are colinear, the proportion of the incident light absorbed for an ordered system with polarized i.r. light will vary as the plane of polarization is rotated about axes of symmetry.

For investigations of crystalline or molecularly oriented substances, two spectra are recorded, one with E parallel and one with E perpendicular to the crystal axis (or direction of elongation in stretched specimens). The two spectra differ particularly in the regions of peak maxima where, in one direction, the absorbance will be higher than in the other, depending on the directional alignment of the bond concerned. A measure of this effect is given by the *Dichroic Ratio*, R, the ratio of the optical density, D, with E parallel to the crystal axis, $D_\parallel$, and perpendicular to the crystal axis, $D_\perp$, i.e. $R = D_\parallel/D_\perp$. Hence, when R of a band is greater than unity, the corresponding bond tends to be aligned parallel to the axis and, when less than unity, perpendicular alignment is indicated.

Infra-red light can be polarized conveniently in three ways. Transmission polarizers have been widely used and consist of a pile of transparent plates inclined at the Brewster or polarizing angle to the beam, so that the angle of incidence θ is related to the refractive index, n, of the plate material by the relation

$$n = \tan \theta$$

Selenium (Elliott, 1969) with a refractive index of about 2·5 and silver chloride (Fraser and Price, 1952) with a refractive index of about 2·0 have been used in the construction of such devices; in Fig. 2.4, there are six plates. As the light is incident on a given plate, a fraction of the radiation with E parallel to the plane

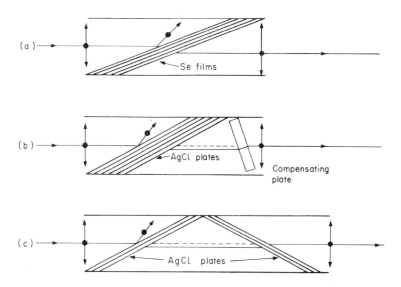

FIG. 2.4. Transmission i.r. polarizers used in dichroism measurements. Lateral displacement of incident beam in (a) can be corrected by compensating plate (b) or equal numbers of polarizing plates mounted in opposition (c). Double arrowheads represent planes of E vibration (Alexander and Block, 1960).

of the plates is reflected away while the remainder is transmitted. After transmission through six selenium plates, the fractional polarization is 0·98 and the percentage of the original beam transmitted is 44%. Slightly lower values are obtained with the more robust silver-chloride sheets.

Reflection polarizers produce a high degree of polarization (0·995) but only at the cost of light intensity (37% of the incident-beam intensity). Selenium or germanium have been used alone but more useful types incorporate a deposit of the element on a glass support. In Fig. 2.5, a single polarizing reflection only is utilized; the remainder occur at aluminized surfaces to maintain the intensity of the i.r. beam.

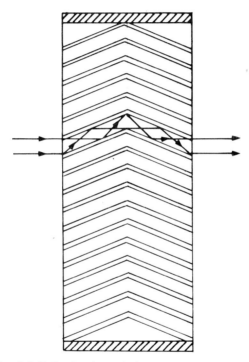

FIG. 2.5. Reflection i.r. polarizer (Takahashi, 1961).

A new commercial polarizer, with good performance, consists of uniformly spaced parallel strips of metallic gold or aluminium deposited on a silver-chloride or KRS-5 (thallium-iodobromide) disc substrate. This system acts as an efficient polarizer and, when used at normal incidence, produces no lateral displacement.

Since all spectrometer grating dispersion systems polarize the (i.r.) light entering the instrument, the measured absorbances in a double-beam instrument will be minimized if the polarizer is mounted either before sample and reference beams are split or after their recombination. Further, for efficient polarization, the beam must be parallel, so that the polarizer site must be one of small beam divergence or convergence. In Fig. 2.1, a grid polarizer is sited between slit *r* and plane mirror *t*.

2.2.2.4. *Differential techniques*

In principle, a small quantity of impurity, either in solid films or in solutions of biological and other materials, should be detectable by placing a sample of the pure material in the reference beam of the double-beam instrument. With similar amounts of the absorbing substance in both beams, then in theory the spectrum should be a horizontal line, due to compensation of the absorptions; alternatively, when an impurity is present in the sample located in the sample beam, the spectrum of the impurity should be superimposed on the compensation line.

The technique is extremely delicate and, for solutions, a variable-path-length cell containing pure solvent is inserted in the reference beam. For solid specimens, a wedge shaped reference sample is mounted laterally and adjusted in and out of the i.r. beam by means of a fine thread screw.

Differential spectroscopy is also useful for detection of a small change in absorption frequency resulting from chemical or physical modification of structure or from interaction with a liquid used as a solvent or as an immersion medium (to reduce the amount of light scattered by solid samples). Figure 2.6

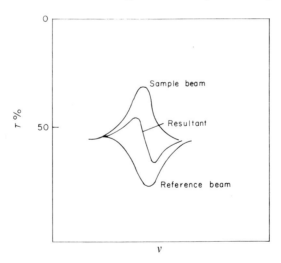

FIG. 2.6. Frequency displacement of absorption in sample and reference beams.

shows spectrometer plots of the resultant of two identical but frequency-displaced bands in the sample and reference beams; since this resultant is similar to the first derivative of an absorption (Section 2.2.2.5), the frequency displacement can be clearly identified.

2.2.2.5. *Double-wavelength and derivative spectroscopy*

Infra-red spectra of polymers, and particularly of biological polymers, arise predominantly from the repeating unit of the polymer molecule (Section 2.2.1). Absorptions from end-of-chain groups, side chains and absorbed water are not generally observed, owing to their low intensity and to the fact that such absorptions often overlap to produce an unresolved absorption which frequently broadens a fundamental mode without resolved band structure. Further, a few broad peaks may overlap to produce a single broad unresolved absorption. Problems such as these, together with those concerned with the detection of component impurities, have led to the development of derivative techniques (Keighley and Rhodes, 1971, 1972) applicable to biological materials. While the complex shape of an i.r. absorption curve cannot be represented exactly by a single mathematical model, two functions approach this, namely (a) Gaussian and (b) Lorentzian:

$$(a) \quad D_1 = a_1 \exp\{-b\tilde{v}^2\}$$

and

$$(b) \quad D_2 = \frac{a_2}{1 + a\tilde{v}^2}$$

where a is the amplitude, $\tilde{v}$ is the wavenumber, D is absorbance, $b = \dfrac{4 \log_e 2}{t^2}$ and $a = 4/t^2$, where $2t$ is the absorption width at half height.

Then

$$\frac{dD_1}{d\tilde{v}} = -2\tilde{v}a_1 b \exp(-b\tilde{v}^2)$$

$$\frac{d^2 D_1}{d\tilde{v}^2} = 2a_1 b(2b\tilde{v}^2 - 1)\exp(-b\tilde{v}^2)$$

$$\frac{dD_2}{d\tilde{v}} = -\frac{2a_2 a\tilde{v}}{(1 + a\tilde{v}^2)^2}$$

and

$$\frac{d^2 D_2}{d\tilde{v}^2} = \frac{2aa_2(3a\tilde{v}^2 - 1)}{(1 - a\tilde{v}^2)^3}$$

For both Gaussian and Lorentzian functions, graphs of $\dfrac{d^2 D}{d\tilde{v}^2}$ approximate sufficiently closely to the second derivative of an absorption curve (Fig. 2.7), especially in the central region, for either mathematical model to be useable. Changes in the precise shape of an experimental absorption vary with the resolution of the spectrometer as a consequence of, for example,

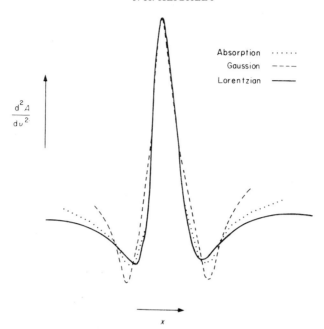

FIG. 2.7. Second derivative of theoretical (Gaussian and Lorentzian) and experimental band shapes (Keighley and Rhodes, 1971).

changes in slit width. Hence differences between experimental curves and those derived from mathematical functions can be ignored.

The experimental curve has been expressed (Fraser and Suzuki, 1966) in terms of the functions D_1 and D_2 (p. 41):

$$D = (1-F)D_1 + FD_2$$

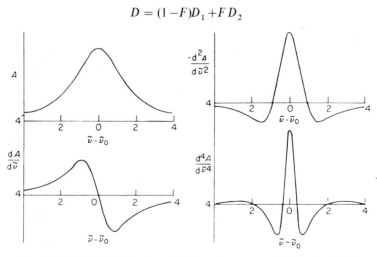

FIG. 2.8. Multiple derivatives of absorption curve, A (Martin, 1959).

where F is the fraction of the Lorentzian component contributing to the total absorption; changes in F will have little effect on the shape of the second derivative curve.

From Fig. 2.8, it is clear that even derivatives producing maxima which correspond to the absorption peaks are more useful than odd derivatives. Despite narrower lines, the fourth derivative is relatively complex and little used. The improved resolution of the second-derivative system results from the reduction of the width at half height to approximately one-third of that of the

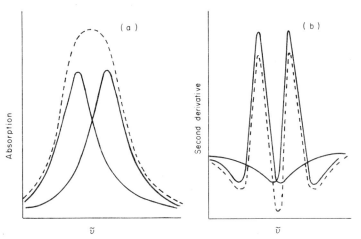

FIG. 2.9. Resolution of doublet by differentiation: (a) apparent singlet in absorption curve yields resolved doublet, (b) in derivative mode. Resultants are shown as dashed lines (Keighley and Rhodes, 1971).

parent absorption. Figure 2.9 shows the effective resolution by differentiation of an apparent singlet formed from two unresolved components.

While the shape of an absorption curve (D or T) will alter slightly if it is changed from linear frequency to linear wavelength, such distortions are not usually significant. Accordingly, the conclusions reached below for frequency apply equally to wavelength.

When an absorption D' arises from two overlapping bands,

i.e. $\quad D' = f(\tilde{v}) + f(\tilde{v} + \Delta \tilde{v})$

where $\Delta \tilde{v}$ is the frequency separation of the two bands,

then $\quad \dfrac{d^2 D'}{d\tilde{v}^2} = f''(\tilde{v}) + f''(\tilde{v} + \Delta \tilde{v}).$

Thus each second-derivative peak appears near the frequency of the absorption maximum, and the frequency separation of the two maxima which arise on differentiation is close to the true separation of the absorbing species. While this conclusion is generally valid, factors which lead to deviations will be discussed later in this section.

Spectra from overlapping absorptions exhibit at a given frequency an absorbance D,

$$D = d_1 + d_2 + d_3 + d_4 + \ldots d_n,$$

where d_1 to d_n are the optical densities of the component bands 1 to n. Hence

$$\frac{d^2 D}{d\tilde{v}^2} = \frac{d^2 d_1}{d\tilde{v}^2} + \frac{d^2 d_2}{d\tilde{v}^2} + \frac{d^2 d_3}{d\tilde{v}^2} + \ldots \frac{d^2 d_n}{d\tilde{v}^2},$$

i.e. the height of a second-derivative curve is equal to the sum of the heights of the individual components provided measurements are made in terms of optical density. When, as in many instruments, the spectrometer is calibrated in terms of percentage transmission, T, different conditions apply.

Now $T = I/I_0$ and $D = \log_{10} 1/T$ so that $T = 10^{-D} = \exp\{-2 \cdot 303 D\}$ and

$$\frac{dT}{dD} = -2 \cdot 303 \exp(-2 \cdot 303 D) = -2 \cdot 303 T.$$

For the Lorentzian function,

$$\frac{dD_2}{d\tilde{v}} = \frac{-2 a_2 a \tilde{v}}{(1 + a \tilde{v}^2)^2}$$

However

$$\frac{dT}{dv} = \frac{dT}{dD} \cdot \frac{dD}{dv}$$

$$= +(2 \cdot 303 T) \left[\frac{2 a_2 a \tilde{v}}{(1 + a \tilde{v}^2)^2} \right]$$

$$= \frac{4 \cdot 606 a_2 a \tilde{v}}{(1 + a \tilde{v}^2)^2} T$$

From this it can be shown that, in order to use $\dfrac{dT}{d\tilde{v}}$ in place of $\dfrac{dD}{d\tilde{v}}$, it is necessary to limit the application to absorbance regions in which these are proportional, i.e. about 15%–75% T. Similar results are obtained when second derivatives are considered. Compliance often necessitates examination at several concentrations or cell-path lengths for derivatives based on transmission curves.

Differentiation can be accomplished either electronically (early valve instruments were noisy) or optically. In the latter case, two identical sources mounted close together are alternately covered by a rotating commutator so that the exit slit receives two signals from two close spectral frequencies. The alternating component which arrives at the detector is proportional to $\dfrac{dD}{d\tilde{v}}$, the momentary gradient of the spectrum, much as indicated in Fig. 2.6. Such spectra are thus only first derivative; other methods are required for second-derivative spectra. Commercial visible-u.v. spectrometers based on the above system to produce the first derivative of the absorption are now available.

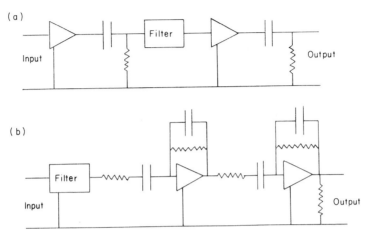

FIG. 2.10. Differentiating circuits.

An alternative system, which can be used as an addition to most commercial instruments, is based on differentiating electronic circuitry. Of the two circuits shown in Fig. 2.10, (a) utilizes the CR circuit whilst (b) utilizes a well-known property of integrated circuits.

Figure 2.11 shows how electronic differentiation methods accentuate the small absorption differences between dry and wet samples of fibrous proteins. While no substantial information has previously been obtained to show the presence of amino groups at the ends of polypeptide chains and amino-acid residues, the presence of non-H-bonded NH and NH_2 groups is clearly indicated in the 3400 cm^{-1} region, in addition to the presence of strongly bound water molecules. Changes in the absorption of absorbed water molecules are also evident. While this subject cannot be dealt with here, (Keighley and Rhodes, 1971), the use of derivative techniques has led to other problems associated with band overlap; this applies equally to the absorption and the derivative modes, and is particularly important in studies of polymers, whose absorption bands are frequently composed of several overlapped absorptions.

Figure 2.12 shows the components and resultant of an unresolved composite band; it is clear from the marked positions of the components that the resultant maxima do not correspond to the frequencies of the maxima of the individual components. In such an absorption band, therefore, the recorded maxima are closer together than the peaks from which they are derived.

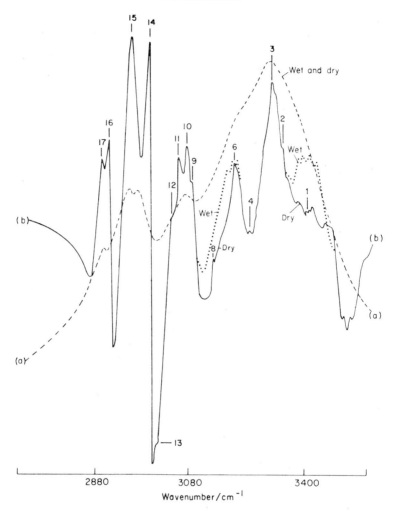

FIG. 2.11. Absorption (a) and second-derivative (b) spectra of keratin.

The family of curves, Keighley and Rhodes (1971) based on the Lorentzian function (Fig. 2.13) compares values of amplitude (a) and half width (c) at different band separations; subscripts refer to band numbers.

Curve	a_1	c_1	a_2	c_2
1	5	0·5	1	0·5
2	5	0·5	1	0·3
3	5	0·5	1	0·1
4	5	0·5	1	0·01

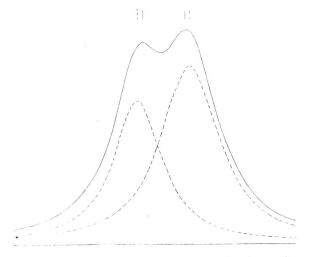

Fig. 2.12. Relative band shift of two overlapping peaks.

From the above table, the shift in frequency of the main peak (Fig. 2.13) is evidently greatest when the peaks are of equal half-width and are near to coincidence.

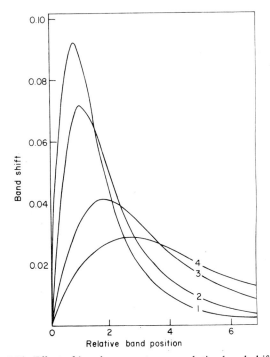

Fig. 2.13. Effect of band parameters on relative band shift.

The derivative-peak positions thus represent the positions of the parent absorptions, provided that there is no major overlap with a neighbouring derivative peak. While this phenomenon is understood in other branches of spectroscopy, rather little attention has been paid to it in the i.r. region. In many studies of biological materials, band shifts which have been ascribed to structural changes consequential to chemical treatment have actually arisen as a consequence of the removal of an overlapping band.

2.2.2.6. Specimen preparation

Biological materials may be examined in liquid, solution or solid form. Highly polar solvents such as water are unsuitable, in view of their intense absorption of incident i.r. light over much of the spectral region. Sample cells are constructed with windows made from materials such as alkali halides (NaCl and KBr are the most widely used), calcium fluoride, silver chloride and KRS-5 (made from thallium iodobromide). The path length of the cell is determined by a spacer of PTFE or lead; in most cases, inlet and outlet ports suitable for connection to a syringe are provided. Variable-path-length cells are also available and in all cases care of the window material is necessary.

Solid samples often present problems in preparation for examination. If the solid can be dissolved in a suitable solvent, a film may be cast from the solution. Often, modification of the structure of the polymeric film prepared results from inclusion of small quantities of residual solvent and this leads to anomalies in the recorded spectrum. In order to circumvent this, sections of some solids can be cut with a simple rocking microtome using a 10° or 90° blade with the sample embedded in epoxy resin, but artefacts, in the form of regions of compression due to the sectioning process (as a consequence of blade "chatter"), may give rise to erroneous results as, for example, when dichroic measurements are being made.

Powdered samples can be examined in one of two ways. In the older technique, the sample is ground to a fine powder, of particle size less than 0·5 μm diameter, to ensure good resolution and low light scatter, and is mixed to a thick paste or mull with a liquid medium which does not absorb significantly in the frequency range to be studied. The mull is placed between two rock-salt windows and compressed to a suitable thickness. Liquid paraffin and carbon tetrachloride are often used; these exhibit only C—H and C—Cl vibrations, respectively, which do not usually interfere significantly.

In the alternative KBr-disc technique, a small quantity of the powdered sample is mixed intimately with approximately 150 mg of potassium bromide. After thorough mixing in a high-speed ball mill, the mixture is placed in a die (Fig. 2.14), evacuated and pressed at approximately 10^8 Nm^{-2} to produce a clear transparent disc. Although this method is convenient, interactions can occur between the sample and matrix, which produce variations in the spectrum, e.g. proteins in the 3300 cm^{-1} region. In spite of this, however, the

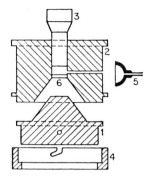

FIG. 2.14. KBr-disc die: (1) anvil, (2) body, (3) piston, (4) clamping ring, (5) suction cup, (6) pressed disc (Kopf, 1964).

KBr-disc technique is perhaps the most widely used in i.r. spectroscopy.

Fibrous polymers have also been studied (Baddiel, 1968) using attenuated total reflection (A.T.R.), a technique whereby the i.r. beam is made to reflect on the inner face of a transparent highly refractive prism (Fig. 2.15a), while the sample is placed in contact with the prism surface at which reflection takes place. The angle of incidence, θ, of the beam is chosen so that complete reflection occurs and so that $\sin \theta > n_1/n_2$ where n_2 is the refractive index of the prism and n_1 is the real component of the sample index. The reflected beam

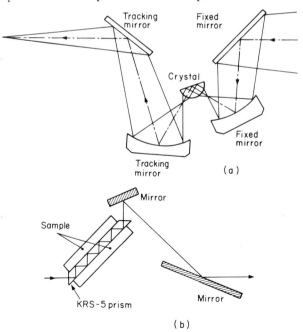

FIG. 2.15. ATR units: (a) single-reflection crystal; (b) multi-reflection prism.

suffers little intensity loss with a transparent sample but, in the i.r. region, the sample will absorb some of the incident light at specific frequencies. The beam only penetrates the surface of the sample, to an extent controlled by variations in θ. While the recorded A.T.R. spectrum closely resembles an absorption spectrum, shifts in specific characteristic frequencies and variations in intensity from the absorption mode are often found.

A.T.R. spectra have an advantage in that they can be recorded without destruction or alteration of a sample; good contact of the sample with the crystal surface is desirable and is best achieved from a film of material. The technique is also useful for the study of samples in which the surface structure differs from that of the bulk of the sample. Single (Fig. 2.15a) or multi-reflection crystals giving up to 25 reflections are available (Fig. 2.15b).

2.3. General factors relevant to the investigation of molecular structure by i.r. spectroscopy

2.3.1. DEUTERIUM EXCHANGE

If an atom present in a molecule is replaced by one of its isotopes, the electronic-charge distribution within the molecule is unaltered, so that the frequencies of vibrations are affected only by the change in mass of the atom. From Section 2.1.5, the stretching wavenumber $\tilde{v} = \dfrac{1}{2\pi c}\sqrt{\left(\dfrac{k}{\mu}\right)}$, where μ is the reduced mass of the system. If the vibrations are restricted to H atoms and the atoms to which they are bonded, then the frequency will depend on the inverse of the square root of the reduced mass. Hence for an N—H group deuteriated to form N—D, the reduced mass will change from μ_H to μ_D, where

$$\mu_H = \frac{\text{mass }(N \times H)}{\text{mass }(N+H)} = \frac{14 \times 1}{14+1} = \frac{14}{15} \quad \text{and} \quad \mu_D = \frac{\text{mass }(N \times D)}{\text{mass }(N+D)} = \frac{14 \times 2}{14+2} = \frac{28}{16}$$

$$\therefore \frac{\tilde{v}_H}{\tilde{v}_D} = \frac{1}{2\pi c}\sqrt{\left(\frac{k}{\mu_H}\right)} \cdot \frac{2\pi c}{1}\sqrt{\left(\frac{\mu_D}{k}\right)} = \sqrt{\left(\frac{\mu_D}{\mu_H}\right)} = \sqrt{\left(\frac{28}{16} \times \frac{15}{14}\right)} = \sqrt{\left(\frac{30}{16}\right)}$$

so that the wavenumber change will theoretically be 1·369. Similar calculations for C and O give changes of 1·363 and 1·374, respectively; since other neighbouring atoms are slightly involved, the measured value is usually near 1·33. Analogous changes may be determined for deformation modes but, in many cases, such vibrations are coupled to other modes of similar frequencies, so that deuterium exchange has a smaller effect than expected.

In biological polymers, the exchange of H atoms of OH, NH and SH groups can be effected by exposure of the polymers to deuterium-oxide vapour (to the exclusion of water vapour) at room temperature or at elevated temperatures, or by immersion directly in deuterium oxide or other deuteriated compounds, such as deuteriated fatty acids, e.g. CH_3COOD. Since CH groups do not readily exchange with deuterium, C—D vibrations can only be examined from specifically synthesized materials; conversely, CD groups are stable in presence of water vapour.

Deuterium substitution is a powerful technique for checking the assignments of bands associated with a particular vibration, e.g. in studying molecular order and disorder in both polypeptides and fibrous proteins (Section 2.4).

In the same way, tritium may be used to produce a bigger frequency shift, but it is radio-active and little additional information has been obtained from the few studies carried out, e.g. on wool (Keighley and McKinley, unpublished measurements).

2.3.2. HYDROGEN BONDING

In a hydrogen bond, $C=O \ldots H—N$, the reduced effective restoring force to the stretching of the N—H bond decreases its frequency of vibration and, conversely, the deformation frequency is increased.

Spectra recorded for a range of concentrations of N-substituted and unsubstituted amides (Richards and Thompson, 1947) in a non- or weakly-

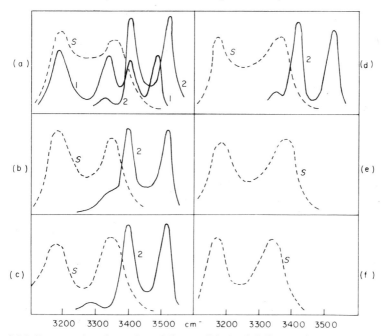

FIG. 2.16. Spectra of amines: dotted lines solid (S); full line (1) in (a) concentrated solution $CHCl_3$; full lines (2) dilute solution in $CHCl_3$. (a) hexoamide; (b) butyramide; (c) phenylacetamide; (d) benzamine; (e) oxamide; (f) succinamide. (Richards and Thomson, 1947.)

polar solvent, such as chloroform, showed four different bands (Fig. 2.16) attributed to the various states of the NH group. In the solid state, strong absorptions at 3180 and 3350 cm^{-1} were assigned to H-bonded NH groups, whilst in dilute solution, peaks at 3400 and 3520 cm^{-1} arose from antisymmetrical and symmetrical "free" NH groups, respectively. All four bands were found in concentrated solution, indicating the presence of both free and H-bonded species.

Since the change in frequency of a stretching vibration is proportional to the strength of the H bond formed, the presence of different structural conformations in a natural material such as protein will be indicated by the number and frequency of H-bonded groups absorbing in a given region of the spectrum. (This characteristic has been particularly useful in studies of proteins.)

2.3.3. DICHROISM AND CRYSTALLINITY MEASUREMENTS

Quantitative measurements of dichroism using polarized light produced by polarizers (see Section 2.2.2.3 and Figs 2.4 and 2.5) can help to confirm or invalidate peak assignments. Even if biological polymers were perfectly oriented, the charge displacements or *transition moments* (Section 1.2.3) responsible for i.r. absorption would not coincide exactly with bond directions, as a consequence of vibrations of neighbouring groups. Furthermore, real polymers involve molecules in differing degrees of molecular order so that, from the point of view of X-ray diffraction, some regions appear "ordered" whilst other "disordered" regions appear to consist of randomly ordered molecules. The way in which such regions are combined is not always known, but the regions of differing molecular order often absorb i.r. light at slightly different frequencies owing to differing molecular environments. Consequently, the observed dichroic ratio will be the sum of the high dichroism of the ordered regions and the lack of dichroism of the disordered regions.

In this way, information can be obtained about degree of crystallinity and spatial arrangements of the chains. Clearly, however, it must be assumed either that the transition moment is oriented at a given angle to the bond (and hence the extent of order within the structure of a polymer must be calculated) or one must assume perfect order and calculate the angle between the bond and the transition moment.

For the second possibility, with the transition moment inclined at an angle α to the polymer-chain direction and distributed evenly around the chain axis, then a dichroic ratio can be defined by $R = 2\cot^2\alpha$. Hence, if $\alpha = 54° 44'$, there is no dichroism (since $R = 1$); for smaller angles, there is parallel dichroism ($R > 1$) and, for larger angles, there is perpendicular dichroism ($R < 1$). Thus, from measurements of dichroic ratio, R, for an absorption peak, α, can be calculated.

If an orientation factor f is defined as the fraction of the polymer which is perfectly oriented (the remainder is randomly oriented), then the dichroic ratio R is given by (Fraser, 1953)

$$R = \frac{f\cos^2 a + \frac{1}{3}(1-f)}{\frac{1}{2}f\sin^2 a + \frac{1}{3}(1-f)}$$

Thus, for a known value of a and a measured value of R, the calculated value of f can be used to determine a for the transition moment of another absorption in the same spectrum. However, it has been shown (Fraser and Suzuki, 1964) that $f = \frac{(R-1)(R_0+2)}{(R+2)(R_0-1)}$, where R is the measured value of the dichroic ratio, and R_0 is that of a perfectly aligned chain. Since $R_0 = 2\cot^2 a$, f may be calculated from the slope of R_0 of a graph of f plotted against $(R-1)/(R+2)$.

Since C—H vibrations are relatively unaffected by neighbouring groups, the transition moment lies in the same direction as the C—H bond. Thus, for symmetric and antisymmetric vibrational modes of the CH_2 group (Fig. 2.17), the resultant change in transition moment bisects the H—C—H angle for the

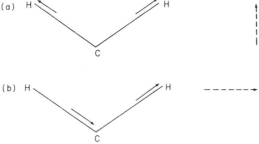

FIG. 2.17. Symmetric (a) and antisymmetric (b) modes of single CH_2 group. Broken arrows indicate resultant direction of transition moment.

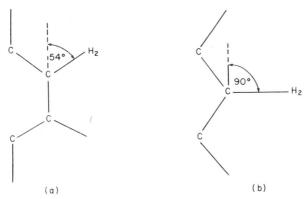

FIG. 2.18. Inclination of transition-moment direction (CH_2 plane) with chain axis for two chain conformations (a) and (b).

symmetric vibration and is at right angles to this for the antisymmetric vibration, so that the dichroic properties of each absorption differ. Figure 2.18 shows two different aliphatic chains. In (a), the symmetric C—H transition moment is at 54° to the chain axis and the antisymmetric is at 90° (i.e. $\alpha = 90°$); experimentally $R = 1·04$, from which $\alpha = 54°$, whereas the observed value of D, which corresponds to $\alpha = 90°$, is 0·72. This discrepancy arises from twisting and wagging vibrations of the CH_2 group which alter the effective dichroic ratio.

In the second chain (Fig 2.18b), with C—H bonds aligned perpendicularly to the chain axis, both symmetric and antisymmetric vibrations are oriented perpendicularly to the chain axis. Theoretically $R = 0$ in both cases but participation of other vibrations accounts for the experimental values of 0·33 and 0·39.

2.4. Application to polypeptides and proteins

2.4.1. AMIDE BANDS

Polypeptides are chains of amino acids linked by amide groups to give the general formula

$$H_2N—CH—CO—NH—CH—\ldots CH \ldots CH_2COOH$$
$$\quad\quad\quad | \quad\quad\quad\quad\quad\quad | \quad\quad\quad |$$
$$\quad\quad\quad R_1 \quad\quad\quad\quad\quad R_2 \quad\quad R_3$$

Many synthetic polypeptides are homopolymers, where $R_1=R_2=R_3$, whereas in proteins the nature of the side group R varies and so may characterize the function of the protein.

The *trans* amide group in peptides

is generally planar and gives rise to a number of amide bands characteristic of both simple peptides (Miyazawa *et al.*, 1958) and fibrous proteins (Bendit, 1966). The frequencies and probable assignments of these bands are shown in the table on p. 55.

In the amide structure, many N—H and C=O groups are H-bonded. Fermi resonance, associated with Amide A and B bands, arises when two interacting vibrations of similar frequencies and symmetries are displaced to give an increased frequency separation. While such peak splitting is common in polymer spectra, it is often difficult to prove unless at least one absorption frequency can be determined under non-resonant conditions.

The assignments given represent simplified origins of the Amide bands; for I and II, in particular, involvement of other vibrations has been suggested

	Approx. wavenumber cm^{-1}	Assignment	
Amide A	3300	N—H stretch	
			in Fermi
Amide B	3100	Amide II first overton	resonance
Amide I	1660	C=O stretch	
Amide II	1570	N—H in-plane bend + C—N stretch	
Amide III	1300	C—N stretch + N—H in-plane bend.	
Amide IV	630	O=C—N in-plane bend	
Amide V	730	N—H out-of-plane bend	
Amide VI	600	C=O out-of-plane bend	

(Miyazawa, et al. 1958; Fraser and Price, 1952). In addition, Amides II and III are both associated with N—H bending and C—N stretching modes, but the phase relations between these modes are thought to differ in the two cases.

Synthetic polypeptides and fibrous proteins illustrate the effective use of polarized i.r. light. Fibrous proteins have their polypeptide chains (Fig. 2.19) in either (a) a helical configuration (α-helix) or (b) in an extended state (β-structure, considered to exist as a pleated-sheet configuration). Elliott and coworkers (Ambrose, et al., 1949; Ambrose and Elliott, 1951) showed that, for proteins in the extended-chain configuration, such as feather, silk or stretched

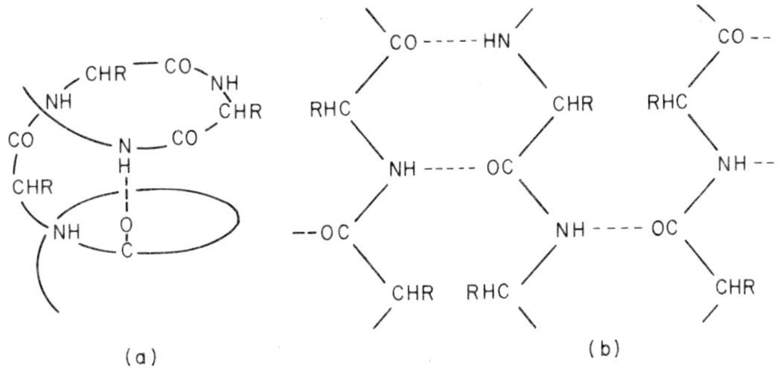

FIG. 2.19. Protein structures: (a) α-helix; (b) extended chain.

hair, Amide A and I bands were more intense when the plane of polarization was perpendicular to the fibre axis, while for proteins in the helical form, the absorptions were strongest when parallel to the fibre axis. This indicated that the N—H and C=O bonds were preferentially aligned along the helix axis for α-proteins and aligned perpendicular to the chain or fibre axis in β-proteins.

In Fig. 2.20, the spectra of hair (b) and (d) and silk gut (a) are compared with the spectra of poly-L-alanine (PLA) predominately in α- and β-conformations (c). (The Raman spectra from α-PLA are discussed in Section 3.3.1). The α-polypeptide film was cast from a solution in dichloroacetic acid and then rolled, and the β-structure was then produced by stretching the film in steam. The lower values of dichroism exhibited by the fibrous proteins arise from the presence of regions of structural disorder in the specimens which, since the absorptions are broad and overlap with absorptions of the ordered regions, decrease the dichroism exhibited by the regions of high structural order.

Deuteriation was found to decrease the intensity of the Amide-A absorption and to increase the dichroic ratio simultaneously (Bendit, 1966; Parker, 1955) by rapid exchange of NH and OH groups present in the side chains of the disordered regions (Fig. 2.21). Calculations based on deuteriation and transition-moment studies (Fraser and Price, 1952; Fraser and Suzuki, 1964; Sandeman, 1955) have shown that the dichroism of polypeptides could be accounted for in terms of the α-helix, whilst in proteins the dichroism was shown to arise from 80% of the NH groups in the α-helical structure.

Among additional absorptions detected (Bendit, 1966) are the 3286 cm^{-1} band associated with helical material and a peak at 3310 cm^{-1} assigned to the disordered phase; in proteins, the ordered region was found to absorb at 3274 cm^{-1}. The far-i.r. spectrum of PLA is shown on p. 136.

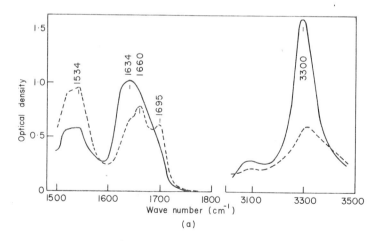

Fig. 2.20. Infrared spectra: full line, E-vector $\perp$ to fibre, broken line, E-vector parallel to fibre. (a) silk gut (Ambrose and Eliott, 1951; see pp. 57 and 58).

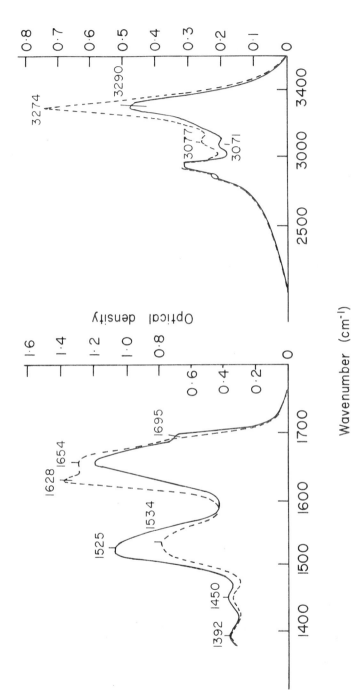

Fig. 2.20(b). Stretched hair (Bendit, 1966).

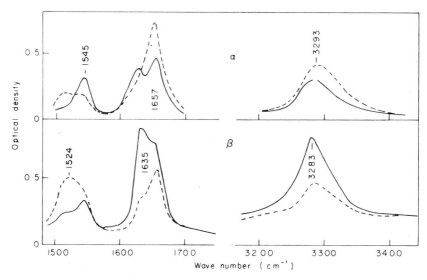

FIG. 2.20(c). Poly-L-alanine (Elliott).

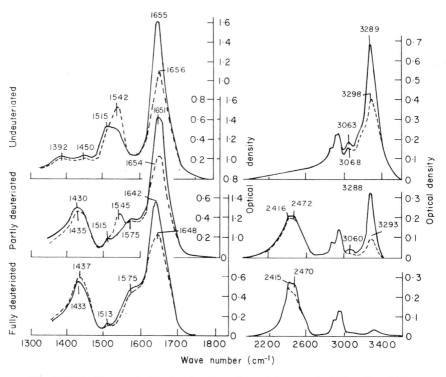

FIG. 2.20(d). Unstretched hair (showing effect of deuteriation) (Bendit, 1966).

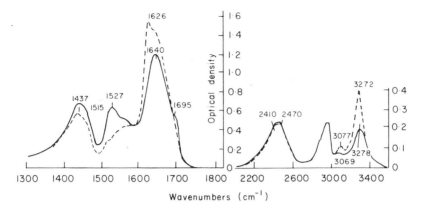

Fig. 2.21. Infrared spectrum of deuteriated stretched hair (Bendit, 1966).

2.4.2. VIBRATIONAL INTERACTION

At this point it is relevant to examine the vibrational interactions which are associated with polypeptide structure. From consideration of the relative phase differences between vibrational interactions, it has been shown (Miyazawa, 1960; Miyazawa and Blout, 1961), that the frequency observed depends on chain conformation; a theoretical basis was derived, which related observed Amide I and II vibrations and dichroisms with α-helical, parallel-chain pleated sheet, anti-parallel-chain pleated sheet and random conformations of polypeptides. It was assumed that each amide group vibration associated with a H bond has a characteristic frequency, v_0, which would be exhibited but for the perturbing influence of adjacent amide groups. For chains of infinite lengths, the frequency of a vibration is given by

$$v(\delta,\delta') = v_0 + \sum_s D_s \cos(s\delta) + \sum_s D_{s'} \cos(s'\delta')$$
$$\text{(intra)} \qquad \text{(inter)}$$

Here the second and third terms are due to intrachain and interchain interactions respectively, δ is the phase angle between vibrations of adjacent groups, and D_s is a coefficient for the interaction between the sth neighbours in the chain.

In the α-helical form of a polypeptide chain, with intrachain H bonds at every fourth group, D_1 and D_3 must be considered. Since there are 3·6 residues/helix turn, δ is either zero or $\dfrac{2\pi}{3\cdot 6}$, which correspond to parallel and perpendicular vibrations, respectively.

The frequency of the parallel band is given by $v(0) = v_0 + D_1 + D_3$ and for the perpendicular band by $v(\phi) = v_0 + D_1 \cos\phi + D_3 \cos 3\phi$. Experimentally, the difference in frequency between the two corresponding Amide-I bands is approximately 3 cm^{-1} and about 30 cm^{-1} for Amide II bands.

Hydrogen bonds may link polypeptide chains with two-fold screw axes, in either parallel or antiparallel alignment, so that the angle ϕ is π and the phase angle δ is either 0 or π. Taking the simpler (in terms of nearest-group interactions) parallel-pleated-sheet case first, then the vibrational frequency of the parallel band is

$$v(0, 0) = v_0 + D_1 + D_1'$$

and the perpendicular band

$$v(\pi, 0) = v_0 - D_1 + D_1'$$

Here (0, 0) and (π, 0) represent the phase difference between the next similar group in the chain and the H-bonded group in the adjacent chain, respectively. The vibrational modes which correspond to the above frequencies are shown in Fig. 2.22.

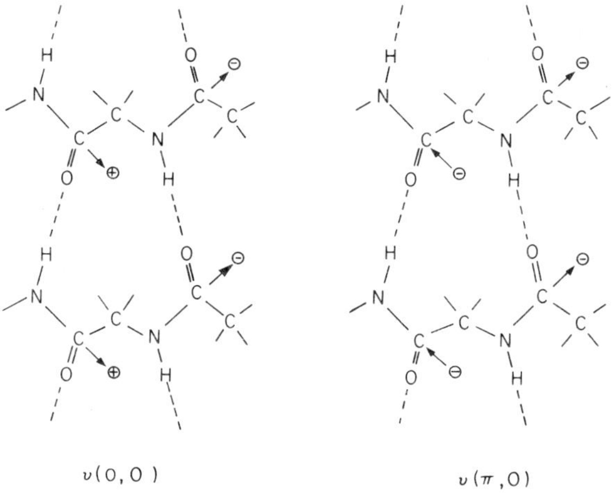

FIG. 2.22. Vibrational modes of peptide groups in parallel pleated sheet. ⊕ indicates direction below plane of paper, ⊖ above.

For a mode of vibration to be i.r.-active, the phase difference between vibrations in adjacent polypeptide chains, and hence unit cells, must be zero, that is, vibrations are in phase. For the parallel polypeptide-chain structure, therefore, both modes satisfy this condition and so are i.r.-active. For the antiparallel structure, however, with modes as shown in Fig. 2.23, the $v(0, 0)$ and $v(\pi, \pi)$ vibrations do not satisfy this condition and so, in theory, are inactive. Since the ($v(\pi, \pi)$) case is inactive only for a *fully* extended chain, a weak band is usually observed.

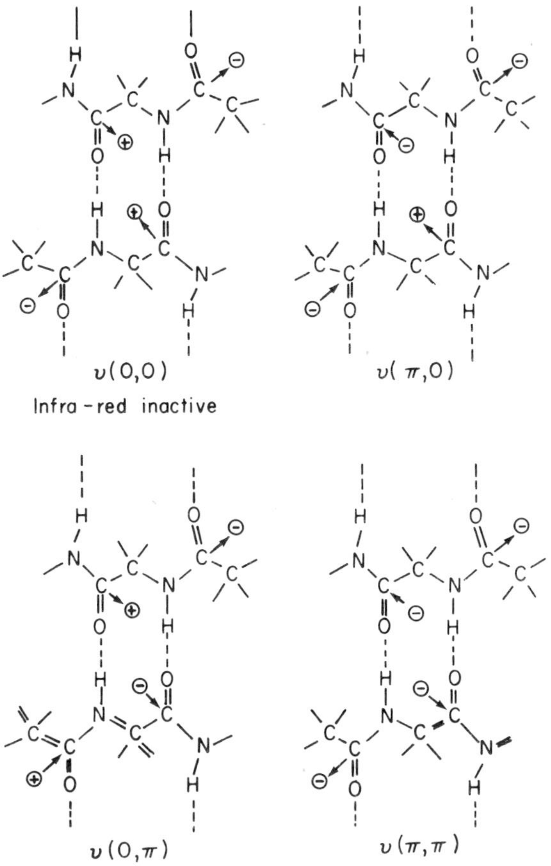

Fig. 2.23. Vibrational modes of peptide groups in anti-parallel pleated sheet.

The three vibrational modes are given by,

$v(0, \pi) = v_0 + D_1 - D'_1$ (parallel);
$v(\pi, 0) = v_0 - D_1 + D'_1$ (perpendicular); and
$v(\pi, \pi) = v_0 - D_1 - D'_1$ (perpendicular).

The direction of the dichroic effects can be obtained from analysis of the axes of symmetry in each case. Each unit cell of the anti-parallel chain structure consists of four peptide groups: two from one chain and two from the adjacent chain, and the four vibrations are shown. The $v(0, 0)$ vibration is i.r.-inactive. The $v(0, \pi)$ vibration is symmetric with respect to the two-fold screw axis parallel to the z or chain axis and, as can be seen from the resultant change in transition moment for each chain, the band is parallel. The $v(\pi, 0)$ vibration, however, is symmetric with respect to the axis parallel to the y axis, normal to the chain axis in the plane of the paper and again, as can be seen from the resultant change in transition moment for each chain, the band is perpendicular. Similarly, the $v(\pi, \pi)$ mode is symmetric about the z axis, normal to the plane of the paper, and again gives rise to a perpendicular band. In a similar way, the parallel-chain structure can be analysed; the $v(0, 0)$ mode is symmetric about the z axis, parallel to the chain axis, and the net change in transition moment is along the chain axis to give a parallel band, while the $v(\pi, 0)$ vibration is symmetric about the y axis to give a perpendicular mode.

Analysis of the spectra of proteins, polypeptides and some polyamides has enabled estimations of the values of D to be made; an analysis of the bands in terms of measured and calculated frequencies is given in Table 2.1. Random-coil samples, studied from solid films of polypeptides have shown that the Amide frequencies are close to the corresponding unperturbed vibrations.

Originally, a weak perpendicular band at 1630 cm^{-1} in α-protein spectra was associated with the presence of a small amount of β material. However, as shown above, calculated frequencies of Amide I and Amide II absorptions for differing polypeptide configurations showed that the presence of additional bands of opposite dichroism was to be expected, owing to symmetrical and antisymmetrical vibrations of neighbouring H-bonded peptide groups (Miyazawa, 1960; Miyazawa and Blout, 1961).

TABLE 2.1. Correlation between amide wavenumbers in proteins and vibrational modes

Conformation	Designation	Amide/cm^{-1}		Amide II/cm^{-1}	
		$\tilde{v}_{obs}$[a]	$\tilde{v}_{calc}$	$\tilde{v}_{obs}$	$\tilde{v}_{calc}$
Random coil		1656^s	1658	1535^s	1535
α-helix	$v(0)$	1650^s	1650	1516^s	1516
	$v(2\pi/n)$[b]	1652^m	1647	1546^w	1540
Parallel-chain pleated sheet	$v(0, 0)$	1645^w	1648	1530^s	1530
	$v(\pi, 0)$	1630^s	1632	1550^w	1550
Antiparallel-chain pleated sheet	$v(0, \pi)$	1685^w	1684	1530^s	1530
	$v(\pi, 0)$	1632^s	1632		1540
	$v(\pi, \pi)$		1668	1550^w	1550
Nylon 66		1640^s	1640	1540^s	1540

[a] Observed intensities indicated by superscript letters: s, strong; m, medium; w, weak.
[b] n is the number of peptide groups/turn of helix.

Since the Amide I band is not associated with replaceable protons, deuteriation of polypeptides has no effect on it. The Amide II absorption is detected in the range 1530–1550 cm^{-1} for α-structures and 1515–1525 cm^{-1} in β structures. The absorption for fibrous proteins near 1540 cm^{-1} shows perpendicular dichroism while that near 1515 cm^{-1} is parallel; these have been found to be consistent with the α-helical structure. Deuteriation of proteins has yielded, in addition to the expected N—D band at 1100 cm^{-1}, a new absorption at 1430 cm^{-1}. The Amide II absorption (Miyazawa et al., 1958) arises from 60% NH deformation coupled with 40% C—N stretching modes; deuteriation breaks the coupling between these modes to give a C—N stretching band at 1430 cm^{-1}.

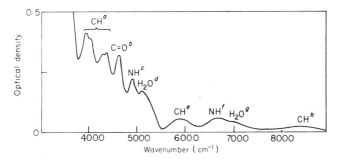

FIG. 2.24. Near-i.r. spectrum of keratin. ^{a}CH: stretching + CH deformation 2900 cm^{-1} + 1300 to 1460 cm^{-1}. ^{b}C=O: 2 × stretching + peptide group mode 2 × 1650 cm^{-1} + 1250 cm^{-1}. ^{c}NH: stretching + peptide group mode 3280 cm^{-1} + 1550 cm^{-1}. ^{d}H$_2$O: OH stretching + deformation 3400 cm^{-1} + 1645 cm^{-1}. ^{e}CH: 2 × stretching 2 × 2900 cm^{-1}. ^{f}NH: 2 × stretching 2 × 3280 cm^{-1}. ^{g}H$_2$O: 2 × stretching + deformation 2 × 1645 cm^{-1} + 3400 cm^{-1}. ^{h}CH: 3 × stretching 3 × 2900 cm^{-1}. (Ambrose, 1951.)

2.4.3. OVERTONE BANDS

Possible assignments in a typical spectrum of fibrous proteins in the overtone i.r. region (∼ 5000 cm^{-1}) are shown in Fig. 2.24.

2.4.4. DERIVATIVE TECHNIQUES

Derivative spectral analysis (Section 2.2.2.5) reveals many more peaks in the 4300–5200 cm^{-1} region, resolving apparently simple bands into multiple components in Fig. 2.25. Peaks 1–5 were clearly associated with the broad absorption band and are considered to arise from water in differing states of bonding in the protein structure; they reflect some of the complexity of the important protein property of absorption of large quantities of water. Peaks 1 and 2 were removed during extensive drying and arose from non-bonded or weakly-bound water, while peaks 3–5 could not be removed and clearly indicate the presence of strongly bound water in the protein. Peaks 6–8 and 9–12 (Fig. 2.25) arise from overtone and combination bands of NH and CO peaks, respectively. In view of the large number of vibrations present in the fundamental region of the spectrum, the multiplicity of derivative peaks is considered to reflect differences in fundamental band origin and peptide

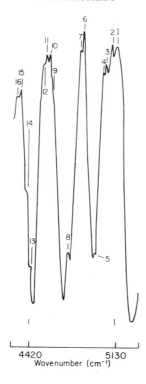

FIG. 2.25. Derivative spectrum of overtone region of α-keratin.

environments. The complex group of bands 13–16 have also been assigned to overtones of various C—H and CH_2 vibrations.

While the use of derivative techniques assists in the analysis of overtone bands which are both broad and of low intensity, in the fundamental region, they also assist in the separation of broad unresolved peaks typical of polymers with low crystallinity. In Fig. 2.11, which shows the absorption and derivative modes of a fibrous protein spectrum in the 3000 cm^{-1} region, the broken lines indicate differences in the spectrum due to further absorption of water. (The peaks in region 1 will be discussed later.) Peak 2 at 3311 cm^{-1} and peak 3 at 3280 cm^{-1} are associated with non-helical protein (Bendit, 1966) and the Amide A band, respectively, while peaks 4, 5, 7 and 8 are associated (Keighley and Rhodes) with absorbed water. While peak 6 has not hitherto been recorded at 3189 cm^{-1}, it is probably a combination band arising from Amide I and Amide II bands of the $v(\pi/3\cdot6)$ vibration. Peaks 9 to 13 at 3080, 3065, 3030 and 2995 cm^{-1} are assigned to various Amide B modes, three of which have been reported before, and reflect the different frequencies of the Amide II bands calculated by Miyazawa and Blout (1961). In addition, peaks 14–17 arise from CH modes.

2.4.5. WATER ABSORPTION BY PROTEINS

Using higher differentiator resolution, the number-1 group of peaks of Fig. 2.11 was analysed with a lower differentiator time-constant and a lower rate of frequency sweep. The results for samples conditioned at relative humidities from 0% to 72% (Fig. 2.26) contrast with diffferent spectra of hydrated proteins referred to dry proteins; the latter gave broad difference spectra with poor resolution (Bendit, 1966).

Of the 28 peaks recorded in the hydrated-protein spectrum, two bands at 3530 cm^{-1} and 3420 cm^{-1} are characteristic of free "non-H-bonded" NH$_2$ groups, being the antisymmetric and symmetric modes, respectively, and one peak at 3470 cm^{-1} arises from free NH groups. Since a large proportion of the protein structure is in disordered and non-helical forms, it is probable that the H bonding is incomplete within such regions, especially in the vicinity of

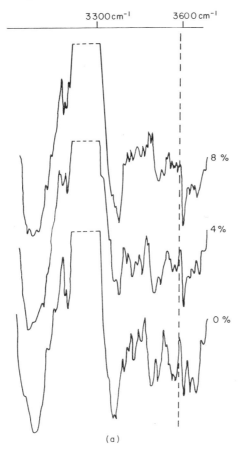

FIG. 2.26. Second-derivative spectra of water in keratin: (a) 0, 4 and 8% R. H. (Whitaker and Keighley, 1976; see also p. 66.)

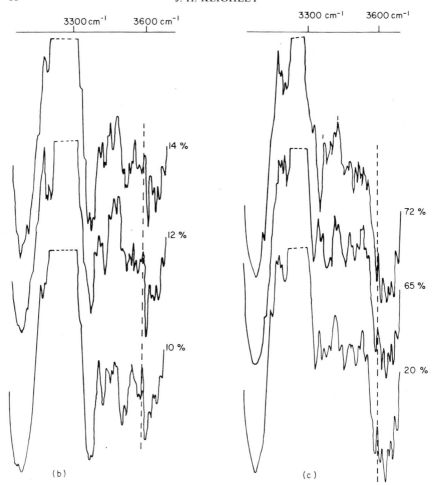

FIG. 2.26. Second-derivative spectra of water in keratin: (b) 10, 12 and 14% R.H.; (c) 20, 65 and 72% R.H.

proline residues which cannot be incorporated into helical regions. Therefore full H-bonding is unlikely in side-chain environments, which precludes the formation of completely H-bonded structures. Although there is no evidence in the absorption spectrum for the presence of free NH and NH_2 groups, such bands would be weak, partly because of their low concentration and partly because the electrostatic nature of H-bonding increases the polarity, and hence dipole moments and absorption coefficients, of H-bonded groups. Consequently, the ratio of band intensities of free and H-bonded NH groups is not proportional to the numbers of absorbing species.

Bands in the regions 3550–3470 cm^{-1} have been assigned to

$$>\!\!C\!\!=\!\!O \ldots H\!\!-\!\!O_{\diagdown \atop H}$$

in varying configurations and those in the range 3390–3470 cm^{-1} to

$$\diagdown\!\!\!\!\!\!\text{N}\diagup \ldots \text{H—O}\diagdown\!\!\!\!\!\!\text{H}$$

groups. Because of the complexity of the system, much band analysis remains to be completed.

2.4.6. CHAIN CONFORMATION IN PROTEINS

When suitably treated by heating or subjection to chemical treatments, many polymers form ordered regions with characteristic X-ray diffraction patterns. Infra-red absorptions characteristic of ordered and disordered zones differ for several reasons (Elliott, 1969):

1. The conformation of the chain in crystallites will be fully extended or folded in some regular manner, usually with only one stereo-isomer present. In disordered regions, however, a variety of stereo-isomers will give their characteristic vibrational frequencies, e.g. *gauche* and *trans* forms.

2. In the disordered regions, molecular symmetry may induce some inactive modes to become weakly active.

3. Vibrations which are inactive in a single regular, but isolated, chain may become weakly active if the symmetry of the structure of the chain is lowered. In the disordered polymeric form, bands of this type will not appear.

4. When several chains constitute the unit cell of the ordered structure, single modes of the isolated chain may become multiple in the ordered form.

5. Since, in general, the ordered regions are more highly oriented than disordered regions, dichroism is usually higher for bands from ordered regions.

Often, the distinction between regions of molecular order and disorder is not clear-cut. Some workers prefer the term "paracrystalline" to "crystalline" and "amorphous" since, in many cases, the regions of high molecular order are not continuous as in true crystals and since difficulties arise in the differentiation between different states of order. This approach is particularly valuable when samples exhibit high dichroism, indicating high molecular orientation without the presence of a significant proportion of regions which are crystalline to X-rays.

Helical bands near 3300, 1650 and 1550 cm^{-1} are associated with ordered regions in protein structure and differ from the bands of random-coil absorption (Table 2.1). In addition, the absorption at 3311 cm^{-1} has been associated with non-helical protein and that at 3280 cm^{-1} with the helix Amide A vibration.

The spectral region 1100–800 cm^{-1} has been called the "backbone" region, since vibrations are due to groups of atoms in the main chain, which may or may not have aggregated to leave ordered or disordered structures in differing conformations. Crystallization of polymers into differing conformations may occur (Bernal, 1958) because the atoms or groups of atoms are held together by forces of differing strengths, so that, when a polymer is able to retain its covalent bonds, it may be found in a large number of conformations and configurations; freedom of rotation of a group around a single bond, together with bond deformations, leads to a number of states of differing molecular order. Biological polymers can initiate the formation of ordered regions by linear polymerization and subsequent formation of a more complex structure by coiling and folding into a more stable structure. The extent of the coiling and folding is determined by the balance between the energy of deformation of the flexible chain and the energy gained by the process of chain folding. Such a process is enhanced by H-bond formation.

A section of the fibrous protein keratin, 7 μm thick, exhibits few resolved absorptions in the 1100–800 cm^{-1} region of the spectrum (Fig. 2.27), but there is some evidence from inflexion points of overlapping bands. The derivative spectrum (Fig. 2.27) also indicates the large number of absorptions present in

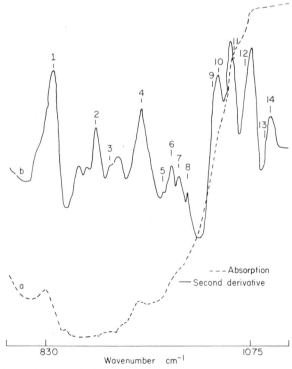

FIG. 2.27. Absorption and derivative spectra of keratin in 1100–800 cm^{-1} region.

this region. As an aid to the analysis of this complex spectrum (Keighley and Rhodes, 1971; Hallos and Keighley), the polyamide nylon 6 provides a suitable model compound (polyglycine is an alternative name for nylon 2).

Early work by Sandeman and Keller (1956) on nylon 6 indicated that bands could be assigned to ordered and disordered regions of the structure and to intermediate states of "crystallization", which corresponded to states in between completely random packing and full structural order. These intermediate states have been associated with chain folding (Michaels and Hausslein, 1965), so that observed bands can be described in terms of crystalline, crystallizable, non-crystallizable and chain-folded regions. Crystalline bands were shown to have an intensity proportional to the degree of crystallinity; these decreased to zero above the melting point, were absent in the amorphous polymer, increased on crystallization and on annealing and were unaffected by chemical reagents unable to penetrate the ordered regions of the polymer.

The intensity of the crystallizable bands was shown to be inversely proportional to the degree of order; it decreased during crystallization, or annealing and when the disordered region of the polymer are attacked chemically. Non-crystallizable bands were shown to be little affected by melting, crystallization, annealing or chemical treatment. Chain-folding bands, which have an intensity proportional to the degree of folding, disappear on melting, increase when crystallization by a chain-folding mechanism occurs, and rapidly disappear on chemical treatment.

These criteria enabled derivative keratin bands, identical in frequency with those in the polyamide spectra, to be classified. Peaks 1 and 10 (Fig. 2.27) were associated with crystalline regions of the structure, peaks 2, 3, 11, 12, 13, 14 with the crystallizable regions, peaks 7 and 8 with the non-crystallizable regions and peaks 4, 5, 6 and 9 with chain folding. Experiments on silk and stretched hair showed similar characteristics; chemical treatment induced the reduction of peaks numbered 4, 5, 6 and 9, substantiating the above assignments.

2.4.7. THE CHEMICAL BONDING OF METAL IONS TO PROTEINS

Biological systems in which metals are bonded to proteins are of wide importance, e.g. in studies of disorders in organs of humans and other mammals, and especially in connection with tumour formation. Accordingly, it is desirable that the changes which occur in the i.r. spectra of such complexes are understood. Molecularly bound metal ions induce profound changes in the structure and function of many polypeptides and proteins with concomitant conformational changes, while the binding of the cation often induces denaturation (Tanford, 1968). Such changes in group environment will induce changes in the i.r. spectrum.

Detailed vibrational assignments are available for many component bands in the i.r. spectra of metal ethylene diamine (en) complexes (Nakamoto, 1970). Conclusive evidence of the presence of chelate bonds in the metal complexes of amino-acids and peptides has also been obtained. The formation of N-to-metal coordinate bonds changes the polarity of the N-bonds (Quagliano et al., 1954–1960; Rosenberg, 1956; Kim and Martell, 1964; Kackowitz et al., 1967) and induces a decrease in the frequency of the N—H stretching vibration, together with an increase in the intensity. Similarly, coordination at carboxyl and peptide carbonyl groups increases; this decreases the deformation and increases the stretching-mode frequencies (Martell, 1961, 1966; Kothari and Bush, 1969; McAuliffe and Perry, 1969).

Spectra of ethylene diamine, Zn-ethylene diamine (zincoxen), keratin and keratin treated with Zn-ethylene diamine (Habib et al., 1975) are shown in Figs 2.28 and 2.29. Grinding the fibrous protein to produce a KBr disc evidently degrades the structure to such an extent that no information is obtained about structural configuration. The strong broad absorption centred near 3425 cm^{-1} contrasts with the sharper absorptions near 3300 cm^{-1} from a section of

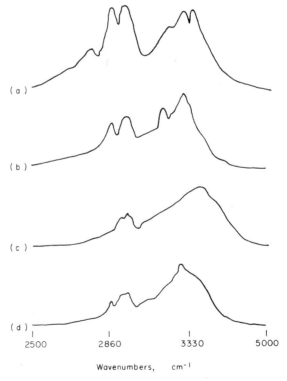

FIG. 2.28. Infra-red spectra in 5000–2500 cm^{-1} region for (a) ethylene diamine; (b) zincoxen; (c) keratin; and (d) zincoxen-treated keratin. (Habib, 1975; see also Fig. 2.29.)

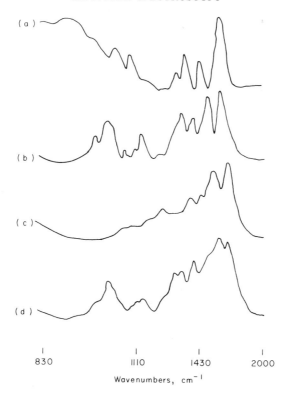

FIG. 2.29. Infrared spectrum in 2000–830 cm^{-1} region for (a) ethylene diamine; (b) zincozen; (c) keratin; and (d) zincoxen-treated keratin (Habib, 1975).

keratin (Bendit, 1956), probably as a result of an interaction between the peptide N—H groups and the halide ions in the KBr matrix.

The frequencies of the Zn-keratin spectra are characterized by a broad absorption in the 3300 cm^{-1} region; as the reaction between the keratin and the Zn complex proceeds, the frequency of the absorption maximum decreases. It does not necessarily follow (see Section 2.2.2.5) that the frequency of the N—H vibration is decreasing, since the shape of the resultant absorption envelope will depend on the relative intensities of the component absorptions. Hence, in keratins treated with the Zn complex, the total absorption (of the Zn complex) at 3270 cm^{-1} will increase in intensity so that the resultant absorption-maximum wavenumber decreases as the uptake increases. Thus Fig. 2.12 exemplifies the effect illustrated on p. 47.

Since any change in frequency of the N—H band cannot be detected, measurements of optical density for the Zn complex, keratin and the Zn-keratin were made at 3270 and 2940 cm^{-1}; these corresponded to the N—H absorption of the complex and a C—H mode of both keratin and complex which is unaffected by the reaction. The ratio of absorbance at 3270 cm^{-1} to that at 2940 cm^{-1} was 1·68 for the Zn complex, 1·70 for the keratin, and 2·00 for the Zn-keratin. This indicated a new absorption at 3270 cm^{-1} ascribed to a shift in Amide A band to a lower frequency. In the light of previous results (Quagliano et al., 1954–1960; Rosenberg, 1956; Kim and Martell, 1964; Kackowitz et al., 1967), this can be interpreted in terms of coordination of peptide N to the Zn reagent.

Similar results were obtained with Cu and Ni ethylene-diamine complexes (Habib and Keighley), but the peak maxima of the resultant keratin-complex spectra were determined at 3270 and 3290 cm^{-1} for the Cu and Ni-keratin samples, respectively; these compare with 3280 cm^{-1} for Zn. Optical-density ratios were determined as 2·05 and 1·80, respectively; comparisons with 1·70 for keratin again indicates increased absorbance at 3270 cm^{-1}.

Examination of Amide I bands of the Zn-, Cu- and Ni-keratin complexes indicated that an essential change in the absorption frequency had occured. The Amide II band at 1510 cm^{-1} of the keratin sample (band frequencies obtained from KBr-disc samples often differ from those obtained from sections of the same material) is obscured by the Zn ethylene-diamine absorption in the same region (Fig. 2.29), but the difference spectrum (Fig. 2.30) with Zn ethylene-diamine in the reference beam of the spectrometer

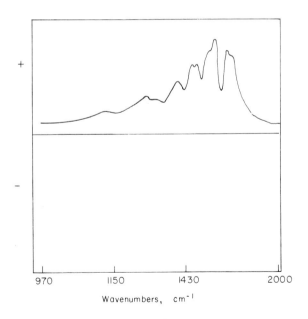

FIG. 2.30. Difference spectrum obtained with zincoxen-treated keratin in the sample beam and zincoxen in the reference beam (Habib, 1975).

indicates a new absorption at 1543 cm^{-1}. The intensities of the Amide I and II bands were reduced by the treatment, indicating that there is an interaction with the amide group and that the new peak is associated with this interaction. The new 1543 cm^{-1} peak may arise from (a) a decrease in frequency of the keratin Amide I band; (b) an increase in frequency of the Amide II band, due to coordination of the N→M type; or (c) absorption of the NH$_2$ groups of Zn diethylene-diamine molecules. Of these, (a) and (c) are unlikely, since the frequency changes involved are too great to be explained in terms of coordinate bond strength (Krishnan and Plane, 1966); an (a) shift has been shown to have a maximum value of 40 cm^{-1} in this region of the spectrum. Hence the new absorption is attributed to an increase in frequency of the Amide II band, in keeping with the behaviour observed in polypeptides (Kackowitz et al., 1967). For Cu- and Ni-keratin complexes, the new peak was similarly detected at 1562 and 1538 cm^{-1}, compared with 1543 cm^{-1} for Zn, and it was concluded that the extent of the shift was dependent on the strength of the N→M coordinate bond formed. On this basis, therefore, the sequence of strengths of the coordinate bonds formed between peptide N and the metal ion is N→Cu > N→Zn > N→Ni. The results in the 3300 cm^{-1} region substantiate this.

Studies of the Amide B band, 3075 cm^{-1} in native keratin, showed that chemical treatment increased the frequency to 3155 cm^{-1} for Cu-keratin, and to near 3125 cm^{-1} for Zn- and Ni-keratin. These changes are in accordance with the origins of the Amide B band. Since increases in frequency of the Amide II have been observed, the maximum increase of 80 cm^{-1} is in line with the increase of 50 cm^{-1} observed for the Cu-complex-treated sample, in view of the Fermi resonance involved. If Amide A and B arise from a resonance effect, then any change in one will influence the other; whichever change is cause and which is effect, it is clear that there is some relation between the Amide A, B and II bands.

In line with changes in the Amide II band, studies of peptides which coordinate with metals have shown that, while the Amide II band increases in frequency, the Amide III decreases. The difference spectrum (Fig. 2.31), in which the treated sample is in the sample beam and the keratin and zincoxen sample is in the reference beam, indicates a decrease in intensity of the Amide III band at 1205 cm^{-1}, together with decreases at 1315 and 1350 cm^{-1}. Simultaneously, however, three new bands are recorded at 1000 cm^{-1}, 1150 cm^{-1} and 1100 cm^{-1} which correspond, respectively, to shifts in the native keratin bands by a frequency factor of $\frac{1}{1\cdot2}$. Since the new bands at 1315 and 1350 cm^{-1} behave in the same way as the Amide III band, their origins must be closely related; they have been designated Amide IIIa and Amide IIIb, respectively. Analogous results were obtained for the Cu- and Ni-triethylene-diamine-treated samples.

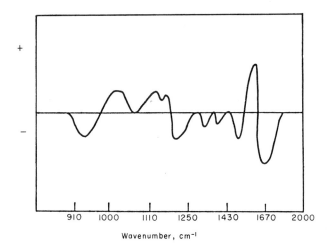

Fig. 2.31. Difference spectrum obtained with zincoxen-treated keratin in the sample beam and (keratin + zincoxen) in the reference beam (Habib, 1975).

From this study of reaction between metal amine and keratin, it is apparent that coordinate bonds are formed between the amide-N atom and the metal ion.

2.5. Polysaccharides

Infra-red spectroscopy has been widely used in studies of the structure of cellulose, chitin and other polysaccharides. Of these, the most widely distributed are homopolymers of monomers such as glucose, xylose, etc. and their derivatives. The spectra of these compounds, as with proteins, are often very similar and small differences are due to chain conformation and packing or anomeric configurations.

2.5.1. α-D-GLUCOSE

The combined use of i.r. and Raman spectroscopy has enabled the majority of the predicted vibrational modes of α-D-glucose (Vasco et al., 1972) to be accounted for. Greatest differences in carbohydrate spectra are found in the OH-stretching region of the spectra and, while five OH stretching bands were predicted, six bands have been detected, indicating that a coupling exists between motions in the crystalline phase. While non-H-bonded OH groups are absent (there is no band at 3650 cm^{-1}), all bands observed are in the region

3200–3500 cm^{-1}, showing that all OH bands are H-bonded. Differences arise from the nature (i.e. a linear or bent O—Ĥ—O, and length) of the H-bond.

2.5.2. CELLULOSE

The distinct polymorphic forms of cellulose can be recognized from analysis of their i.r. spectra in the 4000–2500 cm^{-1} region. Figure 2.32 indicates the differences in the dichroic properties of native (cellulose I) and mercerized cellulose (cellulose II) in the OH- and CH-stretching region (Liang and Marchessault, 1959–1960); the band frequencies, intensities and dichroic properties are shown in Tables 2.2 and 2.3. Cellulose II gives two strong OH bands at 3480 and 3447 cm^{-1} with parallel dichroism, whereas no bands are

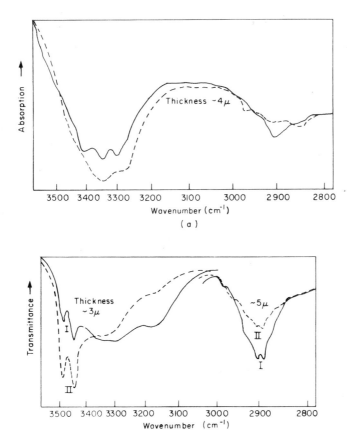

Fig. 2.32. Polarized i.r. spectra of films of (a) cellulose I; (b) cellulose II in 3000 cm^{-1} region (continuous line $E\perp$, broken line $E\parallel$ fibre axis). See Tables 2.2 and 2.3 (Liang and Marchessault, 1959).

TABLE 2.2. Band Assignments for the 2000–4000 cm^{-1} region of the i.r. spectrum of Cellulose I (see Fig. 2.32)

Wavenumber (cm^{-1})		Polarization	Assignment
Valonia and bacterial cellulose	Ramie cellulose		
2853	2853	∥	CH$_2$ Symmetric stretching
2870	2870	⊥	
2897		⊥	
	2910	⊥	
2914		⊥	
2945	2945	⊥	CH$_2$ Antisymmetric stretching
2970	2970	∥	
3245		∥	
3275	3275	∥	
3305	3305	⊥	O—H Stretching (intermolecular hydrogen bonding in the 101 plane)
3350	3350	∥	O—H Stretching (intramolecular O-3—H···O-5 hydrogen bonding)
3375	3375	∥	
3405	3405	⊥	O—H Stretching (intermolecular hydrogen bonding in the 101 plane)
—	3450	∥	Cellulose II impurity in ramie?

TABLE 2.3. Band Assignments for the 2000–4000 cm^{-1} region of the i.r. spectrum of mercerized Ramie cellulose (Cellulose II) (see Fig. 2.32)

Wavenumber (cm^{-1})	Polarization	Assignment
2850	∥	CH$_2$ Symmetric stretching
2874	⊥	
2891	⊥	
2904	⊥	
2933	⊥	CH$_2$ Antisymmetric stretching
2955	∥	
2968	∥	
2981	⊥	
3175	⊥	O—H Stretching (intermolecular hydrogen bonding)
3305	⊥	
3350	⊥	
3447	∥	O—H Stretching (intramolecular hydrogen bonding)
3448	∥	

observed in this region for cellulose I, where the dominant band is at 3350 cm^{-1}, with parallel dichroism. Celluloses III and IV exhibit analogous differences. Studies of the highly crystalline samples of *Valonia ventricosa* have given the most detailed results (Fig.2.33) from a mat of oriented fibres. Since the bands observed in the 1500–800 cm^{-1} region were like to those of α-D-glucose, similar origins were suggested for them.

The glucose residue in cellulose has three OH groups but, since nine OH-stretching bands have been observed in the i.r. and Raman spectra, it is likely that the glucose residues are not arranged in the same way. The unit cell for cellulose I contains two residues, each of two separate chains, which probably (Jones, 1971) has a two-fold screw axis along its length, so that the environment of successive residues is expected to be analogous but not identical. Although such structures can yield up to six bands, two of these will be identical for the intramolecular H-bonds, so that only five will be observed. The four remaining OH bands are thought to arise from H-bonding between elementary fibrils of the cellulose structure. The extensive i.r. spectroscopic studies of cellulose have been surveyed by Blackwell and Marchessault (1971) and successive review articles in the same volume.

2.5.3. CHITIN

Replacement of the C_2 hydroxyl groups in cellulose with the $NHCOCH_3$ group transforms cellulose into chitin. The chain structure resembles that of cellulose, so that i.r. spectra are similar (Pearson *et al.*, 1960).

Chitin exist in at least three polymorphic forms (α-, β- and γ-) distinguished by the wavenumbers of the amide bands:

Amide band	α cm^{-1}	β cm^{-1}	γ cm^{-1}
A	3264	3292	3263
B	3106	3102	3100
I	1656	1656	1656
	1631	1631	1626
	1621		
II	1555	1556	1550

Wavenumbers differences among the Amide A bands arise from differences in the strength of the NH $\cdots$ OĊ H-bond (predicted by Nakamoto (1955) as 0.1Å longer for the β structure) between α-, β- and γ-structures. While such arguments can account for the low-frequency component of the Amide I band, the higher frequency band near 1656 cm^{-1} is thought to arise from hydrate formation, since this band is almost completely absent in β-chitin.

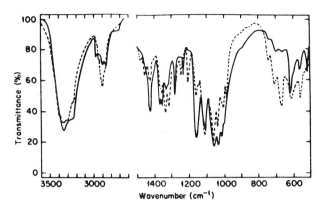

FIG. 2.33. Polarized i.r. spectra of oriented celluose I fibres over range 3500–500 cm^{-1}: continuous line $E\perp$ fibre axis, broken line $E\|$.

2.6. Conclusion

In addition to its application for organic identification, i.r. spectroscopy can be used for structural studies of biological polymers. In this case, it is perhaps especially necessary to appreciate the underlying theory of the subject before the recorded absorptions can be analysed; some attempt has been made in this chapter to outline this process. At the same time, the use with i.r. of several complementary techniques such as Raman spectroscopy (Chapter 3) or X-ray diffraction for analysis of molecular structure, assists in the identification of vibrational modes, an identification which frequently would be impossible without such an association of techniques.

References

Alexander, P. and Block, R. J. (1960) (eds). "A Laboratory Manual of Analytical Methods of Protein Chemistry." Pergman, Oxford.
Ambrose, E. J. and Elliott, A. (1951). *Proc. Roy. Soc.* **A206,** 206.
Ambrose, E. J. and Hanby, W. E. (1949), *Nature* **163,** 859.
Baddiel, C. B. (1968). *J. Mol. Biol.* **38,** 181.
Bendit, E. G. (1966). *Biopolymers* **4,** 539, 561.
Bernal, J. D. (1958). *Disc. Far. Soc.* **25,** 7.
Blackwell, J., Vasco, P. D. and Koenig, J. L. (1970). *J. Appl. Phys.* **41,** 4375.
Blackwell, J. and Marchessault, R. H. (1971). *In* "Cellulose and Cellulose Derivatives" (N. M. Bikales and L. Segal, eds), Part IV, p. 1. Wiley-Interscience, New York.
Elliott, A. *Proc. Roy. Soc.* **A26,** 408.
Elliott, A., Ambrose, E. J. and Temple, R. B. (1947). *Nature* **159,** 641.
 J. Opt. Soc. Am. **38,** 212.
Elliott, A. (1969). "Infra-red Spectra and Structure of Long-Chain Polymers." Edward Arnold, London.

Fraser, R. D. B. and Price, W. C. (1952). *Nature* **170,** 490.
Fraser, R. D. B. (1953). *J. Chem. Phys.* **21,** 1511.
Fraser, R. D. B. and Suzuki, E. (1964). *J. Mol. Biol.* **9,** 829.
Fraser, R. D. B. and Suzuki, E. (1966). *Anal Chem.* **38,** 1770.
Fraser, R. D. B. and Suzuki, E. (1970). *Spectrochim. Acta* **26A,** 423.
Habib, F. K., Keighley, J. H., Rhodes, P. and Whewell, C. S. (1975). *Spectrochim. Acta* **31A,** 1.
Habib, F. K. and Keighley, J. H. Unpublished.
Hallos, R. and Keighley, J. H. (1975). *J. Appl. Polym. Sci.* **19,** 2309.
Kackowitz, J. F., Durkin, J. A. and Walter, J. L. (1967). *Spectrochim. Acta* **23A,** 67.
Keighley, J. H. and Rhodes, P. (1971). *Proc. 22nd C. Elec. Radio Eng.* **21,** 397.
Keighley, J. H. and Rhodes, P. (1972). *Infra-red Physics* **12,** 277.
Keighley, J. H. and McKinley, J. Unpublished.
Keighley, J. H. and Rhodes, P. Unpublished.
Kim, M. K. and Martell, A. E. (1964). *Biochemistry* **3,** 1169.
Koening, J. L. and Hannon, M. J. (1967). *J. Macromol. Sci. (Phys)* **B1(1),** 119.
Kopf, R. (1960). *Bull. Soc. Chim. France*, 1964.
Kothari, V. M. and Busch, D. H. (1969). *Inorganic Chem.* **8,** 2276.
Krishnan, K. and Plane, R. A. (1966). *Inorganic Chem.* **5,** 852.
Liang, C. Y. and Marchessault, R. H. (1959). *J. Polym. Sci.* **37,** 385.
Liang, C. Y. and Marchessault, R. H. (1959). *J. Polym. Sci.* **39,** 269.
Liang, C. Y. and Marchessault, R. H. (1960). *J. Polym. Sci.* **43,** 31.
Martell, A. E., Nakamoto, K. and Morimoto, Y. (1961). *J. Amer. Chem. Soc.* **83,** 4528.
Martell, A. E. and Kim, M. K. (1966). *J. Amer. Chem. Soc.* **88,** 914.
Martin, A. E. (1959). *Spectral Chimica Acta* **4,** 97.
McAuliffe, C. A. and Perry, W. D. (1969). *J. Chem. Soc.* p. 634.
Michaels, A. S. and Hausslein, R. W. (1965). *J. Polym. Sci.* **C9,** 61.
Miyazawa, T., Shimanouchi, T. and Mizoshima, S. (1958). *J. Chem. Phys.* **29,** 3, 611.
Miyazawa, T. J. (1960). *J. Chem. Phys.* **32,** 1647.
Miyazawa, T. and Blout, E. R. (1961). *J. Amer. Chem. Soc.* **83,** 712.
Nakamoto, K., Margashes, M. and Rundle, R. E. (1955). *J. Amer. Chem. Soc.* **77,** 6480.
Nakamoto, K. (1970). "Infra-red Spectra of Inorganic and Coordination Compounds", p. 150. Wiley Interscience. New York.
Newman, R. and Halford, R. S. (1948). *Rev. Sci. Inst.* **19,** 270.
Parker, K. D. (1955). *Biochim. Biophys. Acta* **17,** 148.
Pearson, F. G., Marchessault, R. H. and Liang, C. Y. (1960). *J. Polym. Sci.* **43,** 101.
Quagliano, J. V., Svatsos, S. F. and Curran, C. (1954). *Anal. Chem.* **26,** 429.
Quagliano, J. V., Svatsos, S. F. and Curran, C. (1955). *J. Amer. Chem. Soc.* **77,** 6159.
Quagliano, J. V., Saracen, A. J., Nakagawa, I., Mikushima, S. and Curran, C. (1958). *J. Amer. Chem. Soc.* **80,** 5018.
Quagliano, J. V., Kennelly, M. M., Mizushuma, S. and Tsuboi, M. (1959). *Spectrochim. Acta* **15,** 296.
Quagliano, J. V., Segnini, D. and Curran, C. (1960). *Spectrochim. Acta* **16,** 540.
Richards, R. E. and Thompson, H. (1947). *J. Chem. Soc.*, p. 1248.
Rosenberg, A. (1956). *Acta Chem. Scand.* **10,** 840.
Samuels, R. J. (1965). *J. Polym. Sci.* **A3,** 1741.
Sandeman, I. (1955). *Proc. Roy. Soc.* **A232,** 105.

Sandeman, I. and Keller, A. (1956). *J. Pol. Sci.* **19,** 401.
Singleton, F. and Collier, G. F. (1956), *J. Appl. Chem.* **6,** 494.
Takahashi, S. (1961). *Opt. Soc. Amer.* **51,** 441.
Tanford, C. (1968). *Adv. Protein Chemistry* **23,** 121.
Vasco, P. D. *et al.* (1972), *Carbohydr. Res.* **23,** 407.
Whitaker, A. J. and Keighley, J. H. (1976). *Shriffenreiche Deutsches Wollforschungsinstitut, Aachen,* in press.
Woodward, L. A. (1972). "Introduction to the Theory of Molecular Vibrations and Vibrational Spectroscopy." Oxford.

3. Raman Spectroscopy
J. L. Koenig

3.1. Introduction

While most biologists and biochemists are well aware of the use of i.r. spectroscopy (Chapter 2) for the study of the structure and conformation of biological molecules, relatively few are familiar with Raman spectroscopy (Sections 1.3.4 and 2.1.1), the "other half" of the vibrational spectra. However, in the last few years, progress in the Raman spectroscopic study of biological molecules has reached the point where biologists and biochemists should consider the use of Raman spectroscopy for the solution of research problems.

3.1.1. WHAT IS RAMAN SPECTROSCOPY?

The Raman effect provides a further probe for examining the vibrational energy levels of molecules, which are a sensitive function of the chemical nature of the constituent groups and of the relative positions and interactions between atoms. Figure 3.1 shows how the incident light beam of frequency v_0 interacts with the vibrational energy levels, E_0, E_1, etc. of the molecule. (Elastic scattering gives rise to so-called Rayleigh lines (at v_0); these provide no information of interest in this context and must be separated from the weaker Raman lines scattered inelastically.) The lines occurring as a result of the molecule giving up quantum energy from the photon appear at higher frequency ($v_0 + v$) and are called anti-Stokes lines, while the lines resulting from the sample accepting energy from the photon appear at lower frequencies ($v_0 - v$) and are called Stokes lines. Since a larger number of molecules always exists in the lower energy or ground state at normal temperatures, the Stokes lines are more intense and are usually the lines measured experimentally; the weaker anti-Stokes lines merely duplicate the information contained in the Stokes lines. The scattered photons are characteristic of the scattering molecule in the same way that the i.r. absorption spectrum is a characteristic "fingerprint" of the molecule. In fact, many of the frequencies observed in the i.r. spectrum of a molecule may be observed in the Raman spectrum as well. Experimentally, the Raman effect can be measured from near $10\,cm^{-1}$ to $4000\,cm^{-1}$ on a single instrument. However, Raman and i.r. represent two different

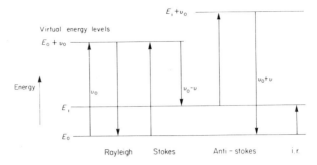

FIG. 3.1. Rayleigh and Raman scattering involving vibrational energy levels.

physical processes. The i.r. absorption arises when the vibrational motion produces a change in the dipole moment, while Raman scattering occurs when the vibrational motion produces a change in the polarizability of the molecule; interaction with the radiation causes an induced dipole moment. Polar chemical groupings (those with large dipole moments) are usually strong i.r. absorbers; if the polarizability can also change, they can also be Raman scatterers.

In addition to the intensity and frequency of the Raman lines, the polarization character of the lines can also be measured. Indeed, it was the unique polarization properties of this "new radiation" which led C. V. Raman to believe he was observing a new phenomenon, rather than normal emission, such as fluorescence. In the usual Raman experiment, the observations are made perpendicular to the direction of the incident beam, which is plane-polarized (as at (a) in Fig. 3.2). The "depolarization ratio" is defined as the

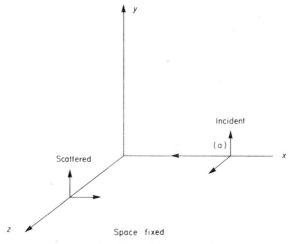

FIG. 3.2. Polarization directions of beams in Raman spectroscopy.

intensity ratio of the two polarized components of the scattered light which are parallel and perpendicular to the direction of propagation of the (polarized) incident light when the polarization of the incident beam is perpendicular to the plane of propagation and observation. The polarization of the Raman lines of CCl_4 is shown in Fig. 3.3. Since the laser beam (the most commonly used radiation source) is polarized and highly directional, polarization measurements are easily made with accuracy. Theoretically, the depolarization ratio thus measured can have values ranging from zero to 3/4, according to the nature and symmetry of the vibration. Non-symmetric vibrations give depolarization ratios of 3/4; symmetric vibrations have depolarization ratios ranging from 0 to 3/4, depending on the polarizability changes and symmetry of the bonds making up the molecule. Accurate values of the depolarization ratios are important for determining the assignment of a Raman line to a totally symmetric or asymmetric vibration. Potentially, depolarization measurements can yield information about the nature of chemical bonds.

Raman spectroscopy has the same general uses as i.r. spectroscopy. Raman spectroscopy can be used for identifying substances, pure or in mixtures, without recourse to assumptions about their structure; it can also be used to demonstrate structural differences between similar compounds or isomers. Raman effects, which arise from inter- or intra-molecular forces, can be used for following structural changes in molecules as a result of induced changes in environment, such as pH, ionic strength or temperature. Finally, Raman spectroscopy can facilitate the structural determination of biological molecules in solid, solution or liquid state.

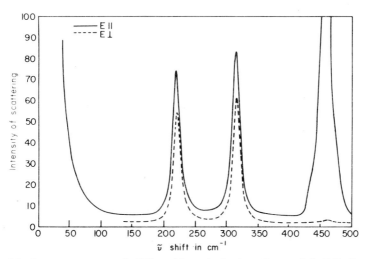

FIG. 3.3. Raman spectra of CCl_4 with polarization E parallel (full line) and perpendicular (broken line).

3.1.2. ADVANTAGES OF RAMAN SPECTROSCOPY OVER I.R. SPECTROSCOPY

Raman spectroscopy has several advantages over i.r. spectroscopy. Raman spectra are obtained from a very small quantity of sample, since the amount required need only fill a focused laser beam; this is an advantage for samples that can not be powdered. Glass sample containers can be used for Raman spectroscopy since glass is a weak Raman scatterer. The sample need not be transparent for Raman spectroscopy since the Raman effect is a light *scattering* technique rather than a light *transmission* technique. The size and shape of the sample can be variable for Raman spectroscopy. The sample need not be pulverized or mulled to be examined by Raman spectroscopy.

For biological applications, the primary advantage of Raman spectroscopy over i.r. spectroscopy (see Section 2.4.5) is the low interference of liquid water; the 2000–200 cm^{-1} region of the vibrational spectrum is completely accessible with either D_2O or H_2O. Thus aqueous solutions of biological systems can be studied for isomeric changes, H-bonding, base-pairing and tautomeric equilibria. In many cases, it may be possible to examine compounds *in vivo* and, unless the sample absorbs the laser beam, spectral measurements can be made without fear of a change in structure.

In summary, the advantages of Raman spectroscopy for biological applications are primarily the low interference of water, the accessibility of the low-frequency modes which are sensitive to conformational changes and the sensitivity to homonuclear bonds, such as S—S, N—N, C—C, C=C and aromatic groupings.

The primary limitation of Raman spectroscopy rests in the appearance of a strong luminesence in the background for some samples. In some cases, this background radiation may be stronger than the Raman signal and completely obsecures the spectrum. The source of this background is not understood but purification procedures are usually successful in removing the interfering background. On occasion, there may be some danger of damage to a sample at the focus of a laser beam.

In general, the signal-to-noise ratio for the Raman spectra decreases as the molecular weight of the sample increases. This is due to the sensitivity of the Raman effect to the density of the sample and the Rayleigh (or elastically scattered light) intensity to the size of the molecule. In addition to this effect, biological molecules are large and complex, with many modes which cannot be separated; the maxima observed can involve the merging of several neighbouring maxima. The Raman scattering intensity is a linear function of the concentration of the scattering species, so low concentrations of impurities or minor constituents do not generally interfere with the spectra of the major components.

Raman spectroscopy has been found useful for studying a variety of biological molecules and these systems will be discussed separately.

3.1.3. EXPERIMENTAL RAMAN SPECTROSCOPY

The experimental apparatus for Raman spectroscopy, outlined in block form in Fig. 3.4, includes: (a) a powerful light source (currently a laser); (b) an illuminating chamber for the sample; (c) a high-performance light-dispersion system to resolve the more intense, elastically scattered, light from the weak, inelastically scattered, Raman signal; (d) a light detection and amplification system capable of detecting weak light levels; and (e) a recorder.

The main assets of a laser as a light source are directionality, coherency, intensity, monochromaticity and polarization. Since the light waves of the coherent light of a laser are all in phase, the additive effect produces a focused light intensity that is 10^4 times as great as for normal coherent light of the same power. The coherent nature of light also yields a collimated beam which spreads out only fractionally with distance. A beam from an Hg lamp, like a flashlight beam, on the other hand, travels only a limited distance before its luminosity is lost, since the beam loses its intensity through spreading. A laser beam can travel large distances and the beam size does not increase substantially. In fact, a laser beam shone from the earth to the moon, as a part of the U.S. space research, spread to a diameter of only one half mile and the reflected beam was detected in Houston, Texas. The laser beam is monochromatic in frequency with a very narrow bandwidth, which is vital to its application as a Raman source. Additionally, the laser beam is linearly polarized; this yields an additional gain factor of two, since one-half of normal light is lost in passing through a polarizer. Since a continuous source is

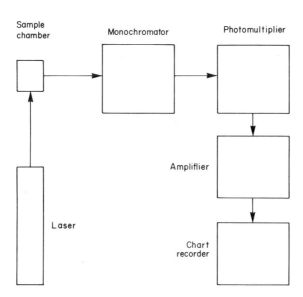

FIG. 3.4. Block diagram indicating essential components of a Raman spectrometer.

desirable for the purposes of spectroscopy, gas lasers such as He-Ne and Ar-ion or Kr-ion lasers are used. The He-Ne lasers are more stable, have a longer life and are cheaper than the Ar- and Kr-ion lasers, but they have a much lower power output. Since the Raman signal increases with the power of the incident beam, high power is desirable. Pulsed lasers can be used and offer some advantages at the detection end of the system, but they are not normally used except for special purposes such as remote Raman spectroscopy.

The illuminating chamber is simple and is built to facilitate the focusing of the beam on the sample, whether it is a gas, liquid, or solid. The beam can be passed through a liquid or gas several times to enhance the Raman signal. The scattered light is condensed using a lens system and is focused on the slits of the light dispersion system.

Special monochromator systems are used in order to reduce the scattered light to a level for the Raman signal to be detected. Since the Raman signal is only $10^{-6}-10^{-9}$ as strong as the Rayleigh line, two and three grating monochromators in tandem have been used in order to reduce the Rayleigh scattered light to an acceptable level. Various configurations of the monochromator are possible, but the requirement of frequency matching in the double or triple monochromators presents a challenging coupling problem for frequency-scanning systems. A mismatch of the frequency viewed by the monochromators is disastrous since all the apparent advantages of using the tandem systems are lost.

For detection of the small number of scattered photons, high-performance photomultiplier tubes are used. Modern tubes have low internal noise and high gain. The amplification methods have evolved from lock-in amplification and direct-current amplification; currently most instruments are equipped with a pulse-counting system. In this, the individual photoelectron pulses from the phototube are detected and processed individually through an adjustable gating system and subsequently integrated over a predetermined time period; the output is recorded on a strip chart recorder as a function of wavelength or frequency.

A number of companies supply Raman instruments as a complete integrated package along with a broad range of sampling devices; these allow spectra to be obtained easily and under special conditions, such as at high and low temperatures and under high pressure.

3.2. Raman spectral identification of amino acids

Raman spectroscopy has played a role in the understanding of the structure of amino acids in aqueous solutions. The classic work of Edsall, starting in 1936, established much of the basis for interpreting the spectra of amino acids. The Raman spectra of the various forms of glycine are given in Table 3.1 (Krishnan and Plane, 1967).

TABLE 3.1. Raman spectra of various forms of glycine

Wavenumbers (cm^{-1})			Assignments
$H_2NCH_2COO^-$	$H_3^+NCH_2COO^-$	$H_3^+NCH_2COOH$	
507 m	502 m	494 m	CO_2^- rock; NH_3^+ torsion
575 w	577 w	—	CO_2^- wag
—	665 w	644 w	CO_2^- scissor
899 m	896 s	869 s	C—C str
968 w	—	—	CH_2 rock?
—	1027 m	1042 m	C—N str
1100 m	1118 m	1123 w	N^+H_2 twist; NH_2^+ rock
1169 w	—	—	CH_2 twist
1315 m	—	—	NH_2 wag?
1343 s	1327 s	1308 s	CH_2 wag
1407 vs	1410 s	—	CO_2^- sym str
1440 sh	1440 m	1433 m	CH_2 scissor
1611 vs	1615 s	1644 vs	CO_2^- antisym str
—	—	1744 s	C=O str
—	2879 w	—	CH_2 str
2935 vs	2968 s	2975 vs	CH_2 str
—	3011 m	3011 s	CH_2 str
3320 m	—	—	NH_2 str
3385 m	—	—	NH_2 str

s, strong; m, medium; w, weak; v, very; b, broad; sh, shoulder.

Edsall laid the foundations for the understanding of the vibrations of the amino group and was the first to utilize Raman polarization spectra as an aid to the interpretation of the spectra of amino acids. With model amine compounds, he established that the symmetric stretching mode of the NH_2 group gives rise to a strong polarized line at 3320 cm^{-1} and the asymmetric stretching mode is weak and depolarized at 3380 cm^{-1} (Edsall and Scheinberg, 1940). The charged species of $-NH_3^+$, $-NH_2^+$ or $-NH^+$ have no Raman lines above 3300 cm^{-1} (Edsall, 1937). When the free amino acids of glycine and alanine were studied, no lines were observed above 3300 cm^{-1}, so it was confirmed that the amino group in the isoelectronic amino acid carries a positive charge and that these molecules exist as dipolar ions (or zwitterions) $H_3^+NRCOO^-$ (Edsall, 1937).

The uncharged amino group of the amino-acid anions gives rise to two Raman lines, one of moderate intensity near 3370 cm^{-1} (the antisymmetric stretching mode) and the other of high intensity near 3305 cm^{-1} (the symmetric stretching mode) (Takeda et al., 1958). When H is replaced by D, these frequencies are displaced to 2500 and 2435 cm^{-1}, respectively. The frequency ratio of the N—H to the corresponding N—D vibrations is consistently 1·35–1·36 (Edsall and Scheinberg, 1940; Takeda et al., 1958). For the zwitterions and cations containing the $-\text{ND}_3^+$ group, there is a broad weak line occurring in the 2175–2200 cm^{-1} region, and the line for the NH_3^+ group is expected near 2970 cm^{-1}, but is masked by the sharper and more intense lines due to the C—H vibrations.

The ionized carboxyl group gives rise to two principal stretching frequencies: an asymmetrical stretching mode in the region 1570–1600 cm^{-1}, and a symmetrical stretching mode near 1400 cm^{-1}. The 1400 cm^{-1} line is very intense in the anionic and dipolar ion forms of all the amino acids and is essentially unaffected in position or intensity by ionization or deuteration of the amino group. The asymmetrical stretching frequency of the ionized (COO$^-$) group is particularly easy to recognize in D_2O and occurs near 1565 cm^{-1} for all anionic amino acids (Takeda et al., 1958).

It is important to determine whether particular amino acids can be detected unambiguously in aqueous solutions of proteins. A simplified first step in this direction is a spectroscopic analysis of the individual amino acids; the Raman spectra of 20 amino acids at various pH have been obtained by Lord and Yu (1970a). Laser-excited spectra are of sufficient quality to allow the detection of most of the residues, so that these spectra have been used for the analysis of the Raman spectra of proteins.

The Raman spectrum of lysozyme was interpreted by Lord and Yu (1970a) by superimposing the Raman spectra of the 20 different kinds of amino acids making up the molecule. They have tabulated the spectra of the amino acids and have given peak intensities relative to an internal standard and corrected for the frequency dependence of the instrument response. Amino-acid mixtures were prepared and the resultant spectra were superimposed to simulate the spectra of the completely hydrolyzed lysozyme in acidic solution. The correspondence between the Raman spectrum of lysozyme and superposition spectra of the constituent amino acids (Fig. 3.5) is excellent. The amino acids, trytophan, phenylalanine and tyrosine, with aromatic side chains, give very intense and highly characteristic frequencies which are readily discernible in the Raman spectra of proteins.

3.3. Raman spectra of chain conformations

3.3.1. VIBRATIONAL SPECTROSCOPY OF THE α-HELICAL CONFORMATION

It has been established that many polypeptides and proteins exist in the α-helical conformation. In the α-helix, there are 18 peptide residues in five turns and, as mentioned in Section 2.4.2, interchain H bonds are found between each pair of every third neighbours. The factor or point group (see Section 2.1.9) for the helix is a $C_{18/5}$ group. The character table (Section 2.1.11) for the $C_{18/5}$ factor group is shown in Table 3.2. The i.r.-active molecular vibrations are

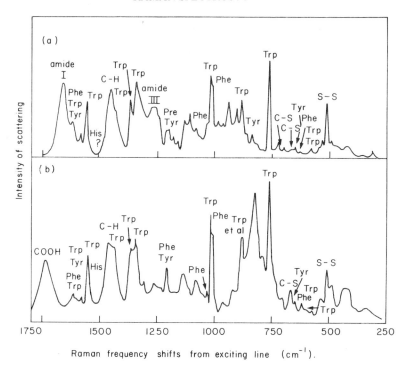

FIG. 3.5. Raman spectrum of lysozyme (a) obtained experimentally in water, and (b) synthesized by superposition of spectra from constituent amino acids.

found in the A and E_1 representations, corresponding to a phase difference of 0 and $2\pi/3\cdot6$ respectively. The A and E_1 vibrations are parallel and perpendicularly polarized with respect to the helix axes, respectively, in the i.r.

In crystal studies, where the molecular axes are fixed, values of the individual tensor components of the polarizability tensor can be measured.

The Raman scattering tensors for the α-helix have been derived (Bamford et al., 1956). The Raman-active modes are found in the A_1, E_1, and E_2 representations; the A modes are polarized in the Raman while the E_1 and E_2 modes are depolarized. For an oriented helix and with polarized radiation, it is possible to isolate each of the A_1, E_1 and E_2 contributions.

Raman polarization measurements are designated by the following notation, $X(zz)Y$. The first capital letter before the parentheses (in this case X) designates the direction of the light propagation; the last capital letter indicates the direction of viewing of the scattered light. The first letter inside the parentheses, which is written in lower case (in this case z), denotes the polarization direction of the incident laser light, while the second lower-case letter in the parentheses denotes the polarization direction of the scattered light. With the $X(zz)Y$ spectral measurement, the A modes are isolated, while the $X(zx)Y$ spectral measurement isolates the E_1 modes. The $Z(xx)Y$ and the $Z(yx)Y$ spectral measurements give rise to both A and E_2 vibrations. These measurements have been made for the Raman spectrum of poly-L-alanine (Fanconi et al., 1969).

Since polyglycine assumes other helical conformations, poly-L-alanine (PLA) is the simplest polypeptide having the α-helical conformation. It is appropriate to examine the vibrational properties of this α-helical polypeptide in some detail in order to understand the spectroscopic implications. The vibrational properties of α-poly-L-alanine have been investigated by i.r. spectroscopy (as discussed in Section 2.4.1) (Bamford et al., 1956; Itoh et al., 1968), Raman spectroscopy (Fanconi et al., 1969; Koenig and Sutton, 1969; Fanconi et al., 1971) and normal coordinate analysis (Fanconi et al., 1971; Itoh and Shimanouchi, 1970; Miyazawa et al., 1967); for far i.r. see p. 136.

TABLE 3.2. Group character table for $C_{18/5}$ factor group

$C_{18/5}$	E	$C_{18/5}^1$	$C_{18/5}^2$	$C_{18/5}^3$	...	$C_{18/5}^{16}$	$C_{18/5}^{17}$	ir	Raman
A	1	1	1	1		1	1	z	$z^2; x^2+y^2-i(xy-yx),$ $x^2+y^2+i(xy-yx)$
ε_1	1	$e^{i5\pi/9}$	$e^{i10\pi/9}$	$e^{i15\pi/9}$	...	$e^{i80\pi/9}$	$e^{i85\pi/9}$	$x+iy$	$(xz-iyz, zx+izy)$
ε_2	1	$e^{i10\pi/9}$	$e^{i20\pi/9}$	$e^{i30\pi/9}$	...	$e^{i180\pi/9}$	$e^{i170\pi/9}$		$[x^2-y^2+i(xy+yx)]$
ε_3	1	$e^{i15\pi/9}$	$e^{i30\pi/9}$	$e^{i45\pi/9}$	...	$e^{i240\pi/9}$	$e^{i255\pi/9}$		
$\vdots$		$\vdots$	$\vdots$	$\vdots$		$\vdots$	$\vdots$		
$B = \varepsilon_9$	1	$e^{i45\pi/9}$	$e^{i90\pi/9}$	$e^{i135\pi/9}$	...				
$\vdots$		$\vdots$	$\vdots$	$\vdots$					
$\varepsilon_3^* = \varepsilon_{15}$	1	$e^{i75\pi/9}$	$e^{i150\pi/9}$	$e^{i225\pi/9}$	...				
$\varepsilon_2^* = \varepsilon_{16}$	1	$e^{i80\pi/9}$	$e^{i160\pi/9}$	$e^{i240\pi/9}$	...				$[x^2-y^2-i(xy+yx)]$
$\varepsilon_1^* = \varepsilon_{17}$	1	$e^{i85\pi/9}$	$e^{i170\pi/9}$	$e^{i255\pi/9}$	...			$x-iy$	$(xz-iyz, zx-izy)$
A	1	1	1	1				z	$z^2; (x^2+y^2, xy-yx)$
E_1	2	$2\cos(5\pi/9)$	$2\cos(10\pi/9)$	$2\cos(15\pi/9)$	...			x, y	$(xz, zx; yz, zy)$
E_2	2	$2\cos(10\pi/9)$	$2\cos(20\pi/9)$	...					$(x^2-y^2; xy+yx)$
E_3	2	$2\cos(15\pi/9)$	$2\cos(30\pi/9)$	...					
$\vdots$		$\vdots$	$\vdots$						
E_8	2	$2\cos(40\pi/9)$	$2\cos(80\pi/9)$	...					

The optical frequencies represent frequencies for particular values of θ on the phonon (lattice vibration) *dispersion curve* (the real component corresponding to the normal absorption), where θ is the phase difference between vibrations of equivalent atoms in adjacent repeat units. The frequencies of the active modes for the α-helix occur at $\theta = 0°$ (A species), $\theta = 100°$ (E_1 species) and $\theta = 200°$ (E_2 species). The E_2 frequencies are at $\theta = 160°$ because of periodicity of the Brillouin zone. In the phonon dispersion curves for α-PLA (Figs. 3.6 and 3.7; Fanconi et al., 1971), the intercepts on the vertical lines at the values of $\theta = 0$, 100, and 160° give the theoretical frequencies which should correspond to the i.r. and Raman data.

The dispersiveness of the curves is an indication of the coupling of the particular normal mode within adjacent repeat units. When the frequency

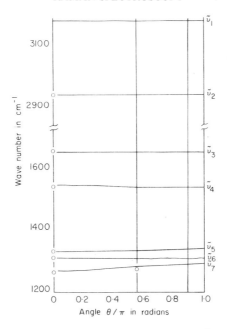

FIG. 3.6. High-frequency range (wavenumber above 1200 cm^{-1}) of dispersion curves for poly-L-alanine (continued in Fig. 3.7; see also Table 3.3).

curves are flat as a function of θ, the vibrations are localized within the chemical repeat unit, indicating that the motion of the adjacent unit has no effect on the frequency. Usually, the internal motions correspond to bond stretching or angle deformations of substituents on the backbone of the polymer chain; these modes correspond to the higher frequency portions of the spectrum. The external modes, i.e. those showing change in frequency with phase angle, usually involve skeletal motions and occur at lower frequencies.

It has been well established that the *trans* -CONH group undergoes highly localized vibrations and exhibits characteristic bands at: 3280 (amide A), 3090 (amide B), *ca.* 1650 cm^{-1} (amide I), 1550 cm^{-1} (amide II) and 1300–1250 cm^{-1} (amide III), 627 (amide IV), 725 (amide V), 600 (amide VI) and 206 (amide VII). Note that these are similar to, but not identical with, the corresponding i.r. frequencies (Section 2.4.1). The amide A and B frequencies are believed to arise from Fermi resonance between the first overtone of the amide II vibration and the N—H stretching vibration. In the Raman spectra, the amide A and B lines are usually marked by water lines and aromatic C—H stretching lines. Lord and Yu (1970a) suggest that the amide A line can be observed for aqueous solutions of proteins if the spectrum is obtained immediately after dissolution in D$_2$O, since the exchange rate of the peptide hydrogens is slow. The amide I vibration is primarily C—O stretching (80%), with some contributions from the C—N stretching and the N—H in-plane

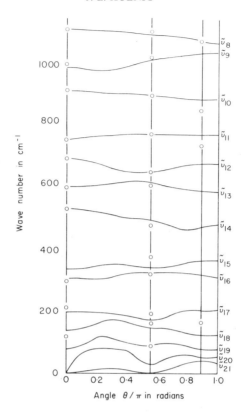

FIG. 3.7. Low-frequency range (wavenumber below 1200 cm^{-1}) of dispersion curves for poly-L-alanine (continued from Fig. 3.6).

bending. The amide I in the α-helix shows very little splitting of the A and E_1 components; experimental resolution of the two components has not been made in the Raman so only a single line is observed at 1651 cm^{-1}. The polarized i.r. of poly-γ-benzyl L-glutamate shows a parallel band at 1650 cm^{-1} and a perpendicular band at 1652 cm^{-1}.

The amide II mode is primarily the N—H in-plane deformation (65%) with a C—N stretching contribution. This mode in the α-helix gives an intense perpendicular i.r. band (1545 cm^{-1}) but only a weak broad Raman line at 1549 cm^{-1}. The calculated frequencies of the amide II modes show very little splitting but it has been suggested that the parallel i.r. mode at 1515 cm^{-1} in poly-benzyl-L-glutamate is the A component of the amide II mode.

The amide III mode is calculated at 1305 cm^{-1} for N-methyl acetamide and at 1328 (A), 1333 (E_1) and 1340 (E_2) mode for α-PLA. Fanconi et al. (1971) assign the line at 1339 cm^{-1} in the Raman for poly-L-alanine to this mode, while Koenig and Sutton (1969) suggest the 1310 cm^{-1} line. Both lines disappear on N-deuteriation. This frequency range is further complicated by the presence of the C_α-H deformation mode.

The amide IV line was assigned to O=C—N in-plane bending, amide V to N—H out-of-plane bending, amide VI to C=O out-of-plane bending and amide VII to torsion about the C—N bond. All of these amide modes are expected to be weak in the Raman effect.

The remainder of the vibrational modes are not easily classified because of the large interaction of several different internal coordinates. The Raman lines at 909 and 889 cm^{-1} may be the C-methyl stretching frequency, which should be fairly localized; most peptides usually have a strong Raman line in this frequency range.

The low-frequency normal modes (below 500 cm^{-1}) are mainly skeletal motions and are highly phase-dependent. These frequencies are expected to be highly sensitive to polymer conformation but they will also depend on the residue, so clear-cut conformational assignments seem unlikely. Fanconi et al. (1971) assign the lines at 371 and 378 cm^{-1} of α-PLA as characteristic of the α-helix, but no Raman lines are found in this region for poly-L-leucine. This difference between alanine and leucine illustrates the sensitivity of these modes to their environment.

3.3.2. RAMAN SPECTRA OF α-HELICAL POLYPEPTIDES (AMIDE MODES)

The Raman spectra of four polypeptides in the α-helical form have been reported—poly-L-alanine (Koenig and Sutton, 1969) (α-PLA); poly-γ-benzyl-L-glutamate (Koenig and Sutton, 1971) (α-PBLG); poly-L-lysine hydrochloride (Koenig and Sutton, 1970) (α-PLL ·HCl); poly-L-leucine (Koenig and Sutton, 1971) (α-PLL). A strong Raman line at 1652 cm^{-1} for α-PLL, α-PBLG, α-PLL ·HCl and α-PLA is the amide I mode. Since this mode is not experimentally resolvable in the Raman spectra, the A, E_1 and E_2 modes are apparently common at 1652 cm^{-1}.

In the amide II region, a weak, broad Raman line is found at 1547 cm^{-1} for α-PLL, 1549 cm^{-1} for α-PLA, with only broad diffuse absorption in this region in α-PBLG. This weak line probably corresponds to the A mode since a strong perpendicular i.r. band is found near 1540 cm^{-1} for these polymers. One would be hard pressed to determine that the i.r. and Raman frequencies are experimentally different in this region.

In the amide III region, a Raman line is observed at 1320 cm^{-1} for α-PLL and α-PLL ·HCl, at 1336 cm^{-1} for α-PBLG and two lines at 1339 and 1310 cm^{-1} for α-PLA. As mentioned in the Section 3.3.1, the C_α—H deformation is also expected in this frequency region of 1200–1300 cm^{-1}.

The remaining lower-frequency amide modes (IV, V, VI, VII) are difficult to assign from the Raman effect, due to their inherent weakness and the presence of strong C—C motions from the skeleton of the α-helix. Assignments for the Raman spectrum of poly-L-alanine (Fig. 3.8) are given in Table 3.3 (Fanconi et

TABLE 3.3 Observed and calculated wavenumbers of α-poly-L-alanine (see Fig. 3.8)

Phonon[b] curve no.	Wavenumbers (cm^{-1})			Assignments[a]
	Raman	i.r.	Calc.	
1	3308		$3184(A, E_1, E_2)$	$r_{NH}(96)$
	2998	2988		Methyl symmetric stretch
	2950	2938		Methyl symmetric stretch
2	2892	2883	$2939(A, E_1, E_2)$	$r_{C'H}(100)$
3			$1657(A)$	$r_{CO}(71), r_{CN}(17)$
	1659	1657($\parallel$)		
3			$1655(E_1, E_2)$	$r_{CO}(72), r_{CN}(18)$
	1587			
	1563			
4			$1545(A)$	$\phi_{NH}(65), r_{CN}(38)$
4	1543	1545	$1542(E_1)$	$\phi_{NH}(69), r_{CN}(32)$
4			$1541(E_2)$	$\phi_{NH}(68), r_{CN}(33)$
	1528			
	1462	1457($\parallel$)		Methyl asymmetric deformation
	1405			
	1390	1384($\perp$)		Methyl symmetric deformation
5			$1340(E_2)$	$r_{CC'}(34), r_{CN}(25), \phi_{NH}(20)$
5	1339	1330($\parallel$)	$1333(E_1)$	$r_{CC'}(21), r_{CN}(17), \phi_{HC'C}(16), \phi_{NH}(17)$
5			$1328(A)$	$\phi_{HC'C}(37), \phi_{HC'N}(16), r_{CO}(14), \phi_{NH}(10)$
6	1309	1308($\parallel$)	$1309(A, E_1, E_2)$	$\phi_{HC'M}(57), \phi_{HC'N}(41)$
7			$1290(E_2)$	$\phi_{HC'C}(57), \phi_{HC'M}(35)$
7	1275	1274($\perp$)	$1282(E_1)$	$\phi_{HC'C}(46), \phi_{HC'M}(20), r_{CC'}(15)$
7	1262	1262($\parallel$)	$1266(A)$	$r_{CC'}(32), \phi_{HC'C}(28), r_{CN}(20)$
	1167	1167($\parallel$)		Methyl rocking
8			$1112(A)$	$r_{C'N}(61), r_{C'C}(20)$
8	1106	1107($\perp$)	$1093(E_1)$	$r_{C'N}(48), r_{C'C}(12)$
8	1072		$1076(E_2)$	$r_{C'N}(45), r_{CN}(10)$
	1058	1050($\perp$)		Methyl mode
9			$1034(E_2)$	$r_{CN}(17), r_{C'C}(15), r_{CO}(10)$
9	1017		$1012(E_1)$	$r_{CC'}(18), r_{C'N}(16)$
9	994		$984(A)$	$\phi_{CO}(19), r_{CC'}(13), \phi_{C'NC}(12), r_{CN}(10)$
10	909	906($\parallel$)	$911(A)$	$r_{C'M}(64)$
10	889	893($\perp$)	$896(E_1)$	$r_{C'M}(50), \pi_{CO}(14), \phi_{MC'C}(10)$
10	842		$883(E_2)$	$r_{C'M}(57), \phi_{MC'C}(15)$
11		772	$771(E_1)$	$\pi_{NH}(28), r_{C'M}(24), \pi_{CO}(20)$
11	757		$752(A)$	$\pi_{NH}(44), \pi_{CO}(18)$
11	728		$769(E_2)$	$\pi_{NH}(34), \pi_{CO}(30), r_{C'M}(10)$
12	695	686	$691(A)$	$\pi_{CO}(29), \phi_{CC'N}(17), \phi_{NCC'}(13)$
12			$673(E_2)$	$\phi_{CO}(38), \pi_{NH}(11)$

TABLE 3.3—continued

Phonon[b] curve no.	Wavenumbers (cm^{-1})			Assignments[a]
	Raman	i.r.	Calc.	
12	662	653	652(E_1)	$\phi_{CO}(28),\pi_{NH}(12)$
13	610	610	611(E_1)	$\pi_{NH}(30),\pi_{CO}(29)$
13		595	599(A)	$\pi_{NH}(31),\pi_{CO}(15),\tau_{CN}(13)$
13			583(E_2)	$\pi_{CO}(26),\pi_{NH}(24),\tau_{CN}(11)$
14	531	526	530(A)	$\phi_{CO}(36),r_{C'M}(15),\phi_{NC'M}(10)$
14	469		493(E_1)	$\phi_{CO}(21),\phi_{NC'M}(15),\tau_{CN}(10)$
14			472(E_2)	$\pi_{CO}(25),\pi_{NH}(20)$
15			363(E_2)	$\phi_{NC'M}(62)$
15	378	371	339(E_1)	$\phi_{C'O}(19),\phi_{MC'C}(16),\phi_{C'NC}(12)$
15			334(A)	$\phi_{MC'C}(35),\phi_{C'NC}(12),\pi_{CO}(12)$
16	314	325	324(E_1)	$\phi_{NC,M}(45),\phi_{MC'C}(16)$
16			313(E_2)	$\phi_{C'NC}(20),\phi_{CO}(21),\phi_{CC'N}(17)$
16	294	287	304(A)	$\phi_{NC'M}(35),\phi_{MC'C}(22),\phi_{CC'N}(11)$
17	214		201(E_2)	$\phi_{MC'}(41),\phi_{NCC'}(16)$
17			200(A)	$\phi_{CC'N}(28),\tau_{CN}(10),\tau_{CC'}(15)$
17	194	190	173(E_1)	$\phi_{MC'C}(20),\phi_{NCC'}(13)$
18	165	167	147(E_1)	$\phi_{C'NC}(19),\tau_{CC'}(15),\phi_{NCC'}(12)$
18	159		125(E_2)	$\tau_{C'N}(34),\tau_{CN}(16),\pi_{NH}(13)$
18		120	137(A)	$\phi_{NCC'}(20),\phi_{C'NC}(19),\phi_{NC'M}(15)$
19			90(E_1)	$\tau_{CC'}(26),\tau_{C'N}(23),\phi_{CC'N}(14)$
19			80(A)	$\tau_{CC'}(31),\tau_{C'N}(23),\phi_{CC'N}(14)$
19			79(E_2)	$\tau_{CC'}(19),\phi_{CC'N}(12)$
20			53(E_2)	$\tau_{CC'}(47),r_{N..O}(10)$
21			40(E_2)	$r_{N..O}(34),\tau_{CN}(21),\tau_{CC'}$
20			26(E_1)	$\tau_{CC'}(33),\tau_{CN}(18),\tau_{C'N}(14)$

[a] Potential-energy contribution (% in parentheses) among diagonal elements of the force matrix: r, bond stretch; ϕ, angle bend; π, out-of-peptide-plane bend; τ, torsion. [b]Curve numbers from Figs 3.6 and 3.7.)

al., 1971). The methyl modes are easily identifiable and in several cases coincide with the observations in the i.r. and Raman spectra of poly-L-valine. The expected doubling of the methyl modes due to the A and E species is rarely observed, indicating that intra-chain interactions for the methyl residue are small.

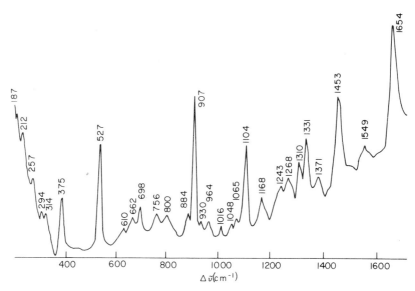

FIG. 3.8. Raman spectrum of poly-L-alanine (PLA) (see Table 3.3).

3.3.3. VIBRATIONAL SPECTROSCOPY OF THE β-CONFORMATION

Another common chain conformation of polypeptides is the so-called β-conformation in which the polypeptide chain is very nearly fully extended; an anti-parallel pleated sheet accommodates the H bonding.

Let us consider polyglycine I as an example of the β-conformation. According to X-ray investigations, the molecule of polyglycine I belongs to a line group whose *factor group* is C_{2v}. As mentioned in Section 2.1.9, the factor group is a combination of symmetry elements associated with all sites within a unit cell. Therefore, the 10-atom (i.e. excluding methylene hydrogens) chain in this structure has $(3 \times 10) - 4 = 26$ normal modes grouped into nine A_1 (polarized Raman and i.r.), four A_2 (Raman), nine B_1 (Raman and i.r.) and four B_2 (Raman and i.r.) modes. The crystal of polyglycine I, with two chains (four chemical residues) per primitive unit cell, has a space group whose factor group is D_2; owing to the interchain H bonding in the crystal, each of the internal modes of the polyglycine I molecule is expected to split into two normal vibrations (Table 3.4)

With the anti-parallel-chain pleated sheet, a unit cell contains four peptide groups; two groups from one chain and two from an adjacent chain. In this case, adjacent chains may move in-phase or out-of-phase ($\delta' = 0, \pi$), as well as the intramolecular phasing ($\delta = 0, \pi$).

Since the i.r. and Raman modes of the crystal arise from in- or out-of-phase motions of corresponding groups in various cells, it is easier and more convenient to understand the splitting of the frequencies in terms of intrachain

TABLE 3.4. Correlation table of the β-conformation of polyglycine

Molecule	Crystal	Activity
C_{2v}	$(\delta, \delta')D_2$	
A_1	$A_1(0, 0)$	R
	$B_1(0, \pi)$	R, i.r.
B_1	$B_2(\pi, 0)$	R, i.r.
	$B_3(\pi, \pi)$	R, i.r.
A_2	$A_1(0, 0)$	R
	$B_1(0, \pi)$	R, i.r.
B_2	$B_2(\pi, 0)$	R, i.r.
	$B_3(\pi, \pi)$	R, i.r.

and interchain interactions. Miyazawa (Miyazawa et al., 1967) has derived the appropriate equations (see Section 2.4.2):

$$v(0, 0) = v_0 + D_1 + D_1'.$$
$$v(0, \pi) = v_0 + D_1 - D_1'.$$
$$v(\pi, 0) = v_0 - D_1 + D_1'.$$
$$v(\pi, \pi) = v_0 - D_1 - D_1'.$$

where D_1 and D_1' are the intra- and interchain interaction terms respectively. These equations are useful in making vibrational assignments for the β-form.

Although all four frequencies are expected in the Raman, only one Raman line at 1665 cm^{-1} is found in the oligomers of polyglycine in the amide I region. A single line at 1674 cm^{-1} is reported for polyglycine I but this frequency appears to lie on the upper edge of the experimental limit when compared with other measurements (Small et al., 1970). Two of the three i.r. lines are observed at 1685 and 1636 cm^{-1}. From these frequencies, the three parameters, v_0, D_1 and D_1' can be calculated for polyglycine I (Smith et al., 1969).

The amide II mode is weak but two lines are observed at 1564 and 1515 cm^{-1} in the Raman spectra of polyglycine I and a single strong line at 1517 cm^{-1} in the i.r. In the Raman spectra of polyglycine I in the amide III region, lines are observed at 1252, 1234, and 1219 cm^{-1} and weak i.r. lines at 1243, 1230 and 1214 cm^{-1}. The lines at 1234 and 1230 and at 1219 and 1214 cm^{-1} are the same within experimental error. Therefore, all four modes are observed. The 1252 cm^{-1} mode is assigned to the i.r.-inactive A_1 $v(0, 0)$ mode as it is the only mode of the four not observed in the i.r. spectrum. This amide III mode will probably have different frequencies for different β structures, since the interaction terms can have a large range of values for this mode. Since the Raman spectra of the α- and β-forms differ appreciably, little difficulty is expected in differentiating between these two conformations.

The Raman spectra have been obtained for several polypeptides in the β-form, including poly-L-valine (β-PLV), poly-L-serine (β-PLS), polyglycine (PGl), and oligomers of poly-L-alanine (β-PLA). In the amide I region, one observes a strong Raman line at 1666 cm^{-1} for β-PLV, at 1665 for PGl, at 1663 for β-PLA and at 1668 cm^{-1} for β-PLS. These modes have the same frequency within the limits of experimental error. It is most likely that this unique Raman line is the A_1 $v(0, 0)$ mode. The i.r. bands at 1685 and 1625 cm^{-1} have been established as characteristic of the β structure and the Raman line at 1666 cm^{-1} falls into the same category.

The amide II region (see Section 2.4.7 for keratin in the i.r.) is rather weak and diffuse in the Raman effect for the β conformation and peak positions are difficult to determine. The characteristic i.r. frequencies in the 1530 cm^{-1} and 1555 cm^{-1} are apparent in these polymers. It would appear that the amide II mode is of little value in the Raman effect for the β-structure because of its broad, weak character.

The amide III region of the Raman spectra is rich with lines in the β-form. For β-PLV, three lines at 1291, 1276 and 1231 cm^{-1} are found; for the oligomers of β-PLA, lines are observed at 1268, 1250 and 1231 cm^{-1}; for β-pentaglycine, lines are observed at 1296, 1252 and 1234 and for PGl at 1295, 1255, 1234 and 1220 cm^{-1}. For PGI, Small, Fanconi and Peticolas assigned the 1255 cm^{-1} mode to the CH_2 twisting mode and the 1234 and 1220 cm^{-1} modes to the amide III. A single line is observed at 1235 cm^{-1} in β-PLS. It is apparent that this region is rich in lines and is highly diagnostic of the β-conformation. Although the frequencies are slightly variable because of the extensive coupling, this frequency range should be inspected when β-conformation is suspected and the line at 1230–1235 cm^{-1} seems most diagnostic.

3.3.4. VIBRATIONAL SPECTROSCOPY OF OTHER HELICAL FORMS OF POLYPEPTIDES

In addition to the α-helical conformation, other helical conformations can be assumed by polypeptides. Polyglycine can exist as a 3_1 helix (PGII) as well as the β-form; in this conformation, the molecule has factor-group symmetry isomorphous with the $C(\frac{2\pi}{3})$ point group and the chain vibrations can be classified as symmetric modes (A: $\delta = 0$) and degenerate asymmetric modes (E: $\delta = 2\pi/3$). Normal coordinate analysis and the phonon dispersion for PGII have been reported (Small et al., 1970). The A modes are polarized in the Raman and have parallel i.r. dichroism, while the E modes are depolarized and have perpendicular i.r. dichroism. The primary differences between the i.r. and Raman spectra of the polypeptide in the 3_1 helix conformation are in the intensities of the modes; i.e. some modes which are weak or absent in the Raman spectra may appear strong in the i.r. spectra and vice versa. The symmetric A modes are expected to be the strongest in the Raman spectra with the E modes strongest in the i.r. spectra.

3.3.5. VIBRATIONAL SPECTROSCOPY OF THE RANDOM-COIL CONFIGURATION
It is possible for polypeptides to exist in the random-coil conformation as well as the extended β and folded helical conformations. Poly-α-L-glutamate and poly-L-lysine have been prepared in the random-coil conformation. In the random coil conformation, the frequency shifts due to the interchain interactions, as well as the intrachain interaction, will average zero. This will result in broad Raman lines indicative of liquid scattering. In addition, the frequency shifts arising from coupling of modes from units having identical structure and energies will be small. While this latter effect may result in little change in frequency, it will cause a substantial increase in the extinction and scattering coefficients in the i.r. and Raman, respectively. Infra-red studies indicate that the amide I band is at 1655 cm^{-1} for the random-coil conformation whose peptide groups are H-bonded with others. The amide II appears at 1535 cm^{-1} in the i.r. for the random coil. Unfortunately, these i.r. frequencies resemble those in the α-helix in the amide I region (1656 cm^{-1} for random and 1650 cm^{-1} for α-helix); the amide II band is weak for the random-coil and is often interfered with by the band due to the ionized carboxyl mode.

The random-coil conformation is apparently as illusive for detection by Raman spectroscopy as with infra-red (p. 62). The major differences involve a small lower-frequency shift in the amide I region and higher-frequency shift and a greatly enhanced intensity of the amide III mode for the coil compared to the α-helix.

3.3.6. RAMAN SPECTRA OF AQUEOUS SOLUTIONS OF POLYPEPTIDES
Although considerable information exists about the structure of polypeptides in the solid state, much less information is available about their conformation in aqueous solutions. The reasons for this are primarily the lack of proper instrumental techniques for the determination of the conformation on a molecular level. The overall size and shape of the molecule can be measured by light and X-ray scattering techniques, but more detailed information is lacking. As a result, many questions remain unanswered. Does a structural

change occur when a polypeptide is dissolved in water? What structural changes are induced by environmental differences such as pH, ionic strength, temperature and mixed solvents?

Raman studies can attempt to answer many questions of this type, due to the sensitivity of the Raman effect to conformational differences and the minimal interference of the Raman scattering of water. Some difficulties arise since the levels of concentration required for quality Raman spectra are generally higher than occur *in vivo*, so that questions of the concentration dependence of the results can justifiably be asked. With the continued increase in laser power and spectrophotometric detection and amplification, appropriately low concentrations of polypeptides can be studied and so the concentration dependence of the structural changes can be studied. At the present time, Raman spectroscopy is one of the techniques offering fast, easy and detailed structural information for the study of biopolymers in aqueous solutions.

The scattering due to water interferes with the measurement of the amide I mode, but studies of D_2O solutions are not encumbered with this difficulty; so when it is desirable to study the amide I region in the Raman effect, D_2O should be used. The other Raman lines due to water at 800, 450 and 175 cm^{-1} are less of a problem because of their low intensities.

The signal-to-noise ratios of the Raman spectra obtained from aqueous solutions are not as high as for solids or liquids, probably because of the optical inhomogeneities of the aqueous solutions. These inhomogeneities act as Rayleigh scattering centres so that the exciting line is broader in frequency than for liquids and the higher level of scattered light in the monochromator results in a lower signal-to-noise ratio. In addition, since the desire to operate at the lowest possible concentration consistent with good Raman spectra is reflected in the lower signal from the solution, a higher amplification factor is generally used.

The first Raman spectrum of an aqueous solution of a polypeptide, poly-L-lysine, was reported by Edsall and coworkers (Fanconi *et al.*, 1969); it was obtained photographically and was extremely weak. For poly-L-lysine in the random coil, a conformational change is reflected by the shift of the amide III mode from 1216 cm^{-1} to 1243 cm^{-1} in going from the solid state to aqueous solution.

At room temperature and pH 10, it is believed that poly-L-glutamic acid assumes a random-coil form with its carboxyl side groups ionized and the polypeptide backbone hydrated. The amide I mode appears at 1665 cm^{-1} and is quite broad (Lord and Yu, 1970b). In the amide III region, Lord and Yu (1970a) report only one broad line at 1250 cm^{-1}.

The Raman spectra of PGA in aqueous solutions of different pH show spectral shifts indicative of conformational transitions. The amide I modes in D_2O at a pD of 4·5 and 10·5 occur at 1656 cm^{-1}. The methylene wagging and twisting modes and the amide III occur at 1353, 1313 and 1249 cm^{-1} at pH

11·4, but shift to 1338, 1305 and 1238 cm^{-1} at lower pH. The C—C stretch of the C—C—N unit shifts from 949 cm^{-1} to 931 cm^{-1} when the pH is lowered.

3.4. Raman spectra of proteins

The quality of the Raman spectra of proteins is significantly lower than that of the amino acids and polypeptides. Edsall attributes the difficulty of recording quality Raman spectra of proteins to the increase in Rayleigh scattering with increasing size of the molecule, while the Raman scattering is roughly proportional to the mass of the molecule/unit volume. Consequently, the signal-to-noise ratio decreases in systems where the molecules become larger and the molecular weight increases as in proteins. In addition, most proteins show fluorescence which means that the spectra are obtained on an ever-changing background. Since there is always the possibility that the laser beam will damage the molecule in solution, additional testing is required to ensure sample integrity.

Finally, the spectra are expected to be very complex because of the large number of residues and the high molecular weights. Because the intensity of the Raman effect is linear with concentration, many weak lines are expected which cannot be distinguished individually; the overall spectra become diffuse with broad, overlapping, weak Raman lines.

Edsall (Garfinkel and Edsall, 1958) obtained the first Raman spectrum of a protein, lysozyme, with an Hg arc source; altogether 14 faint lines were reported. More recently, Raman spectra of crystalline lysozyme, pepsin and α-chrymotrypsin have been obtained with laser excitation (Tobin, 1968). Tobin (1968) reports a total of 21 Raman lines with about half of these frequencies corresponding to the previous observations. Finally, Lord and Yu report the spectra of lysozyme (Small *et al.*, 1970), ribonuclease (Lord and Yu, 1970b) and α-chymotrypsin (Lord and Yu, 1970b).

The peptide groups of lysozyme have the amide I mode at 1660 cm^{-1}, amide II at 1567 cm^{-1} and the amide III appears as a triplet at 1240, 1262 and 1274 cm^{-1}. The amide I mode was characterized by its narrow bandwidth (half-width of about 30 cm^{-1}) compared to the more normal half-width of 50 cm^{-1}. Lord and Yu interpret the narrow bandwidth as reflecting that the majority of the amide groups in the backbone of native lysozyme are not appreciably solvated. They further suggest that the three amide III peaks might possibly arise from the three distinct structural components of crystalline lysozyme, i.e. the α-helix with residues 5–15, 24–34 and 88–96, the anti-parallel pleated sheet and the sequences which are folded in irregular ways (Lord and Yu, 1970a).

In the region 500–725 cm^{-1} in the Raman spectrum of lysozyme, the C—S stretching frequency of the C—S—S—C group is observed as a weak line at 661 cm^{-1}, while in the simulated spectra the line shows up quite strongly at 667 cm^{-1}. Lord and Yu interpret this difference as reflecting that the conformations of the C—S—S—C crosslinks in lysozyme are different from those of the monomer analogues in solution (Lord and Yu, 1970a). It appears that the C—S—S angles in lysozyme are near 114° while they are at 104° for cystine hydrohalides. The differences in angle are reflected in the frequency differences between the symmetric and asymmetric C—S stretch and in the relative intensities of the $v(S-S)$ and $v(C-S)$ modes in the Raman spectra.

The Raman spectrum of native ribonuclease has been reported by Lord and Yu (1970b) in solution and by Koenig and Frushour in the solid and aqueous phase. The Raman spectra of the solution and solid state are shown in Figs 3.9 and 3.10 and the frequencies are tabulated in Table 3.5. The spectrum was

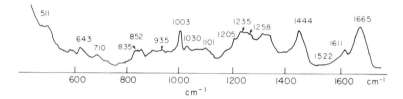

FIG. 3.9. Raman spectrum of ribonuclease in the solid state.

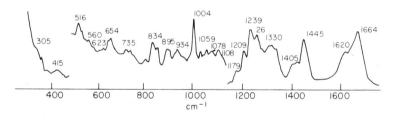

FIG. 3.10. Raman spectrum of ribonuclease in aqueous solution (10% concentration, $pH = 4.0$).

interpreted with the aid of the spectra of the constituent amino acids obtained by the superposition of the spectra of mixtures of amino acids (Lord and Yu, 1970b). The frequencies of the *p*-hydroxyphenyl ring are at 836 and 855 cm^{-1} in the ribonuclease spectrum and these frequencies are usually characteristic of the ionized *p*-hydroxyphenyl ring. However, Lord and Yu (1970b) suggest that these results may not indicate that the *p*-hydroxyphenyl rings of tyrosine are ionized but may be due to tyrosyl-carboxylate-ion side-chain interactions. Six tyrosyl groups are reported to exist as two different types but the Raman spectrum of native ribonuclease does not show the expected line splitting. The S—S and C—S stretching frequencies of the four C—S—S—C crosslinks appear at 516 and 659 cm^{-1}, respectively, with the ratio of their intensities being about one. This suggests that the C—S—S angles around the S—S crosslinks in ribonuclease are near 104° and are, therefore, smaller than in

TABLE 3.5. Raman and i.r. spectra of aqueous and solid ribonuclease (Wavenumbers/cm^{-1})

Raman (aqueous)	Raman (solid)	i.r. (solid)	Assignment
		3300 sb	v(NH)
		3200 mh	
	3052 mh	3060 mh	
	2982 mh	2965 mb	
	2951 sb	2935 mb	
	2902 mh		v(CH)
	2832 wb	2875 mh	
1664 sb	1665 sh	1650 sh	Amide I
1620 mh	1611 wb		
	1522 wb	1525 sh	Amide II
1445 sh	1444 sh	1445 wh	δ(CH), δ(CH$_2$), δ(CH$_3$)
1426 mb			
1405 mb	1396 wb	1395 wh	v(COOO$^-$)s
1330 mb	1317 mb	1340 wb	
1312 mh		1305 wb	
1283 wh			
1261 sh	1258 mb	1260 wh	Amide III
1239 sh	1235 mb	1230 wb	Amide III
1209 mh	1205 mh		Tyr, Phe^a
1179 mh	1175 wh	1180 wh	Tyr, Phe^a
1153 wh			
1109 wh	1101 wb	1115 mh	
1078 wh			skeletal modes
1059 wh			v(C—C), v(C—N)
1033 wh	1030 wb		
1004 sh	1003 mh		Phe^a
956 wh	962 wb		
934	935 wb		v(C—C)
	901 wb		
895 mb			v(C—C)
853 mh	852 wh		Tyr^a
834 mh	835 wh		Tyr^a
748 wb			
737 wh			
725 wb	710 wb		v(C—S) Methionine^a
654 mb			v(C—S) disulfide
647 mb	643 wh		Phe^a
623 wh	618 wh	620 mh	
583 wb			
560 wb			
532 wh			
516 mh	511 wb		v(S—S)
415 wb			
351 wh			
305 wh			

s = strong, m = medium, w = weak, b = broad, h = sharp
[a] Assignments based on work of Lord and Yu (Koenig and Sutton, 1970).

lysozyme. The methionine residue is in the *trans* form as indicated by the line at 725 cm^{-1}.

Differences are noted between the spectra of the aqueous and solid ribonuclease. The amide I line occurs at 1664 and 1665 cm^{-1} in the aqueous and solid phase, respectively, indicating the possible presence of some β-like conformations. Lines are observed in the amide III regions for ribonuclease at 1261 and 1239 cm^{-1} in the spectrum of the aqueous solution and 1258 and 1235 cm^{-1} in the spectrum of the solid. The intensity of the amide III is consistent with 20% α-helical content; upon dissolution, the amide III intensity increases, indicating further loss of order. The Raman lines of the aqueous solution spectra are sharper than the lines of the solid spectra, indicating that the solid state was partially dehydrated.

3.5. Raman spectra of polynucleotides and nucleic acids

Raman spectroscopy can be used for the identification of purine and pyrimidine in the base residues in aqueous solution of nucleic acids. In the Raman spectra of the bases adenine (A), uracil (U), guanine (G) and cytosine (C), intense characteristic Raman lines are present which are not appreciably affected by protonation, deuteriation, deprotonation, or ionizations of the phosphate groups, by chemical substitution at external positions, or by changes in secondary structure (Lord and Thomas, 1967). These characteristic lines (Tables 3.6–3.9) have been used by Thomas (1970) in the analysis of the Raman spectra of ribosomal RNA. The purine rings give very intense lines at 1575 cm^{-1}, arising from the adenine and guanine residues, and at 1484 cm^{-1}, predominantly due to the guanine base. The pyrimidine rings give lines at slightly lower frequencies in the 1400–1300 cm^{-1} region; the line at 1340 cm^{-1} arises from uracil and at 1300 cm^{-1} from cytosine. In the 800–650 cm^{-1} region of the spectrum, intense lines also are indicative of the bases: the line near 720 cm^{-1} is due mainly to the adenine ring, the line at 675 cm^{-1} to guanine and that at 785 cm^{-1} to uracil and cytosine; these lines are only slightly shifted by deuteriation.

The bases can exist in different tautomeric forms in solution. For example, guanine and its 9-N-substituted derivatives can exist either in the keto or enol forms in neural solution, as shown below:

TABLE 3.6. Frequencies cm^{-1} characteristic of the uracil ring[a]

Conditions of observation		Uracil	Uridine	UMP-5'	1-Methyl-uracil	1,3-Dimethyl-uracil	Intensity Raman[b]	i.r.
Solids and all solutions		790±8	786±4	786±4	768±13	691±4	s,PP	vw
		822§	830§	810±10	802±10	804±4	m,PP	s
		1390±10	1392±10	1391±11	1388±2	1382±2	s,P	m·s
All forms except alkaline solutions	Non-deuteriated	1236±1	1233±1	1233±2	1237±5	1273±5	vs,P	vw
	Deuteriated	1268±2	1248±1	1249±1	1271	—	vs,P	vw
Alkaline solutions only		1034±1	1020±2	1018±2	1011±1	—	m,DP	w
		1212±2	1209±1	1209±1	1208±3	—	s,DP	w
		1284±11	1296±2	1296±2	1275±1	—	vs,DP	m·s

[a] In this and Tables 3.7–3.9 the wavenumber entered is a mean value for what is believed to be the same vibration over the several conditions of observation reported at the left.
[b] PP, P and DP denotes strongly polarized, polarized and depolarized lines respectively.

TABLE 3.7. Frequencies characteristic of the adenine ring

Conditions of observation	Adenine	Adenosine	AMP-5'	9-Methyl-adenine	Intensity Raman	i.r.
Solids and all solutions	714±11	726±7	723±5	714±11	s,PP	w·m
	1338±8	1336±12	1337±7	1335±5	vs,P	m·s
All acidic forms	1412±2	1415±2	1414±1	1412±9[a]	vs(B),P	w
Non-acidic forms						
(1) Solids and solutions	1313±2	1308±2	1309±1	1313±2	s·vs,P	s
	1456±6	1485±5	1478±8	1487±3	m·s,PP	m
(2) Solutions only	1388±3	1383±5	1383±3	1388±3	m·s,P	—
	—	1425±2	1426±4	1427±1	w·m,?	—

[a] Resolved into 1398m, 1420 vs and 1432s components in HCl solution.

TABLE 3.8. Frequencies characteristic of the cytosine ring

Conditions of observation	Cytosine	Cytidine	CMP-5' (Disodium)	1-Methylcytosine	Intensity Raman	i.r.
Solids and all solutions	786±9	781±9	780±6	785±10(D)	s·vs,PP	m·s
	1215±5	1218±7	1218±8	1209±9	s,P	w·m
	1282±13	1254±11	1254±11	1278±15	vs,P	vw
		1302±10	1306±6		vs,P	s
	1375±10	1383±8	1383±8	1392±7	w·m,DP	s
Non-deuteriated forms	970±5	995±6	1000±5	985±10	w·m,DP	vw
Deuteriated forms	—	1036	1037	1025±5	m,DP	—
Acidic solutions	1420±10	1430±20	1430±20	1445±5	m·s,P	w

TABLE 3.9. Frequencies characteristic of the guanine ring

Conditions of observation	Guanine[a]	Guanosine	GMP-5'	9-Ethyl-guanine[b]	Raman Intensity	Polarization
All solution forms	638 ± 13	668 ± 8	672 ± 8	622 ± 7	$m(B)$	PP
	1361 ± 11	1367 ± 5	1364 ± 4	1350 ± 2	m·s	P
All acidic forms	1260	1285 ± 10	1285 ± 10	1280 ± 10	m·s	PP
	1410	1405 ± 10	1405 ± 10	1414 ± 4	s·vs	P
All non-acidic forms	1326 ± 3	1315 ± 10	1313 ± 12	—	m	P
	1457 ± 2	1480 ± 10	1481 ± 9	—	vs	PP
Alkaline forms only; OH−, OD−	1326 ± 3	1341 ± 4	1342 ± 3	—	vs	P
Deuteriated forms	1190 ± 2	1170 ± 5	1170 ± 5	1172	w·m	

[a] Neutral solution spectra not available.
[b] Neutral and basic solution spectra not available.

In the Raman spectra of guanine in water, the line at 1680 cm^{-1} (1670 cm^{-1} in D$_2$O) is very broad and strongly polarized (Lord and Thomas, 1967). This line is assigned to the carbonyl stretching vibration. The keto structure predominates because, if the guanine derivatives existed in the enolic form in solution, no carbonyl mode would be expected. If the imino tautomer predominated, this frequency at 1680 cm^{-1} would have to be assigned to the —C=NH stretching mode; this usually does not have such a high frequency and also would be expected to have a greater isotopic shift than the 10 cm^{-1} observed.

The double-bond stretching region (1500–1800 cm^{-1}) of the Raman spectra of these bases is also useful since the lines are very intense. These frequencies are affected by the structural changes induced by pH. Attachment of an H atom to the nucleus causes a shift to higher frequency and, conversely, removal of a proton from the free base causes a shift to lower frequency of the double-bond modes. Raman frequencies can be found which are sensitive to the different structures found in neutral, acidic and basic solutions. Uracil, for example, has three double bonds and their stretching vibrations (not shown in Table 3.6) produce intense Raman scattering. In D$_2$O, the strong lines at 1690 $\pm$ 10 cm^{-1}, 1658 $\pm$ 5 cm^{-1} and near 1620 cm^{-1} must be attributed entirely to double-bond stretching for uracil and its 1—N derivatives. The 1690 cm^{-1} line is attributed mainly to the non-conjugated 2-C-carbonyl stretching mode, the 1658 cm^{-1} line to conjugated 4-C-carbonyl stretching mode, and the lowest frequency to the ring C=C stretching mode. When the solvent pD is lowered for the free base, uracil, the carbonyl frequencies increase since the attachment of a D, apparently to the 1-N-ring position, affects the frequency of both the 2-C carbonyl and the 4-C-carbonyl modes. In water solutions, only two distinct carbonyl modes can be observed: one very broad mode at 1680 cm^{-1} and a shoulder at 1635 cm^{-1}. This result suggests vibrational coupling between the C=O stretching and 3-NH-bending motions. In basic solutions, the deprotonation, which must occur at the 3-N position in uracil, results in an anionic structure of the type

$$\text{[anionic uracil structure]}$$

with a high degree of π-electron delocalization. In NaOH and NaOD, there is a systematic decrease of all Raman double-bond frequencies. Lines are observed, for all derivatives of uracil (except the methyl uracil) at 1640 ± 5 cm^{-1}, 1600 cm^{-1} and 1505 cm^{-1} in alkaline media. These features are assigned to the anionic uracil residue and arise from the C=C and O—C—N—C—O structures. As expected, no similar changes occur in basic solutions of 1,3-dimethyluracil.

Systematic shifts to *lower* frequency of each Raman line in the 1300–1200 cm^{-1} region occurs for the transition from neutral solution to acidic solution for all cytosine derivatives; this is opposite to the behaviour of the Raman lines above 1500 cm^{-1}. The vibrational modes whose frequencies lie in the 1300–1200 cm^{-1} region involve a larger component of single-bond motion and the acidification tends to localize the π-electrons and "fix" the double and single bonds of the ring, so the frequencies in this region should decrease when the single-bond character is increased.

The Raman spectra of the nucleosides are essentially the same as for the bases because of the weak Raman scattering of the ribose residue. The spectra of D(−) ribose (Lord and Thomas, 1967) and deoxyribose (Small and Peticolas, 1971) have been reported to have only very weak lines relative to the ring modes of the bases. The observed frequencies are shown in Table 3.10 and 3.11 (Small and Peticolas, 1971). A weak Raman line appears at 1465 cm^{-1} in the Raman spectra of all nucleosides and nucleotides which apparently arises from the sugar residue.

The Raman spectra of the nucleosides and nucleotides at a given pH or pD are identical except for the 980 cm^{-1} line and a very weak line near 1100 cm^{-1} mode. These frequencies are assignable to the symmetric and degenerate PO$_3$ stretching modes. The (ROPO$_2$OH)$^-$ ion does not appear to give rise to any prominent Raman lines.

A Raman study has been made of the pH dependence of the spectra of the three adenosine phosphates AMP, ADP and ATP (Rimai et al., 1969). The

TABLE 3.10. Raman wavenumbers/cm^{-1} and assignments for ribose-5-phosphate

Ribose-5-phosphate	Ribose	Dimethyl-phosphate	Assignments for ribose-5-phosphate
	421		
	464		
505			
546	548		ribose
575			
	588		
	601		
	650		
	680		
720	729		ribose
	798		
806		759	—O—P—O—symmetric stretch
		816	
	832		
851			
	879		
902			
	918		
930			
	968		
981			
	1012		
1050		1040	
	1052		
		1056	
	1083		
1086		1085	O=P=O$^-$ symmetric stretch
	1125		
1140			
	1159		
		1196	
		1220	
1263	1269		ribose
	1325		
1462	1467		ribose

spectra indicate that the three phosphates can easily be distinguished by the intensity, frequency and shape of lines in the neighbourhood of 1125 cm^{-1}. Additionally, the ADP has a line at 710 cm^{-1}, while ATP has a line of 680 cm^{-1}, but AMP has no frequency in this region.

The pH dependence of the spectra shown in Table 3.12 suggests that the lines at 817 and 1082 cm^{-1} are associated with the protonated form (PO$_2$)OH, whereas the lines at 882, 980 and 1005 cm^{-1} correspond to transitions with the ionized PO$_3$. These highly polarized Raman modes at 980 and 1082 cm^{-1} can be assigned to the PO symmetric stretching vibrations involving a resonant

TABLE 3.11. Raman wavenumbers/cm^{-1} and assignments for deoxyribose-5-phosphate

Deoxyribose-5-phosphate	Deoxribose	Dimethylphosphate	Assignments for deoxyribose-5-phosphate
	392		
	422		
504			
	510		
	560		
	619		
	643		
	676		
	719		
	756		
	802		
804		759	—O—P—O— symmetric stretch
		816	
	817		
845			
	869		
	897		
921	919		deoxyribose
980	984		deoxyribose
1007	1009		deoxyribose
		1040	
	1047		
		1056	
	1081		
1087		1085	O=P=O$^-$ symmetric stretch
	1119		
	1142		
1190	1190	1196	phosphate, deoxyribose
		1220	
1259	1270		
	1325		
1365	1358		deoxyribose
	1388		
	1446		
1463	1459		deoxyribose

double bond in both the ionized and protonated form. In ADP and ATP these lines due to the ionized form occur at 1108 and 1126 cm^{-1}, respectively, and the protonated form at 1090 and 1112 cm^{-1}, respectively.

The conjugate pair, below and above the pK, in AMP at 980 and 1082 cm^{-1}, respectively, are highly polarized and assignable to the PO symmetric stretching vibrations of the ionized PO$_3$ (980) and protonated form (PO$_2$)OH (1082 cm^{-1}). These modes are expected to couple in the polyphosphates and a shift in frequency is predicted. In ADP, these two polarized modes occur at 1108 and 1090 cm^{-1}, with the lower frequency corresponding to the fully ionized variety, while the analogous lines in ATP appear at 1126 and 1112 cm^{-1}, respectively. The shift of the symmetric phosphate line upon ionization decreases from 100 cm^{-1} for the mono-, to 18 cm^{-1} for the di-, to 13 cm^{-1} for the triphosphate. This change is consistent with the change in the degree of vibrational coupling with number of phosphates in sequence for this mode.

TABLE 3.12. Lines assigned to phosphate vibrations (wavenumbers/cm^{-1})

AMP		ADP		ATP	
pH: 5.5	7.5	5.5	7.5	5.5	7.5
817[a]	817[a]	826	828		920[c]
858	850	860			830[c]
	882				
1082[e]	980[d]	1108[e]	1090[e]	1125[e]	1112[e]
1005[b]	1005[b]	1005		1008	
	1123[f]				
	1170[f]				
			710		680[g]

[a]This line is stronger in low pH.
[b]This line is stronger in high pH.
[c]The structure of the lines in this region change in a complicated way with pH.
[d]Symmetric stretch of PO$_3$ group.
[e]Symmetric stretch involving combinations of O=P=O.
[f]Twofold degenerate asymmetric PO$_3$ stretch, split by lower symmetry in the molecule.
[g]A collective mode.

Although synthetic polyribonucleotides have been studied by Raman spectroscopy, only the results for poly(adenylic) acid will be discussed.

Poly(adenylic acid) (poly A) exists in two forms: a two-stranded helix below pH 6.0 and an ordered, single-stranded conformation at a higher pH; it has received study by Raman spectroscopy (Aylward and Koenig, 1970; Tomlinson and Peticolas, 1970). The spectrum of poly A as a solid in the form of a single-stranded helix is shown in Fig. 3.11. It is very similar to spectra of

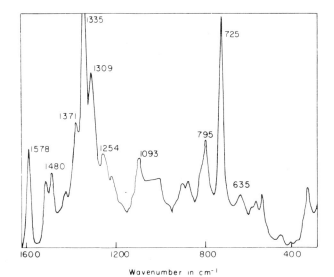

FIG. 3.11. Raman spectrum of the potassium salt of poly-(adenylic acid) in the solid state.

other adenine derivatives and is dominated by the strong lines of the adenine ring at 1335 and 725 cm^{-1}, as indicated in Table 3.13. When poly A is dissolved in water, most of the lines are not appreciably shifted, except that a line at 795 cm^{-1} shifts to 811 cm^{-1}. When the aqueous solution is heated from 20–80°C, the chain goes into an "unstacked" conformation; these changes are reflected in a substantial hypochromic change in the intensity of the 725, 1252, 1303, 1326, 1424 and 1500 cm^{-1} lines (Tomlinson and Peticolas, 1970).

The melting curve of the double-stranded helix indicates that the melting has occurred in a non-cooperative manner (Small et al., 1970). As poly A is tested to 80°C, the spectrum approaches that of AMP at 20°C (Fig. 3.12). These hypochromic effects for poly A have received considerable study. The theoretical basis of the Raman hypochromic effect has been attempted by Peticolas based on a theory of Raman intensities (Peticolas, 1970) but, since the theory involves infinite sums over all the excited states, the theory does not lead to useful quantitative predictions. A simplified treatment making qualitative predictions has been given more recently (Small and Peticolas, 1971).

The spectrum of the double helix of poly A (Fig. 3.13) is similar to that of the protonated form of AMP-5'; the spectral changes are consistent with the suggestion that the protonation occurred at N-1. No spectral features could be associated with interbase H bonding which is consistent with the results found for the nucleotides.

Poly A and poly U form a double-stranded helix which undergoes a transition at 57°C in a 2% solution at neutral pH. The Raman spectrum of this

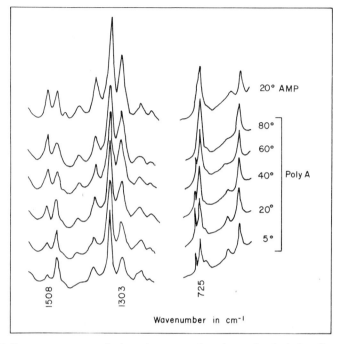

FIG. 3.12. Raman spectrum of adenosine monophosphate of poly-(adenylic acid) as a function of temperature (°C).

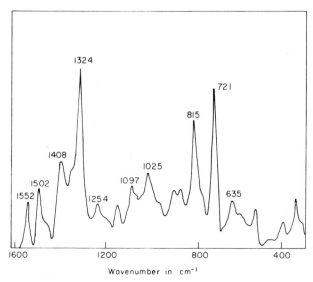

FIG. 3.13. Raman spectrum of the double helix of poly-(adenylic acid) in the solid state.

TABLE 3.13. Raman wavenumbers (cm^{-1}) of poly A and its derivatives

Poly A (solid)	Intensity	Poly A i.r.	Intensity	Adenine (solid)	Adenosine (solid)	Deoxyadenosine (solid)	Intensity	Structural assignment	Vibrational assignment
		1644	v.s.						
		1598	s	1600	1600	1613	v.w.	Ring	
1578	m	1571	s		1575	1570	s	Ring	
				1530					
1506	w	1500	w		1510			Sugar	
1480	w	1472	m	1485	1480	1480	m	Ring	
		1452	w	1462		1448	v.w.		
1415	w	1414	m		1420	1415	w	Sugar	
					1388	1390	m		
1371	m	1366	w	1370	1375	1379	m	Ring	
						1349	v.s.		
1335	v.s.	1326	m	1332	1335			Ring	
						1324	w		
1309	s			1312	1306	1302	m	Ring	
		1294	m						
					1275				
1254	m			1252	1248	1250	m	Ring	v(CNH$_2$)
		1238	s					Phosphate	v_a(O$\cdots$P$\cdots$O$^-$)
				1225					
1213	w				1215			Sugar	
						1198	v.w.		
						1191	v.w.		
					1180	1176	v.w.		
1172	v.w.	1166	w						
		1125	m	1130	1132				
						1112	v.w.		
1093	m							Phosphate	v_s(O—P—O$^-$)
						1067	w	Sugar	
1033	v.w.	1000–1100	s		1037				v(CO)
				1025					
1006	w				1013	1014	w	Sugar	
		990	m		988				
964	v.w.								
		952	w	945		949	w		
		924	w						
					912	911	w		
907	w	905	w	900					
876	w								
853	v.w.	858	w		846	849	w		
815	w	814	w	820	825	811	w		
795	m	790	w			795	w		
					763	756	s		
						741	s		
725	v.s.	717	w	725	723			Ring	
						698	v.w.		
						662	w		

TABLE 3.13—continued

Poly A (solid)	Intensity	Poly A i.r.	Intensity	Adenine (solid)	Adenosine (solid)	Deoxyadenosine (solid)	Intensity	Structural assignment	Vibrational assignment
635	w	638	w		640	639	v.w.	Sugar	
				620					
583	v.w.				587				
564	w					568	v.w.		
532	w	530	w	535	537	536	v.w.		
						500	v.w.		
456	v.w.								
					414	416	w		
						379	w		
						351	m		
				334		335	s		
318	w				320				
					290	298	w		
					275				
248	w								
					230	237	w		
						150	w		
				130					
					113				

w = weak; m = medium; s = strong; v = very; v = stretch, v_a = asymmetric stretch, v_s = symmetric stretch

helical complex (Fig. 3.14) has been reported at temperatures above and below the melting point (Small and Peticolas, 1971). The adenine lines at 730 cm^{-1}, 1300 cm^{-1} and 1510 cm^{-1} show a decrease with helical formation with the poly A–poly U complex. However, the decrease in the adenine line at 1510 cm^{-1} is not as large as in single-chain poly A solutions. The uracil lines at 781, 1236 and 1403 cm^{-1} show Raman hypochromism in the poly A–poly U complex. The frequencies reflecting the unstacking are different for adenine and uracil in the complex, and their separate behaviour can be followed with increase in temperature. The poly A–poly U complex has a strongly polarized Raman line at 814 cm^{-1}, which is assigned to the symmetric stretch of the phosphate diester. This 814 cm^{-1} line may arise from a specific conformation of the sugar phosphate backbone; Small and Peticolas (1971) suggest the *gauche-gauche* configuration of the R—O—P—O—R linkage. In the C=O stretching region, the double helix has one line at 1681 cm^{-1}, which disappears on melting, and two new lines appear at 1698 and 1660 cm^{-1}; these two probably correspond to the carbonyl stretches on the 2 and 4 positions of the uracil. The appearance of the 1660 cm^{-1} line indicates the break-up of interbase H bonding of the double helix.

The Raman spectra have been reported (Tobin, 1969) of calf thymus DNA

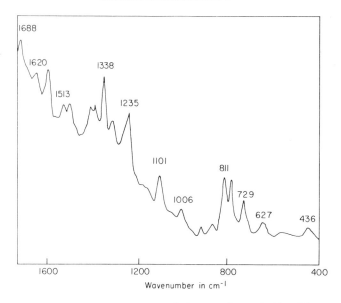

Fig. 3.14. Raman spectrum of the complex Poly (A+U).

as a solid, aqueous gel and D_2O gel, and also as dilute solutions (Small and Peticolas, 1971). The Raman spectra of salmon testes DNA as a solid, aqueous and D_2O gel and as an aqueous gel of $pH = 00$ have also been observed (Tobin, 1969). The Raman frequencies and assignments for Calf thymus DNA are given in Table 3.14, in which the characteristic ring vibrations of the bases have been labelled. The spectrum is not simply the superposition of the spectra of the bases. The differences between the spectra of the two DNA's are, with the exception of the 1650–1700 cm^{-1} region, relatively minor; changes in the spectra from the solid to the gel are also relatively small. The intensities of the lines at 786 and 800 cm^{-1} reverse as the sample goes from the solid to H_2O gel (Tobin, 1969). Tobin (1969) suggests that a line at 1090 cm^{-1} is characteristic of the DNA helix. Peticolas considers that the decrease in intensity of many of the ring modes in DNA indicates more ordered stacking in DNA than in the more disordered polyribonucleotides. On the basis of a temperature-independent line at 795 cm^{-1} (corresponding to the 814 cm^{-1} line in RNA) exhibited by DNA, Small and Peticolas suggest that the changes between ordered and disordered forms which occur in the backbone of deoxyhelices are different from those for the ribonucleic acids.

Similar studies have been made on RNA from *Escheriochia coli* (Thomas, 1970) and the observations concerning the Raman spectra are similar to those of DNA.

The Raman spectrum of yeast t-RNA has also been reported (Small and Peticolas, 1971). A melting curve for t-RNA as a function of temperature,

TABLE 3.14. Raman frequencies and assignments for calf thymus DNA

Frequencies (cm^{-1})		Assignments[a]
H$_2$O solution	D$_2$O solution	
	500	Deoxyribose-phosphate
	567	Deoxyribose
672	662	T
683	685	G
730	725	A
752	743	T
	774	C, T
787		O—P—O Diester symmetric stretch Overlapping C, T
	792	O—P—O Diester symmetric stretch
835	838	Deoxyribose-phosphate
	871	Deoxyribose-phosphate
895	897	Deoxyribose-phosphate
917	921	Deoxyribose
975	977	Deoxyribose
	1000	Deoxyribose
1017	1015	C—O stretch
1058	1053	C—O stretch
1094	1095	O—P—O$^-$ Symmetric stretch
1144		Deoxyribose-phosphate
1180		Base external C—N stretch
1214		T
1226		A
1242		T
1259		C, A
1304	1307	A
1320		G
1340	1351	A
1378	1382	T, A, G
1423	1424	A, G
1448	1449	Deoxyribose
1462	1465	Deoxyribose
1491	1486	G, A
	1504	A
1514	1524	A
1534		G, C
1580	1580	G, A
	1621	
	1672	C=O stretch

[a] T, C, A, and G indicate vibrations characteristic of the thymine, cytosine, adenine and guanine bases, respectively, listed in order of their relative contributions, with the largest contribution first.

"Deoxyribose-phosphate" indicates probable origin is in the deoxyribose-phosphate chain but cannot be readily assigned specifically to deoxyribose or phosphate.

derived from the change in intensity of the 814 cm^{-1} line, shows that the melting process of sRNA is non-cooperative. From a comparison of the spectra of DNA and RNA Small and Peticolas (1971) surmise that guanine undergoes significantly stronger stacking interactions in the DNA than in RNA. In particular, the guanine-ring mode appears at 670 cm^{-1} in RNA and at 683 cm^{-1} in DNA.

3.6. Conclusion

Raman spectroscopy would appear to be a sensitive tool for the study of biological systems in aqueous solution. The informational content of the Raman effect is high and the sampling requirements are low. Consequently, Raman spectrometers will soon be found in many biological laboratories and can be expected to play an increasingly important role in biological research.

References

Aylward, N. N. and Koenig J. L. (1970). *Macromolecules* **3**, 590.
Bamford, C. H., Elliott, A. and Hanby, W. E. (1956). "Synthetic Polypeptides", Chapter 5. Academic Press, New York and London.
Edsall, J. T. (1936). *J. Chem. Phys.* **4**, 1.
Edsall, J. T. (1937). *J. Chem. Phys.* **5**, 225.
Edsall, J. T., and Scheinberg, H. (1940). *J. Chem. Phys.* **8**, 520.
Fanconi, B., Small, E. W. and Peticolas, W. L. (1971). *Biopolymers* **10**, 1277.
Fanconi, B., Tomlinson, B., Nafie, L. A., Small, W. and Peticolas, W. L. (1969). *J. Chem. Phys.* **51**, 3993.
Garfinkel, D. and Edsall, J. T. (1958). *J. Amer. Chem. Soc.* **80**, 3818.
Itoh, K. and Shimanouchi T. (1970). *Biopolymers* **9**, 383.
Itoh, K., Nakahara, T. and Shimanouchi, T. (1968). *Biopolymers* **6**, 1759.
Koenig, J. L. and Frushour, B. Unpublished.
Koenig, J. L. and Sutton, P. L. (1969). *Biopolymers* **8**, 167.
Koenig, J. L. and Sutton, P. L. (1970). *Biopolymers* **9**, 1229.
Koenig, J. L. and Sutton, P. L. (1971). *Biopolymers* **10**, 89.
Krishman, K. and Plane, R. A. (1967). *Inorganic Chemistry* **6**, 55.
Lord, R. C. and Thomas, G. J. Jr. (1967). *Spectrochimica Acta* **23A**, 2551.
Lord, R. C. and Yu, Nai-Teng (1970a). *J. Mol. Biol.* **50**, 509.
Lord, R. C. and Yu, Nai-Teng (1970b). *J. Mol. Biol.* **51**, 203.
Miyazawa, T., Fukushima, K., Sugano, S. and Masuda, Y. (1967). "Proteins". Vol. II. p. 567. Academic Press, New York and London.
Peticolas, W. L., Nafie, L. A., Fanconi, B. and Stein, P. (1970). *J. Chem. Phys.* **52**, 1576.
Rimai, L., Cole, T., Parsons, J. L., Huiekmott, J. T. Jr. and Carew, E. B. (1969). *Biophys. J.* **9**, 230.
Small, E. W. and Peticolas, W. L. (1971). *Biopolymers* **10**, 69.
Small, E. W., Fanconi, B. and Peticolas, W. L. (1970). *J. Chem. Phys.* **52**, 4369.
Small, E. W., Biggerstaff, A. and Peticolas, W. L. *Biopolymers* (In press).

Smith, M., Walton, A. G. and Koenig, J. L. (1969). *Biopolymers* **8,** 29.
Takeda, M., Iavazzo, R. E. S., Garfinkel, D., Scheinberg, I. H. and Edsall, J. T. (1958). *J. Amer. Chem. Soc.* **80,** 3813.
Thomas, G. J. Jr. (1970). *Biochim. Biophys. Acta* **213,** 417.
Tobin, M. C. (1968). *Science* **161,** 68.
Tobin, M. C. (1969). *Spectrochim. Acta* **25A,** 1855.
Tomlinson, B. L. and Peticolas W. L. (1970). *J. Chem. Phys.* **52,** 2154.

4. Far Infra-red Spectroscopy

T. R. Manley

4.1. Introduction

4.1.1. GENERAL

The i.r. region of the spectrum (Fig. 1.1) lies between the visible and the microwave portions and (Section 1.3.3) is arbitrarily divided into three parts. The section between the visible and 2·5 μm is known as the near i.r. from 2·5 to 25 μm is the middle or conventional region and from 25 to 1000 μm, or 1 mm, is the far i.r. region. These divisions have no fundamental basis but arise mainly from the characteristics of the materials used in the instruments employed to study the spectra. In the near i.r. region, for example, glass is transparent to the radiation and thus it is convenient to use the same sort of sample cells as are used in the visible region. From 2·5 μm onwards is the sodium-chloride region; the limit of the far i.r. region, which in early work was considered to be about 15 μm, gradually was pushed out as better instruments became available that could utilize the much smaller amount of energy in the far i.r. region. The region between 2·5 and 15 μm is usually referred to as the fingerprint region, since most of the characteristic absorptions of chemical groups occur in this range. The far i.r. region is also referred to as the sub-mm region by those whose work is primarily with microwaves. Another term that may be found for far i.r. spectroscopy is Fourier-Transform spectroscopy (Section 1.5.3), although Fourier-Transform spectroscopy is not restricted to this region (Section 7.3.3; Griffiths, 1975).

The techniques for study of the far i.r. spectrum also fall into three groups, namely the use of diffraction-grating spectrometers, interferometers and harmonic generators. The diffraction-grating spectrometers are similar in principle to those used in studying the near and middle i.r. regions; they are the simplest instruments to use and also the cheapest. They have the additional advantage that any i.r. spectroscopist has no difficulty in using them. The disadvantage of a diffraction-grating spectrometer is that the energy available falls off markedly as one goes from the near i.r. to the conventional region, and from the conventional to the far i.r. region; this means that the resolution obtainable is limited and the time to obtain the spectra becomes inordinately long. If one's interest in the far i.r. does not go beyond say 50 µm, however, the grating spectrometer would probably be the most useful instrument.

The second technique is an extension of those for studying the radiowave section of the spectrum, using methods based on the generation of harmonics of microwave sources. These techniques give very high resolution, since coherent sources are used, and, of course, the basic technique has been intensively investigated at wavelengths above 1 mm; here again the available power decreases as one enters the i.r. region, power decreasing with decreasing wavelength in this case. This technique also suffers from the disadvantage that it is not possible to scan a range of frequencies and, therefore, it too can be rather time-consuming.

The third technique is the use of interferometric spectrometers (Section 4.2.3).

4.1.2. HISTORICAL

The far i.r. region of the spectrum was the last to be investigated. Although many experimentalists had tried to study absorptions in this region, the difficulties appeared insurmountable; in fact the region was sometimes referred to as the forbidden section of the spectrum. The first work in the far i.r. by Rubens and Nichols in 1897 used restrahlen techniques. Over the next 14 years they developed other methods for looking at the region, but they did not, of course, consider the use of Michelson's interferometer; although this was being developed contemporaneously, it did not appear to have any conceivable use other than for qualitative measurement in the visible region.

The introduction of echellette gratings by Wood in 1910 opened the way for the construction of useful prism-grating spectrometers for up to 15 µm. The introduction of the KBr prisms in 1930, together with improvements in the sensitivity of detectors and amplifiers, extended this to 30 µm. In the 1940's KRS-5 prisms (p. 39) extended the region even further out to 40 µm. The prism spectrometer could therefore be said to have penetrated the far i.r. region; but in later instruments the prism was abolished and reflection gratings were used to reduce stray radiation. In these instruments the Littrow mirror was replaced by a suitable grating in the spectrometer. Long-wave spectra in the region 20–135 µm were obtained by Barnes and Czerny (1931) on apparatus using wire gratings.

In 1958, Fellgett pointed out that the interferometer has a considerable signal-to-noise advantage over a conventional spectrometer and that this advantage is particularly useful in the far infra-red. In 1956, Gebbie and Vanasse published the first far i.r. spectrum with this system.

4.2. Equipment for far infra-red spectroscopy

4.2.1. MATERIALS

To work in the far i.r. (FIR), as in other parts of the spectrum, materials are needed that are transparent in this region and which are suitable for use as dispersion media and instrument windows. Diamond is transparent between 450 and 40 cm^{-1} and possesses ideal physical and chemical properties for use as a window material in the far i.r., but it can only be used to cover small apertures such as detector windows. Silicon plates about 2 mm thick form convenient windows and germanium is used particularly for work in the region 500–800 cm^{-1}. The most widely useful materials, however, are synthetic polymers that are available as tough sheets or powders and are readily formed into cells or other shapes. Willis *et al.* (1963) suggested the use of the high-density polythene "Rigidex 35" for the construction of absorption cells, and the less brittle "Alkathene" low-density polymers for use as vacuum-chamber windows. High-density polymers are more rigid, have higher melting points than the low-density materials and also have better resistance to chemical attack and organic solvents. Absorption bands of the highly-branched polymers were found to be relatively much more intense than those in similar unbranched materials. Certain commercial products tend to show greater absorption than research specimens, indicating that much of the absorption is due to materials other than the base resin. An examination of various polymers (Manley and Williams, 1965a), in order to test their suitability as window materials and, more especially, their use as transparent dispersion media in the far i.r., confirmed the hydrocarbon polymers as being most suitable. The spectra of certain hydrocarbon polymers in Fig. 4.1 have been corrected for instrument background and appear to agree with the uncorrected curves of Willis *et al.* These materials are very weakly absorbent in the far i.r. and thick specimens are required to obtain good spectra. Conversely, thinner specimens of these materials are useful as windows. The

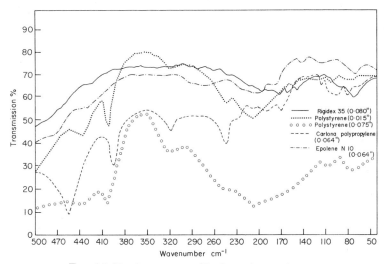

FIG. 4.1. Far i.r. spectra of hydrocarbon polymers.

best window was found to be high-density polythene but polystyrene was a useful complement as it is transparent at 70 cm^{-1}. On the other hand, polystyrene absorbs strongly at 200 cm^{-1} and, at frequencies between 10 000 and 300 cm^{-1}, troublesome interference bands are found in laminated materials.

The polythene wax "Epolene N 10" has transmission properties comparable to "Rigidex 35" but it lacks rigidity and mechanical strength and is partially soluble in solvents transparent in the far i.r., such as CS_2 and CCl_4. However, its low melting point (90°C) enables many thermally unstable materials to be readily dispersed in the matrix without decomposition, and the films resulting on cooling between laminating plates can be readily varied in thickness and in sample concentration; it usually gives rise to well-defined spectra without the need for prior grinding of the specimen. Such grinding may be undesirable as it can lead to phase transitions, readily detectable in the far infra-red. "Rigidex" cannot be readily employed for this purpose and it is better used as a window material where thicknesses of up to 0.5 in. can be tolerated without undue attenuation and in the preparation of pressed polythene discs analogous to KBr discs employed in the conventional i.r. region (Section 2.2.2.6). "Rigidex 35" has now been withdrawn from the selling range of British Petroleum Ltd.; related "Rigidex" homopolymers are "Rigidex 50" and "Rigidex 25". "TPX" (I.C.I. trademark for poly-4-methylpentene-1) is a useful window material because of its high strength; it may be used at temperatures up to 200°C.

Polypropylene is effective as a window at frequencies lower than 200 cm^{-1} and is the preferred alternative to high-density polythene in this range. Absorption maxima occur in the region of 180 cm^{-1} in polythene, polypropylene and polystyrene, which may be attributable to skeletal modes, involving out-of-plane deformation or torsion in the carbon chain. Polypropylene also retains good mechanical strength up to 150°C. Polytetrafluoroethylene, polyoxymethylene and vinyl-acetate polymers were too absorbent for use as windows.

Most of the spectra in Fig. 4.1 have low absorption at frequencies less than 200 cm^{-1}; the absorption is not zero but is spread over a wide range. This is in agreement with the very low power factor in the microwave region, e.g. for polythene $\varepsilon = 0.0013$ at 25 000 MHz (1.2 cm^{-1}), which corresponds to an absorbance α of 0.005. The value for very pure material is probably lower than this. Polar groups which increase the i.r. activity of the torsional modes of the polymer chains, as in the long-chain ketones, are probably introduced as a result of the excitement of impurities in the polymer, thereby increasing the general absorption at frequencies less than 200 cm^{-1}. The impurities are generally present in the form of traces of Zeigler-type catalysts such as VCl_5; partial oxidation of the polymer chain initiated by u.v. radiation, X-rays or oxygen-catalysed reactions during synthesis gives rise to active carboxyl groups. The latter form of impurity is particularly prevalent in polythenes.

Rothschild (1965) studied standard absorption cells for vapours or gases equipped with polythene windows and found that the vapour of many compounds diffuses spontaneously into the polythene. The resulting enrichment of the compound in the polythene windows gives rise to spectra which are superpositions of the spectra of the molecules of the vapour phase. The cells must be used with caution since the effect could give rise to "solvent shifts" resulting in additional spurious bands in the spectrum; band contour may also be affected. Other polymers may well be preferable to polythene for this purpose. In any case, it is advisable to use more than one type of sample cell to guard against this eventuality.

4.2.2. FILTERS

The intensity of far i.r. radiation from hot-body sources is many orders of magnitude less than that in the fingerprint region; consequently, it is essential that the near-i.r. and shorter wavelength radiations are removed. In addition, when grating instruments are used, further filters are necessary in order to remove radiation that would fall in the second and higher orders of the dispersion grating.

Transmission filters are more convenient than reflection filters because their operation is not critical; they do not complicate the optical path of the

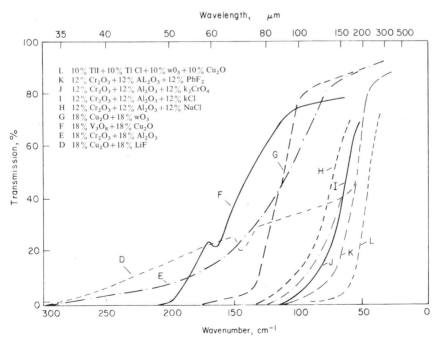

FIG. 4.2. Transmission characteristics of filters between 33 and 160 μm; curve I is truncated.

instrument and they can readily be changed. The most convenient and flexible form of filter is the transmission filter made by adding an inorganic powder to polythene; after rolling and pressing, a uniform dispersion of the powder is obtained in a sheet of polythene (Manley and Williams, 1965b). By varying the inorganic powder, these filters can be made to cut off at various points: Fig. 4.2 shows the curves obtained with some of these filters between 33 and 160 μm. Alternatively, a sheet of polythene may be used on which the rulings of a grating have been impressed; radiation of shorter wavelength than the groove spacing is scattered or diffracted out of the beam. Quartz is a convenient filter; the crystal is opaque between 2000 and 250 cm^{-1}, whilst fused quartz begins to transmit around 100 cm^{-1}.

Four kinds of reflection filters are commonly found. The first is the "scatter plate", made by abrading a metal mirror; this reflects wavelengths greater than the depths of the abrasions and scatters shorter wavelengths out of the beam. Grade 120 carborundum on an alloy reflector cuts off around 150 μm (Martin, 1967). The second is the original "restrahlen reflector"; the third is the use of a grating in zero order; and the fourth, which has a very sharp cut-off and is particularly useful at longer wavelengths, are the metallic wire meshes that reflect long waves but transmit or scatter short waves.

4.2.3. INSTRUMENTS

4.2.3.1. *Grating spectrometers*

There are four main items in a grating spectrometer used in the FIR, viz a broad-band source; a monochromator with the dispersive grating and filter; the sample holder; and the detector. The latter two items are often arranged so that they may be cooled by liquid nitrogen or helium.

Far i.r. detectors must have a high sensitivity because far i.r. sources have limited energy. The Golay pneumatic cell, for which the minimum detectable power is of the order of 5×10^{-11} W for a 1 Hz bandwidth and the time-constant is about 15 ms, is the most useful detector. Work with low-temperature i.r. detectors has led to great improvement in the signal-to-noise ratio of far i.r. spectra and to increased spectral resolution. Some cryogenic detectors, which are stated to be orders of magnitude better than the Golay cell at room temperature, are the carbon bolometer and the gallium-doped Ge bolometer. The superconducting Sn bolometer is very sensitive but requires liquid-He coolant. A simple bolometer may be constructed from a C resistor having a high sensitivity at 1·5 K. The temperature is maintained so that the sensitive element is just over the threshold of superconductivity; hence the absorption of radiation leads to a large change in current which may be amplified.

Other detectors include the antimony-doped Ge photoconductive cell, which has its highest sensitivity in the 70–125 cm^{-1} region, and a time constant of 10^{-5} s. Another photoconductive detector, useful below 100 cm^{-1}, uses an element of high-purity n-type indium antimonide; this detector needs a magnetic field of about 0·6 T, obtained from a superconducting niobium solenoid with a temperature near 1·5 K for efficient operation.

Typical commercial instruments are the Perkin Elmer 577 and 180 and the Grubb Parsons GS 8 and GM 3.

4.2.3.2. Microwave techniques

Far i.r. sources of high intensity, produced from the oscillation of electron beams in cavities, are very suitable for spectra containing hyperfine structure. However, such sources have an extremely narrow bandwidth and may be scanned only over a small frequency range. Very high precision in frequency measurement can be obtained by frequency division, but it is necessary to generate high harmonics from a klystron, owing to the difficulty of decreasing the size of the microwave cavity resonators beyond a few millimetres.

4.2.3.3. Interferometers

Two kinds of interferometer have been employed for far i.r. spectroscopy although, in principle, any double-beam interferometer is suitable. The Michelson instrument is more widely used and is described later. The lamellar grating interferometer is often made by the conversion of a large diffraction-grating monochromator through replacement of the diffraction-grating mount with lamellar-grating plates. Lamellar-grating interferometers are most useful below 20 cm^{-1}, where they can be designed to give satisfactory results that could only be obtained with a Michelson instrument by frequent changes of beam splitter.

The basic principles of interferometry were known at the beginning of this century. Michelson discovered that the interference pattern from a two-beam interferometer, when expressed as a function of the path difference between the beams, is the Fourier-Transform of the light that falls on the interferometer. Michelson himself was unable to gain much benefit from the study of these interference patterns because he could only make visual observations of the fringes and therefore could not measure the absolute intensity of the interference pattern as a function of the path difference. Another and more fundamental restriction was that facilities for the computation of Fourier transforms were not available to him.

A typical far i.r. Michelson interferometer is shown in Fig. 4.3. The source S is a high-pressure Hg lamp which transmits a continuous spectrum into the interferometer; this consists of two plain mirrors at right angles, the right angle being bisected by a beam splitter when the path difference is zero. One of these mirrors M_f is fixed and the other M_m is moveable. The beam splitter B is a thin dielectric film normally of polyethylene terephthalate (I.C.I. "Melinex"), thickness d. The refractive index of the film is n and of the detector is D.

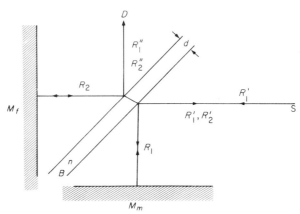

FIG. 4.3. Sketch of a Michelson interferometer (see also Fig. 4.5).

A ray R leaving the source S is partially reflected R_1 and partially transmitted R_2 at the beam splitter. After reflection by the plane mirrors, R_1'' is transmitted and R_1' reflected, whilst R_2'' is reflected and R_2' transmitted. These (R_1'; R_2') second transmissions and reflections are lost. R_1'' and R_2'' then recombine but they may be out of phase with one another, according to their frequency and the position of the moving mirror. These variations are observed on a Golay detector as the moving mirror travels at a constant velocity. When the path difference is zero, all frequencies are exactly in phase and the output from the detector is at a maximum. When the mirror begins to move, however, a path difference is introduced and the different frequencies are no longer in phase, the difference in phase being proportional to the frequency. Therefore, with increasing path difference, there is first a drop in the signal; this then goes through a minimum and afterwards passes through a series of subsidiary maxima and minima, the amplitude of the fluctuations diminishing as the path difference increases. The graph of the relationship between the signal and the path difference is known as an interferogram; a typical example is shown in Fig. 4.4. The interferogram is an intensity function of the path difference $I(x)$. $I(x)$ has a maximum value at $x = 0$, since it is only at the zero path difference that all the spectral elements are in phase. If one takes the Fourier Transform of the interferogram, one obtains a spectral intensity curve for the instrument and, if this is followed by obtaining a second spectrum with a sample in the instrument, then by ratioing the two spectra a transmittance spectrum may be obtained.

The production of an interferogram is more readily understood if we consider radiation of one frequency only (monochromatic) and assume that the moving mirror goes in discrete steps. In the simplified sketch of Fig. 4.5, the paths of lost radiation are omitted for clarity, so that only radiation that reaches the detector is considered. Initially, both mirrors are stationary, equidistant from the beam splitter; R_1 and R_2 are in phase, so that the detector

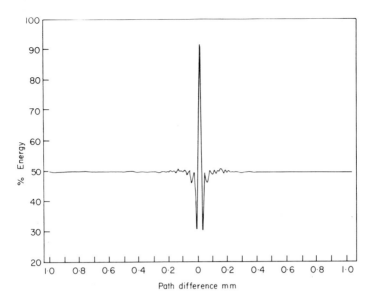

Fig. 4.4. Interferogram of syndiotactic-PMMA.

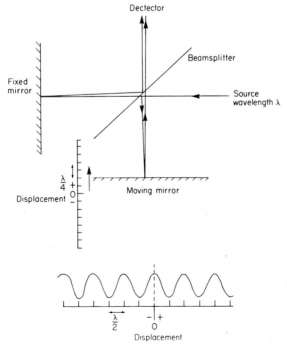

Fig. 4.5. Simplified diagram of mode of production of interferogram.

records a signal proportional to the sum of the energy in each beam. Now let M_m move by a distance $\lambda/4$, where λ is the wavelength of the incident monochromatic radiation; this increases the path of r_1 by $\lambda/2$. Thus, after recombination at the beam splitter, R_1 and R_2 will be 180° out of phase and the detector will record the difference between the radiation from each beam (i.e. zero). If M_m now moves to $2\lambda/4$, the two amplitudes will again be in phase, and the sum of the energies is recorded. When M_m moves to $3\lambda/4$ the detector again reads zero. In other words, a simple cosine-wave interferogram is obtained from incident monochromatic radiation. This cosine wave goes through one cycle as M_m is displaced by $\lambda/2$; if M_m, instead of moving in discrete steps, moves smoothly at a speed S, then the frequency of the detector or signal D is given by

$$D = \left(\frac{S}{\lambda/2}\right) = 2S\tilde{v}$$

where $\tilde{v}$ is the wavenumber of the incident monochromatic radiation. Thus, there is a linear relation between the detector-signal frequency and the incident-radiation frequency when the mirror velocity is constant. The amplitude of the low-frequency signal is proportional to the intensity of the monochromatic radiation. In the case of polychromatic radiation, each frequency undergoes a similar transformation and produces a detector wave of unique frequency. The signal from the detector is the sum of all such waves and is a complex signal; this is the so-called interferogram, e.g. Fig. 4.4. The centre maximum of the interferogram occurs when both mirrors are equidistant from the beam splitter, so that all the various components are in phase. The peak amplitude of the interferogram is proportional to the total energy in the incident radiation. Intensity and frequency information are obtained from the peaks of lower amplitude on both sides of the central peak.

This interferogram is the Fourier transform of the spectrum that was wanted. To recover the spectrum, an inverse Fourier transform has to be performed; this requires a computer and the process is discussed below.

4.2.3.4. Computation of interferograms

If the general case of an interferogram is represented by the intensity function $I(x)$ for path differences in the region $-\infty < x < +\infty$, where x is in cm, and if the power spectrum corresponding to the energy distribution from the interferometer is allowed to be $G(k)$ for frequencies $-\infty < k < +\infty$, where k is in cm^{-1},

then $G(k) = \int_{-\infty}^{+\infty} I(x)\exp(i2\pi kx)\,dx$

and $I(x) = \int_{-\infty}^{+\infty} G(k)\exp(-i2\pi kx)\,dx$

These show the fundamental relationship between the interferogram and the spectrum and may be summarized as

$$I(x) \longrightarrow G(k)$$

i.e. $I(x)$ has the transform $G(k)$.

In practice one cannot plot interferograms to infinity, nor do we require output frequencies to infinity; therefore restrictions are imposed on path differences of

$$-X < x < +X$$

and, for frequencies, of

$$-K < k < +K$$

Finite limitations on an interferogram extending to infinity are applied by multiplying by a truncating function of unit amplitude

$$\left[I(x)\right]_{-X}^{+X} = \left[I(x)\right]_{-\infty}^{+\infty} \left[F(x)\right]$$

where

$$\left[I(x)\right]_{-\infty}^{+\infty} \longrightarrow G(k)$$

$$\left[F(x)\right]_{-\infty}^{+\infty} \longrightarrow F(k)$$

and $F(x) = 1$ where $-X < x < +X$, and elsewhere $F(x) = 0$.
$F(k)$ is a spectral function of the form $(\sin k)/k$, whereas the transform of $\left[I(x)\right]_{-X}^{+X}$ is an ideal spectrum, in which each point has become the $(\sin k)/k$ function $F(k)$ and truncation to finite limits $\pm X$ limits the resolution of the resultant spectrum. The minimum resolved width $\Delta(k)$ is given by

$$\Delta(k) = 1/x$$

a further small error is caused by side lobes on the function $F(k)$, minimized by *apodization* (*vide infra*). The computation of a spectrum from an interferogram is performed by sampling at discrete intervals of path difference

$$\Delta x \text{ from } x = -X \text{ to } x = +X$$

No error is introduced by this technique of sampling provided that

$$\Delta(x) \leq 1/2K$$

where K is the maximum frequency transmitted in the interferogram. The error due to side lobes on the $F(k)$ function is substantially eliminated at the cost of a slight loss in resolution by introducing a convolution that changes the function to the form $(\sin^2 k)/k^2$. This has no negative values and the side lobes are greatly reduced; the central band, however, is made slightly broader, a convolution brought about by multiplying the interferogram by a triangular function. This *apodization* step is incorporated into the computer program.

The imaginary portion of the Fourier transform integral may be eliminated if the interferogram is absolutely symmetrical about $x = 0$ and the sampling points are symmetrical about $x = 0$. In this case, the transform reduces to the simpler form

$$G(k) = 2 \int_0^\infty I(x) \cos 2\pi kx \, dx$$

Absolute symmetry of the interferogram requires perfect optical accuracy and alignment of the interferometer, while symmetrical sampling requires a sampling point exactly at $x = 0$. In practice, these requirements are difficult to meet, leaving the following alternatives.

Both sides of the interferogram are sampled and the complex transform is used, or one side of the interferogram is sampled and an error correction is introduced into the transform program. The latter method is rapid but requires accurate determination of the error of location between the central maximum of the interferogram and the closest sampling point; it is the preferred technique for all preliminary work. The former method takes twice as long to obtain the necessary information but the complex transform is not affected by the relative location of the sampling points; when accurate frequency and intensity measurements are required, this is the preferred method. The problems of sampling and digitizing interferograms have been discussed by Horlick and Malmstadt (1970).

4.2.3.5. *Computer*

It is evident from the above that a computer is a *sine qua non* for Fourier-transform spectroscopy. While analogue computers may be used and some commercial instruments incorporate computers, it is usually more convenient to use a good digital computer. A fast modern instrument such as the IBM 360, the ICL KDF 9 or Atlas is preferred, but smaller machines may be used. The IBM 1130 gives a spectrum at 10 cm^{-1} resolution in 15 min. Such a spectrum required 4 h on Pegasus (1966) or 4 min on Atlas (1972). As computing time increases with the square of the resolution required, it is only with the advent of modern machines that resolution of 1 cm^{-1} has become routine.

A computer should be selected with immediate-access storage of about 8000 12–16 bit words and an access time of 4 μs. A paper-tape reader and line printer or preferably a graph plotter should also be available. Alternatively, a tape punch for paper-tape output with an off-line tape to graph plotter may be used.

As a last resort, if the computer available takes only punched cards, these can be produced from the tape output of the interferometer and used as input to the computer.

4.3. Applications of far infra-red spectroscopy

So far, there has been relatively little study of biopolymers in the far i.r. because the technique is relatively new. The potential value of the technique is considerable, however, as can be seen from the following applications, which include uses in synthetic polymers that are equally applicable to biological materials.

4.3.1. INVESTIGATION OF SKELETAL VIBRATIONS

The longer wavelength region of the spectrum is particularly fruitful as a means of studying skeletal vibrations. As an illustration, we may take the skeletal vibrations of syndiotactic polymethylmethacrylate (s-PMMA) ("Perspex"). The ester side-chains are arranged alternately on either side of the polymer chain and not at random, as in atactic PMMA, nor all on one side of the chain, as in isotactic PMMA. Specimens were polymerized at room temperature with u.v. radiation. Cast film (0·001 in.) and microtomed sections (0·003–0·005 in.) were used. The spectrum in the mid i.r. region (4000–650 cm^{-1}) showed that the specimens were in fact syndiotactic in arrangement.

The far i.r. measurements were made on a Grubb Parsons DM4 (500–200 cm^{-1}) and the IS3 interferometer, using beam splitters (PET) of 0·001, 0·0005 and 0·000025 in. Raman spectra were also obtained on a rod 0·5 in. on a Spex Ar-arc laser (1·5 W, exciting line 20493 cm^{-1}). The DM4 spectrum is shown in Fig. 4.6. The interferogram obtained with the IS3 is given in Fig. 4.4 and the computed spectrum in Fig. 4.7.

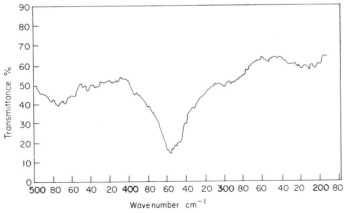

FIG. 4.6. Spectrum of s-PMMA (grating instrument 500–200 cm^{-1}).

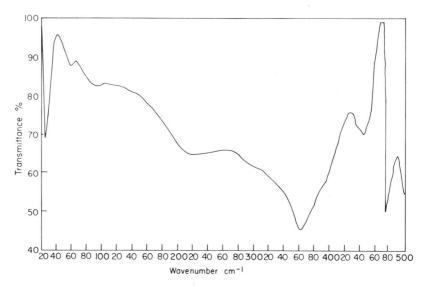

FIG. 4.7. Spectrum of s-PMMA (interferometer).

Several new bands were observed in the longer wavelength region, viz. 35 cm^{-1}, 58 cm^{-1}, 95 cm^{-1}, 225 cm^{-1} (216 cm^{-1} Raman). A schematic model (Fig. 4.8) for the treatment of the skeletal modes may be derived from consideration of a single s-PMMA chain in which the backbone C atoms form an extended planar zig-zag arrangement, whilst the CH$_2$ group, the ester, and α-methyl group are treated as structureless masses in a plane perpendicular to the backbone axis.

From the symmetry elements of this model, it can be shown that the line group (Section 3.3.3) of the polymer chain has a factor group (Section 2.1.9) isomorphous with the point group C_{2v}. The alternating aspect of the syndiotactic chain gives four "masses" in the repeat unit, so that there will be 3N (i.e. 12) skeletal normal vibrations, of which four will be non-genuine, viz. three translations (T) and one rotation (R) about the chain axis. Of the remaining eight genuine vibrations, six will be i.r.-active and all are Raman-active. The symmetry species, characters (Section 2.1.11), number of skeletal normal modes and selection rules for s-PMMA are given in Table 4.1.

In order to calculate the expected frequencies of the skeletal modes, the model in Fig. 4.8 was considered as an infinite planar zig-zag chain of point masses M_1 and M_2, where M_1 is the mass of a (CH$_2$) group and M_2 is the mass of the (CH$_3$—C—COOCH$_3$) group. These calculations (Manley and Martin, 1971) give eight genuine frequencies and four zero frequencies, as predicted by the group-symmetry analysis (Table 4.1).

The calculated frequencies are compared with the observed frequencies and their assignments given in Table 4.2. In the calculations, the (CH$_2$) and

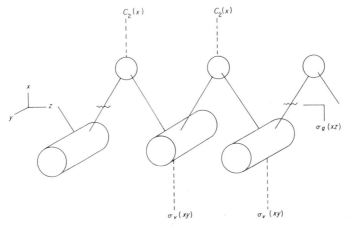

FIG. 4.8. Model of skeletal modes of s-PMMA.

(CH_3—C—$COOCH_3$) groups were replaced by point masses. However, this approximation applies only when there is strong coupling between the chain C atoms and the substituents; for weaker coupling, the mass M_2 must be replaced by a point mass somewhere between the mass of a C atom and the mass of the (CH_3—C—$COOCH_3$) group. Hence the calculated frequencies will be higher for lesser degrees of coupling. Also, since the results are affected by choice of force constants, the calculated frequencies may only be used as guides to assignments.

TABLE 4.1. Symmetry species, number of skeletal normal modes and selection rules for s-PMMA

C_{2v}	E	$C_2(x)$	$\sigma_v(xy)$	$\sigma_g(xz)$	n	T,R	i.r.	Raman
A_1	1	1	1	1	2	T_x	a	a
A_2	1	1	−1	−1	2	—	i	a
B_1	1	−1	1	−1	2	T_y, R_z	a	a
B_2	1	−1	−1	1	2	T_z	a	a

a = active, i = inactive, T = translational mode, R = rotational mode.

Without the far i.r. measurements, it would have been impossible to obtain satisfactory assignments of the skeletal vibrations of this polymer.

4.3.2. PEPTIDES
There has been a good deal of interest in the i.r. spectrum of N–methylacetamide, $CH_3CONHCH_3$, as it provides an opportunity to study the peptide bond in a small molecule. In studies of the far i.r. spectra of

TABLE 4.2. Skeletal modes and assignments of syndiotactic poly(methyl methacrylate)

Observed Infra-red (cm^{-1})	Wavenumber Raman (cm^{-1})	Calculated Wavenumber (cm^{-1})		Assignment		
807 vw	786 w^a	v_2 803	v(C—C) $\parallel$	B$_2$	$v_+(\pi)$	
749 m	732 w	v_4 761	v(C—C) $\perp$	A$_1$	$v_+(0)$	
828 vw	828 msh	v_5 815	v(C—C)	A$_2$ $\}$	$v_+\left(\dfrac{\pi}{2}\right)$ and	$v_+\left(\dfrac{3\pi}{2}\right)$
552 vw	552 w	v_7 695	v(C—C)	B$_1$ $\}$		
320 sh	296 w	v_6 324	δ(C—C—C) in-plane	B$_1$ $\}$	$v_-\left(\dfrac{\pi}{2}\right)$ and	$v_-\left(\dfrac{3\pi}{2}\right)$
	266 sh^b	v_8 274	δ(C—C—C) in-plane	A$_2$ $\}$		
225 w,br	216 w^c	v_{10} 182	δ(C—C—C) out-of-plane	B$_2$ $\}$	$v\left(\dfrac{\pi}{2}\right)$ and	$v\left(\dfrac{3\pi}{2}\right)$
95 w	—	v_{12} 73	δ(C—C—C) out-of-plane	A$_1$ $\}$		
58 vvw	—	—	hindered rotation			
34 vvw	—	—	or translation			

^{a}Observed as broad weak shoulder on strong band $\Delta v = 810$ cm^{-1}.
^{b}Very broad band $\rightleftharpoons 186–246$ cm^{-1}.
^{c}Outside range of instrument without special filters.
Abbreviations: w = weak, v = very, m = medium, sh = shoulder, br = broad, v = stretching, δ = bending.

N–methylacetamide and other monosubstituted amides, Itoh and Shimanouchi (1967) found bands characteristic of the peptide bond that were assigned to H-bonding and to the C—N torsional vibration. They were also able to explain, from a calculation of force constants, the spectral changes observed in the solid-phase transition at 10°C from a disordered to an ordered crystal. Figure 4.9 shows the far i.r. spectra of various N-substituted amides and methyl acetate as liquids (i.e. at 30°C), as obtained by Itoh and Shimanouchi. An absorption band near 100 cm^{-1} was found in all the monosubstituted amides (Fig. 4.9 a–f), but not in dimethylformamide nor in methyl acetate (Fig. 4.9 g, h). All these molecules, except the last two, contain a proton-donor group and give rise to H-bonding. Accordingly, this band near 100 cm^{-1} (102 cm^{-1} for N-methylacetamide) was assigned to the CO ... NH H-bond vibration; its position, as expected, showed little shift on deuteriation.

N-methylacetamide has a band at 201 cm^{-1}, assigned to C—N torsional vibration. The shifts on deuteriation support this assignment (195 cm^{-1} for CD$_3$CONHCH$_3$ and 185 cm^{-1} for CD$_3$CONHCD$_3$).

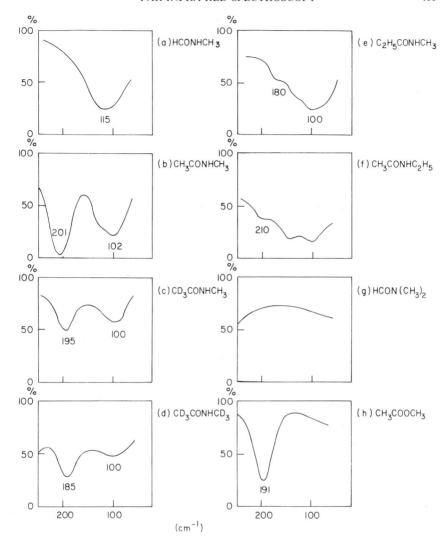

FIG. 4.9. Far i.r. spectra of N-substituted amides and methyl acetate.

At lower temperatures, shifts and splitting occur. At $-10°C$, a new band appears at 90 cm^{-1}, the H band shifts to 130 cm^{-1}, and the internal-rotation band shifts to 210 cm^{-1}. At $-180°C$, the 90 cm^{-1} band intensifies, the H band shifts to 132 cm^{-1}, and the internal-rotation band splits at 22 and 243 cm^{-1}. Using crystal-structure data, the force constants were calculated; from these, the spectral changes, consequent on the transition from an ordered to a disordered crystal at 10°C, could be explained satisfactorily.

4.3.3. POLYALANINES

Itoh et al. (1968) studied the far i.r. spectra of poly L-alanines with the α-helical and β-form structures, a copolymer of glycine-L-alanine, various copolymers of D- and L-alanine, and silk fibroin. Figure 4.10 shows the far i.r. spectra of poly-L-alanine. The first point of interest is that the FIR spectra show a

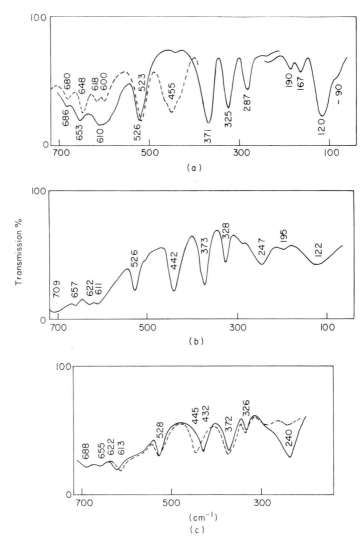

FIG. 4.10. F.i.r. spectra of poly-L-alanine: (a) polymerized in acetonitrile (broken line shows N-deuteriated PLA); (b) polymerized in dimethylsulphoxide (DMSO); (c) as for (a) (hydrogenated PLA) but after stretching with E-vector $\parallel$ (full line) and $\perp$ (broken line) direction of stretch.

difference between the two polymers, depending on whether acetonitrile (Fig. 4.10a) or dimethylsulphoxide is the solvent for the polymerization. In the former solvent, the α-helical conformation is assumed by the polymer, whereas the β-form predominates in the latter instance (Fig. 4.10b). The characteristics of the α-helical conformation are seen in Fig. 4.10(a), whilst the bands due to the β-form appear in Figs 4.10(b) and 4.10(c). The bands at 622, 613, 445, 432 and 240 cm^{-1}, with strong dichroism in the far i.r., were assigned to the β-form; since the last three bands are absent from Fig. 4.10(a), a definite assignment may be made. These bands at 442 and 247 cm^{-1} are found in silk fibroin and the glycine-L-alanine copolymer and they support X-ray evidence that the silk fibroin crystallizes in the β-form; the same deduction is made for the copolymer. (For i.r. and Raman spectra see pp. 56 and 90 et seq.)

4.3.4. COPOLYMERS OF D, L-ALANINES

The far i.r. spectra of the copolymers with different proportions of D- and L-alanine show bands at 580, 478 and 420 cm^{-1} when the proportion of D is 3, 7 or 10%. These bands are absent from the spectra of each homopolymer. They are, therefore, ascribed to the structure of the D-residue when incorporated into the right-hand α-helix (and conversely, as the proportion of D-residue increases, of the L-fraction in the left-handed α-helix).

As the bands in this region are sensitive to changes in configuration, it is likely that the frequencies differ between the structures ($\psi = 123°$, $\phi = 132°$) and ($\psi = 237°$, $\phi = 228°$). Support for this view is provided by the linear relation found between the relative intensities of the 478 and 420 cm^{-1} bands and the percentage of D-residue in the copolymer.

Assignment of the bands to interaction between the right-hand and left-hand α-helix is ruled out because the bands do not appear in a mixture of homopolymers. Evidence from optical-rotation measurements eliminates the possibility of assignment of the bands to a random-coil structure or the joint between the right-handed and left-handed α-helix. As the proportion of the D-residue increases, the spectra illustrated in Fig. 4.11 are obtained. Only the bands of the α-helix and the three bands at 580, 478 and 420 cm^{-1} are seen. No band is found apart from those attributed to the α-helix. The deformation of the α-helix, through the incorporation of amino-acid residues from opposite enantiomorphs, can be followed in the copolymers of D- and L-alanine by the shift in the 325 cm^{-1} band to 319 with broadening of the band, and by the broadening of the three bands at 686, 653 and 610 cm^{-1} until they coalesce at 645 cm^{-1}.

4.3.5. POLARIZED FAR INFRA-RED MEASUREMENTS

As in the conventional i.r. region, the use of polarized radiation (see p. 37) greatly increases the amount of information that may be derived from a specimen. The first results on a polymer obtained with polarized radiation in

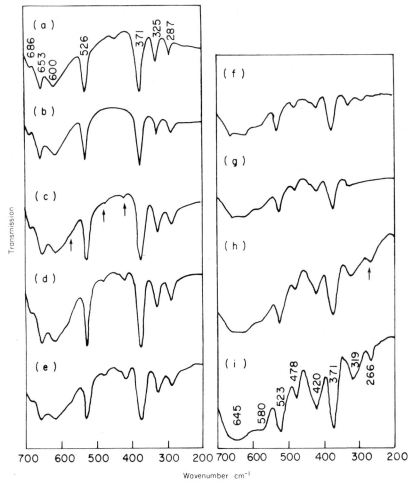

FIG. 4.11. F.i.r. spectra of mixtures poly-D- and poly-L-alanine with % of D increasing from 0 (in b) to 50 (in i); arrows mark peaks characteristic of α-helix.

the far i.r. were presented in 1967 at the I.U.P.A.C. meeting in Brussels (Manley and Williams, 1969). Oriented specimens of polethylene terephthalate (PET) were examined and the work was subsequently extended to poly (cyclohexane 1,4-dimethylene terephthalate), PCHT (Manley and Williams, 1969b, 1971). Taken in conjunction with standard i.r. measurements, the use of polarized far i.r. enabled assignments to be made with greater confidence and to include skeletal-deformation and torsional modes. Hence the changes in the conformation of the polymer induced by orientation and crystallization could be rigorously defined.

The polarizer used was a stainless-steel tube 150 mm long and 42 mm in diameter. An inner sleeve of black polythene had an internal diameter of

30 mm. The refractive index of polythene remains substantially unchanged over the whole of the far i.r. region at 1·46, and hence the Brewster angle (p. 38) is 55°. Twelve polythene sheets were used to make the polarizer. Six of the sheets were 20 μm thick and six were 30 μm; the space between successive sheets was approximately 300 μm (0·3 mm). The use of polythene sheets of two different thicknesses is an attempt to overcome limitations caused by interference. The results obtained are shown in Figs 4.12 and 4.13.

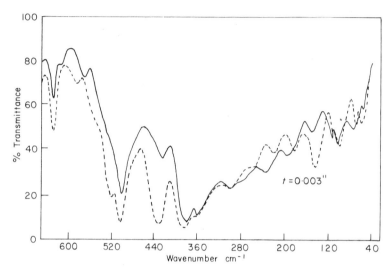

FIG. 4.12. Polarized f.i.r. spectra of heat crystallized PET.

In Fig. 4.12, polarized far i.r. spectra of PET, uniaxially oriented, heat crystallized, for a film thickness of 75 μm (0·003 in.) in the region 640–20 cm^{-1} may be seen. The continuous curve represents the spectrum with the electric vector perpendicular to the stretching direction; the dashed curve represents the spectrum with the electric vector parallel to the stretching direction. Considerable dichroism of the majority of the bands may be observed, as well as appreciable band shifts that occur between the two orientations.

Figure 4.13 shows polarized spectra of uniaxially oriented, non-heat-crystallized PET film of thickness 100 μm (0·004 in.). Many bands occurring in the heat-crystallized specimen are noticeably attenuated or are completely absent in the non-heat-crystallized sample, and a number of bands common to both spectra have been shifted to slightly longer wavelengths. The medium band at 430 cm^{-1} in the heat-treated sample gives rise to a doublet in the non-heat-treated film.

The origin of the so-called crystal and amorphous bands has been attributed by several workers to the rotational isomerism of the —OCH$_2$—CH$_2$O— group. The intensity variations of certain bands in the spectrum which occur

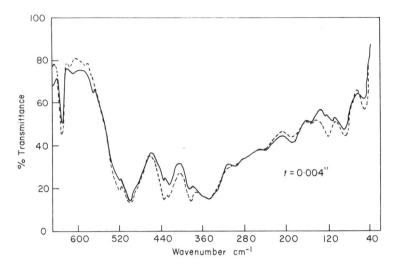

Fig. 4.13. Polarized f.i.r. spectra of non heat-crystallized PET.

upon crystallization were attributed to a *trans* configuration of the glycol residue in the crystal, and to a *trans-gauche* configuration in the amorphous material. However, the *cis* and *trans* isomers of PCHT possess a number of bands of similar frequency, intensity, band shape, polarization and crystal or amorphous origin, as in PET. It is, therefore, likely that these bands are due to the benzene framework in both PET and PCHT; this would support the contention that many of the changes occurring upon crystallization in PET are associated with changes in the symmetry and resonance characteristics of the benzene ring.

It has been predicted that bands showing parallel dichroism would occur at 425, 340 and 162 cm^{-1} in the *gauche* configuration and, in fact, medium bands of amorphous origin at 422, 344 and 155 cm^{-1} were observed. Similarly, bands predicted for the *trans* configuration were observed at 546 and 379 cm^{-1}. All bands were completely absent from the spectrum of PCHT and were, therefore, assigned to skeletal —C—O—C— vibrations in PET.

A weak band at 37 cm^{-1} in the amorphous film was attributed to a pure torsional vibration of the skeleton; its absence in the oriented samples might be due to the relatively higher crystal symmetry in the latter, rendering the mode inactive. A band in the region of 50 cm^{-1} in the uniaxially oriented samples gives rise to two peaks at 44 and 55 cm^{-1} in the amorphous film. Similar bands were observed in PCHT and were thought to be due to out-of-plane twisting of the ring framework.

The medium π bands at 68, 78 and 176 cm^{-1} observed in the biaxially oriented heat crystallized film, and virtually absent from the amorphous film spectrum, are attributed to crystal-lattice modes which are activated by the presence of two mutually perpendicular chain axes. The medium amorphous band at 193 cm^{-1}, showing perpendicular dichroism in the one-way-drawn film, is attributed to one of the principal out-of-ring-plane deformation modes of the benzenoid ring system; a similar band was observed in PCHT. The behaviour of this band in PET tends to support the argument that the orientation of the benzene ring about the fibre axis is random in the uniaxially uncrystallized specimen and parallel to the film surface in the biaxially oriented heat-treated film.

Far i.r. dichroic measurements were subsequently made (Itoh and Shimanouchi, 1970) on the right-handed α-helix of poly L-alanine. These results, taken in conjunction with those mentioned in Section 4.3.3, permitted the assignment of the characteristic bands and also the calculation of the frequencies and normal modes of the poly-L-alanine α-helix. Figure 4.14 shows the far i.r. dichroism of this polymer (in the original paper, the spectra in this figure are transposed). The frequencies of the vibrations of the α-helix with various phase differences were calculated and an estimate was made of the frequencies of the "accordion like" vibrations. The values calculated for the frequencies of the right-handed α-helices of poly-D-alanine (b) and poly (L-α-amino-$n$butyric acid) (a) were useful in interpreting the spectra of these polymers. In the latter case, they provided evidence in support of the presence of rotational isomers in the side chain.

4.3.6. CALCULATION OF YOUNG'S MODULUS FOR HELICAL POLYMERS

If the structure of a polymer crystal is known, it is possible to calculate the theoretical elastic modulus of the crystal. Structural information is obtained from X-ray diffraction and from a knowledge of the complete i.r. spectrum of a polymer. As most of the conventional i.r. information on polymers is readily available, the determination of the far i.r. spectrum permits one, if X-ray data are available, to calculate the force constants of the chemical bonds of the polymer from the vibrational frequencies, and hence the modulus of elasticity in the principal chain direction of the polymer. In the absence of far i.r. data, the calculation of force constants is impossibly difficult.

Helical polymers are of particular biological interest. The calculation of the elastic constants of these polymers based on a Urey–Bradley force field is very complex and, in an attempt to reduce the labour of computation, a modified method has been used (Manley and Martin, 1971) to calculate the elastic modulus of simple helical polymers. The method is an adaptation of the energy method introduced by Jaswon *et al.* (1968) and used by them to calculate the elastic constants of cellulose with respect to the principal areas of elasticity. The first polymer studied was the inorganic elastomer polyphosphonitrilic chloride (PNC).

The values obtained were $1·38 \times 10^9$ dyne cm^{-2}, assuming that the molecule was a uniform helix, and $1·66 \times 10^{10}$ dyne cm^{-2} (1 dyne cm$^{-2} = 0·1$ N m^{-2}), if

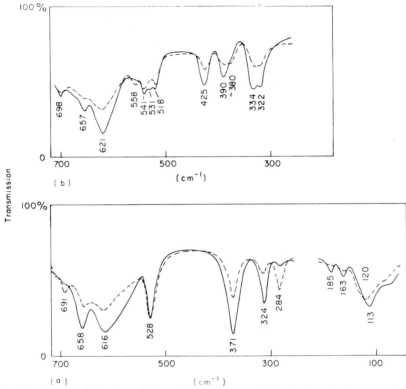

FIG. 4.14. Far i.r. dichroism of the poly-L-alanine α-helix (*E*-vector ⊥ (full line) and ∥ (broken line) to orientation direction): (a) poly-L-alanine; (b) poly-(L-γ-amino-*n*-butyric acid).

a *cis*-planar structure is assumed. Thus the FIR results enable a distinction to be made between possible structures of a helix. In this instance, however, further X-ray data or changes occurring in the crystal-lattice spacing when the crystal is stressed are needed.

The Young's modulus of the α-helix of poly-L-alanine has been calculated (Itoh and Shimanouchi, 1970) as

$$E = 2 \cdot 31 \times 10^{10} \text{ Nm}^{-2}$$

4.4. Conclusions

The advantages of Fourier-transform spectroscopy may be summarized as high speed and high resolution. The resolution in a F.T. Spectrometer is variable at will; there is no restriction to narrow slits for high resolution as in dispersive spectrometry. Resolution of $0 \cdot 1$ cm^{-1} is commonplace but, of course, such fine resolution takes much greater measurement and computational time. There is no stray light problem with an interferometer but, as has been seen, the use of filters enhances the accuracy of results.

The major advantage, which arises from the freedom from the necessity to disperse or filter, is the signal-to-noise ratio (Fellgett's advantage). In a conventional spectrometer, the intensity of each resolution element (r) is proportional to the time spent in scanning it, because each packet of radiation or resolution element is scanned over the detector. The signal (i.e. the intensity of an element of radiation) is therefore proportional to T/r where T is the total scan time. Noise, being random, is proportional to the square-root of the observation time. The signal-to-noise ratio, therefore, varies as the square root of T/r. In an interferometer, however, each resolution element is observed all the time, so that the signal-to-noise ratio now varies as the square root of T. This improvement, by a factor of $r^{\frac{1}{2}}$, is particularly important where high resolution is required.

If high resolution is not essential, Fellgett's advantage enables spectra to be obtained rapidly; a scan at 10 cm^{-1} resolution can be made in two min. Further handling of the information from the interferometer is also facilitated by the fact that it is presented on punched tape, so that data processing or further computation of the spectra are readily performed.

By use of Fourier-transform spectroscopy, the embargo on studying the far i.r. imposed by the lack of energy has been lifted. The way is now open, therefore, for the study of effects that appear in the far i.r. region. These include intra-molecular motions of macromolecules, e.g. hindered rotation of one section of a polymer with respect to another; the vibrational motions of heavy atoms in complex systems; the bending motion of H bonds; skeletal vibrations of polymers; vibrational modes of polymer crystal lattices; modes due to crystallization and orientation of both linear and helical polymers. To date, few biopolymers have been studied but, as the technique becomes better known, it should prove one of the most powerful weapons in the armoury of biological research workers.

References

Barnes, R. B. and Czerny, M. (1931). *Phys. Z.* **72**, 447.
Fellgett, P. (1958). *J. Phys. Radn.* **19**, 187.
Gebbie, H. A. and Vanasse, G. A. (1956). *Nature* **178**, 432.
Griffiths, P. R. (1975). "Chemical Infrared Fourier Transform Spectroscopy". Wiley-Interscience, New York.
Horlick, G. and Malmstadt, H. V. (1970). *Anal. Chem.* **42**, 1361.
Itoh, K. and Shimanouchi, T. (1967). *Biopolymers* **5**, 921.
Itoh, K., Nakahara, T. and Shimanouchi, T. (1968). *Biopolymers* **6**, 1759.
Itoh, K., Shimanouchi, T. and Oya, M. (1969). *Biopolymers* **7**, 649.
Itoh, K. and Shimanouchi, T. (1970). *Biopolymers* **9**, 383.
Jaswon, M. A., Gillis, P. P. and Mark, R. E. (1968). *Proc. Roy. Soc. A* **306**, 389.
Manley, T. R. and Williams, D. A. (1965a). I.U.P.A.C. Macromolecular Conference, Prague, Paper 467, July, 1965.

Manley, T. R. and Williams, D. A. (1965b). *Spectrochim. Acta* **21,** 737.
Manley, T. R. and Williams, D. A. (1969a). *J. Poly. Sci.* **C22,** 1009.
Manley, T. R. and Williams, D. A. (1969b). *Polymer* **10,** 339.
Manley, T. R. and Martin, C. G. (1971). *Polymer* **12,** 524.
Martin, D. H. (ed.) (1967). "Spectroscopic Techniques". North Holland, Amsterdam.
Randall, H. M. (1939). *J. App. Phys.* **10,** 768.
Rothschild, W. G. (1965). *Spectrochim. Acta* **21,** (4), 852.
Richards, P. L. (1963). *J. Appl. Phys.* **34,** 1237.
Rubens, H. and Nichols, E. F. (1897). *Ann. Phys. Chem.* **60.**
Strong, J. and Vanasse, G. A. (1958). *J. Phys. Radium* **19,** 192.
Strong, J. and Vanasse, G. A. (1960). *J. Opt. Soc. Amer.* **50,** 113.
Willis, H. A., Miller, R. G. J., Adams, D. M. and Gebbie, H. A. (1963). *Spectrochim. Acta* **19,** 1457.

5. Electronic Absorption and Emission Spectroscopy

S. Ainsworth

In this chapter, the ways in which molecules absorb and emit light in the visible and u.v. regions of the spectrum are described, and an indication is given as to how these processes may be used to investigate molecules of biological interest.

5.1. The nature of light and its interaction with molecules

The description of light and of its generation and absorption by matter is fundamentally ambiguous. For light may be thought of both as discrete packets of energy and as an electromagnetic wave, the link between the two concepts being provided by the relationship $E = h\nu$, which defines the energy of the packet or quantum of light in terms of Planck's constant, h, and the frequency of the light considered as a wave motion.

The quantum description of light is particularly useful in considering the energetic aspects of its relationship with molecules. Molecules exist in distinct energy states characterized by quantum numbers which differ for each state, so that, if a molecule passes from one such state to another, the transition is accompanied by the uptake or release of energy. When the energy is absorbed or emitted as light, its frequency is defined by the energy change of the transition (Section 1.2.1).

On the other hand, the wave description of light is more helpful when the probability of light interacting with a molecule is considered, particularly so when the spatial aspects of this interaction are of interest. Light, considered as an electromagnetic wave, arises when two electrical charges are accelerated periodically with respect to one another so as to constitute an oscillating dipole. The movement of the charges induces changes in their associated electrical field and, because moving electrical charges correspond to an electrical current, with their associated magnetic field also. The electrical and magnetic disturbances do not reach every point in space instantaneously but spread outward from the dipole with a constant speed (the speed of light), the frequency of the dipole oscillation determining the wavelength of the radiated disturbance. Application of the well-known right-hand rule shows that the directions of the electrical and magnetic fields are mutually perpendicular, with both perpendicular to the direction of propagation. However, because the interactions of light with molecules that are our concern depend exclusively on the electric vector, we shall confine our future attention to this component alone. We may describe its properties most simply by comparing it with the wave propagated in a string by oscillating one end of it. In both, it will be observed that the plane of the wave is confined to the direction of the oscillation which is its cause. Again, in both, the direction of oscillation constitutes an axis of symmetry for the propagated waves, which have a maximum amplitude when the direction of propagation is within the equatorial plane, diminishing to zero as the direction of propagation is moved upwards to the pole.

It has already been noted that light is only generated or absorbed by molecules when its frequency matches the energy difference between two molecular states. It can now be seen intuitively that this condition is insufficient to determine whether the interaction will occur and that another requirement (there are still others) is that a translation of charge must occur during the molecular transition to correspond to the classical oscillator that has been taken as the origin of light. Another conclusion is that interaction will require the transition dipole to be in the same direction as the plane of polarization of the light beam.

5.2. Energy states of molecules

In order of increasing energy, the energy states of a molecule include contributions from molecular rotations, intermolecular vibrations and electrostatic interactions—the last including repulsions between electrons and attractions between nuclei and electrons. Light in the u.v. and visible spectrum is energetic enough to bring about changes in all these molecular states both in absorption and emission; nonetheless, to a first approximation, their energy contributions may be considered separately (Section 1.2.2).

5.2.1. ELECTRONIC ENERGY LEVELS

The electronic state of an atom is described by three spatial quantum numbers which define the size, shape and orientation of the orbitals that can contain electrons, and by the spin quantum number which allows each orbital to be occupied by two electrons with opposed spins. Molecular orbitals are assumed to arise by the combination of atomic orbitals; as before, each orbital can accommodate two electrons with opposed spins.

Light absorption in the u.v. and visible spectrum is generally restricted to molecules with double bonds and, at the longer wavelengths of the range, to molecules with conjugated bond systems. These transitions involve the promotion of an electron from a π orbital (formed by the combination of two p-type atomic orbitals) or from a non-bonding p-type orbital into a higher anti-bonding π^* orbital. The shapes of the different orbitals and the transitions between them are illustrated in Fig. 5.1, using formaldehyde as the example. Figure 5.1 shows that the movement of electron density that takes place during the π–π^* transition is in the plane of the molecule, while that occurring during the n–π^* transition is at right angles to it: in other words, the two transitions are polarized at right angles to one another. It should be emphasized that both these transitions involve a spatial quantum number change only. The spin of the promoted electron remains unchanged in the π^* orbital and opposed to the spin of the single electron left behind, either in the non-bonding or the π orbital. Energy states of the molecule where spin pairing is retained are called "singlets"—a term used in atomic spectroscopy to denote transitions that are not affected by a magnetic field. If, however, both the spatial and spin quantum numbers were to change in forming the excited state of the molecule, the resultant spin would no longer be zero and the energy of transitions to the excited state would be affected by a magnetic field. Such excited states of the molecule are termed "triplets", and always have a lower energy than those of the corresponding singlet states.

Similar transitions occur in molecules possessing conjugated double-bond systems, with the difference that the π-molecular orbitals are now "delocalized" and spread out over the atoms comprising the conjugated structure. The π electrons are thus free to run the whole length of the conjugated chain. In linear conjugated molecules, such as the carotenes, the

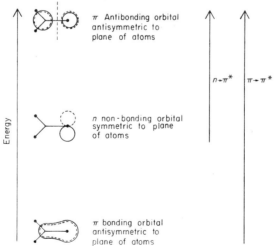

FIG. 5.1. Molecular orbitals of formaldehyde involved in n–π^* and π–π^* transitions. The solid line represents the positive region of the orbital wave function; the broken line, the negative region. The square of the wave function gives the probability distribution for an electron in the corresponding orbital. The non-bonding electrons are only loosely coupled to the rest of the molecule and therefore give rise to the transition with smaller energy.

energy levels may be defined by treating the conjugated chain essentially as a stretched string and calculating the energies of the standing waves of electron density that may be fitted within it. A similar approach may be applied to aromatic molecules, the length of the conjugation path now comprising the molecular perimeter. In both instances, the π-electrons are fitted in pairs with opposed spins into the lowest energy states of the molecule, and light absorption occurs when an electron is promoted into a more energetic state (Platt, 1956; Mason, 1961).

5.3. Processes of absorption and emission

The processes of absorption and emission are summarized in the Jablonski diagram shown in Fig. 5.2. The diagram represents the electronic, vibrational and rotational energy levels associated with the singlet and triplet systems and shows the transitions between them. Particular aspects of the Jablonski diagram will now be considered (see also Fig. 1.1).

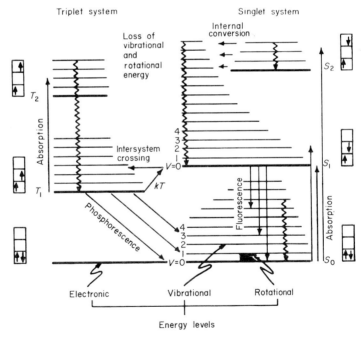

FIG. 5.2. Jablonski diagram showing the energy states of the singlet and triplet systems and the transitions between them.

5.3.1. ABSORPTION

At room temperature and at normal light intensities, the overwhelming majority of molecules are in the lowest electronic state of the molecule, the so-called ground state, and the majority of these are in their lowest vibrational level with quantum number $v = 0$. Transitions to higher singlet states take place with absorption of light and, depending on its frequency, the excited molecules also acquire vibrational and rotational energy. To see how this affects the amount of light absorbed at different frequencies, i.e. the absorption spectrum, an energy diagram must be considered which correlates the energy of the molecule with its nuclear configuration (see Figs 1.1 and 5.3).

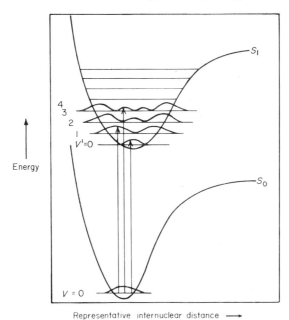

FIG. 5.3. Potential-energy curves for the ground and excited states. The total energy of the molecule in different vibrational levels of the two states is represented by horizontal lines, upon which are drawn curves proportional to the probability of finding the representative atoms at the given internuclear separation. The vertical lines correspond to likely transitions brought about by light absorption.

In Fig. 5.3, lines are drawn representing the total energy of the molecule in different vibrational levels of the first excited state S_1 and in the lowest vibrational level of S_0. The curves represent the potential energy of the molecule in its S_0 and S_1 states as a function of the separation of two of its atoms that are vibrating with respect to one another. It will be noted that the curve for S_1 is displaced, relative to that for S_0, towards greater inter-nuclear separations; this indicates that the structure of the excited state is looser than that of S_0, a feature consistent with the more extended orbital occupied by the promoted electron.

The relative probabilities of the two nuclei being at any given internuclear separation are represented by the curves drawn on the vibrational levels of the excited state. For the lowest vibrational level, the most probable position of the atoms is at the equilibrium extension of the bond, but for all higher levels, the atoms are most likely to be found at the shortest and longest extensions of the bond between them. At these separations, the movement of the nuclei changes direction so that their speed is least and the likelihood of being present greatest: for the same reason, the energy at extremes of bond length is predominantly potential energy.

In discussion of transitions between two electronic states, two points must be considered. First, an electronic transition takes place within the period of a light wave—about 10^{-15} s. This is possible because electrons can move very rapidly—e.g. the traverse of a Bohr orbit takes about the same length of time. On the other hand, nuclei are much heavier than electrons and have a greater inertia; for comparison, the period of a vibrational oscillation is about 10^{-13} s. The Franck–Condon principle, therefore, states that an electronic transition takes place in a molecule so rapidly, compared to the vibrational motion of the nuclei, that the internuclear distance can be regarded as fixed during the transition. It follows from this principle that electronic transitions may be represented on the energy diagram as vertical lines connecting the initial and final states of the molecule at a fixed internuclear distance. Second, a transition is most likely when the molecule has an internuclear distance such that the transition connects probable states of the molecule in the ground and excited states.

Therefore, starting from the lowest vibrational level of the ground state, absorption of light populates the several levels of the first excited state in proportion to the probability of finding the molecule at the same internuclear separation. Several such transitions are represented in Fig. 5.3.

The above treatment is overly simplified for, in polyatomic molecules, many different characteristic vibrations have to be taken into account. In addition, rotational quantum number changes take place with much smaller increments of energy. A large number of overlapping transitions therefore results, the effect of which is to produce a "band" of absorption rather than the "line" which would correspond to the electronic transition itself.

5.3.2. PATHS OF MOLECULAR EXCITATION LOSS
As the Jablonski diagram shows, absorption of light excites a molecule to one of a number of electronically excited states. The return of the molecule to the ground state and the effect of the possession of excitation energy on molecular properties will now be discussed.

5.3.2.1. *Loss of vibrational energy*
Excitation to higher electronic levels is accompanied by vibrational excitation, the excess vibrational energy being lost so quickly that within 10^{-11} s it has all been dispersed. This loss results from collisions, generally with solvent

molecules, the energy being redistributed in such a way as to be consistent with the conservation of energy and quantum restrictions on allowed energy levels. As a result, the molecule ends in the lowest vibrational level of the excited state.

5.3.2.2. *Internal conversion and intersystem crossing*
Internal conversion is a process, not involving the emission of light, whereby the structure of an excited molecule is changed from one singlet state to another with lower energy. Intersystem crossing is a similar process taking place between the lowest excited singlet and triplet levels of a molecule. The first of these decay processes involves a change in a spatial quantum number, the second, a change in the spin quantum number of the excited molecule (Lower and El-Sayed, 1966).

Referring to the potential-energy diagram shown in Fig. 5.4, we see that both processes arise through the crossing of potential-energy surfaces of different states of the same molecule. Initially, the molecule is excited to the vibrational level A of state 2. Loss of vibrational energy leads to the nuclear configuration B which is also a possible configuration of state 1. The equilibrium which then exists between the two states is rapidly displaced by the further loss of vibrational energy, leading the molecule to the potential well of state 1.

Internal conversion is a rapid process, taking place in about 10^{-13} s which, in conjunction with the loss of vibrational energy, leads to all the excited molecules in the singlet system descending to the lowest vibrational levels of

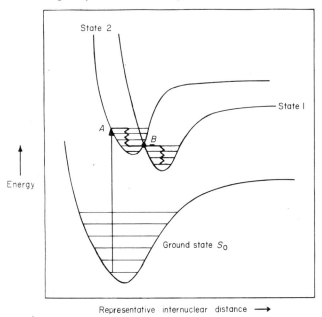

FIG. 5.4. Potential-energy curves leading to internal conversion or intersystem crossing.

the first singlet excited state. The energy gap between these levels and the ground state is generally larger than the gaps between the higher states; in consequence, their potential-energy surfaces do not cross and internal conversion is not an important means of depopulating S_1. Molecules, therefore, persist within S_1 for much longer times, about 10^{-8} s, before returning to the ground state, either by the emission of light or via intersystem crossing to the triplet state.

Intersystem crossing is a much slower process because it involves a change in electron spin and takes about 10^{-7} s to occur. This is just fast enough to compete with the spontaneous emission of light by S_1 but far too slow to compete with internal conversion. Intersystem crossing is, therefore, limited to transfer from the S_1 state.

The lowest triplet state, populated by intersystem crossing, has a long lifetime—10^{-2} s or longer—because its decay to the ground state again involves a spin reversal. The return is accomplished by the emission of light or by collisional quenching by solvent molecules. These decay processes will be dealt with below.

5.3.2.3. Fluorescence and phosphorescence

As previously stated, only the lowest singlet and triplet states are sufficiently stable to emit light in measurable quantities. The light emissions that occur are illustrated in Fig. 5.2. There it will be observed that fluorescence is light emitted in the return of the molecule from the S_1 to the S_0 state, while phosphorescence is limited to the transition between T_1 and S_0. The times these emissions persist after removal of the exciting light is determined by the stability of the S_1 and T_1 states; thus, fluorescence persists for some 10^{-8} s, while phosphorescence may be emitted up to a few seconds after removing the exciting light. Delayed fluorescence is emitted when thermal excitation of molecules in the T_1 state allows them to return to the S_1 state; because of this, delayed fluorescence, like phosphorescence, persists for relatively long times after excitation has ceased. Its intensity is generally very small (Parker, 1968).

The spectral location of the emissions relative to the absorption band of least energy is determined by three facts. First, absorption starts from the lowest vibrational level of S_0 but takes molecules to many vibrationally excited states of S_1. Second, the excited molecules rapidly lose vibrational energy, so that emission occurs from the lowest level of S_1, but returns the molecules to many vibrationally excited states of S_0. Third, T_1 has a lower energy than S_1. Examination of Fig. 5.2 shows that all absorption frequencies are not less than that corresponding to the transition between the zero vibrational energy states of S_0 and S_1 (the so-called zero-zero transition), while all fluorescence emission frequencies are not greater. Assuming that the vibrational structures of the S_0 and S_1 states are identical, the absorption and fluorescence spectra, therefore, present the appearance of mirror images reflected about the frequency of the zero-zero transition. The preceding treatment is strictly applicable only at the absolute zero of temperature. At higher temperatures some transitions do take place from higher vibrational levels; the effect of this is to move the absorption and fluorescence spectra towards one another so that they overlap. This is shown in Fig. 5.5, together with the phosphorescence spectrum, which is similar to the fluorescence spectrum but displaced to lower frequencies.

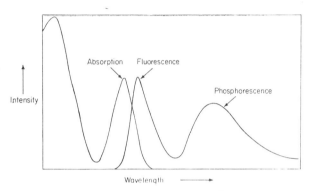

FIG. 5.5. Relative spectral location of absorption, fluorescence and phosphorescence.

5.3.2.4. *Quenching and photochemical reactions*

Quenching is a non-radiative process by which excited molecules lose their excitation energy by collisions with other molecules in their ground state. The triplet state is very subject to this type of energy loss because of its long lifetime; for this reason, phosphorescence is generally observed only in cold, rigid media. Typical quenching agents include O_2, which possesses electrons with unpaired spins, heavy atoms such as I, and paramagnetic ions like Fe^{3+}. All induce intersystem crossing on close approach to the excited molecule without being chemically affected thereby. Because the quencher is present at much higher concentrations than that of excited molecules, the quenching process can be represented by an apparent first-order constant which includes the quencher concentration.

Electronic excitation energy can frequently be used to promote chemical reactions. This capacity may be reflected by a change in the effective acidity or redox potential of the molecule, brought about by excitation, that allows reactions to proceed which would be impossible in the ground state. An example of obvious importance occurs in photosynthesis where an electron is moved against a ground-state electrochemical gradient as a result of light absorption by chlorophyll.

In general, the excited state is predisposed towards chemical reaction, not only because of its increased energy, but also because of its looser structure. These features are common to both the singlet and triplet states, but it is the latter especially that is important in promoting photochemical reactions: first, because its long lifetime greatly enhances the chances of reaction taking place; second because the transition state of the reaction may be very similar to the triplet state itself. Thus, as the latter has already been obtained by irradiation, the reaction is promoted (Reid, 1957).

5.3.2.5. *Energy transfer*

The energy of the excited state may be transferred directly to other molecules without the intervening emission of radiation. As a result, the donor molecule is returned to its ground state at the same time as the acceptor molecule is raised to its first singlet excited state. Energy transfer is an important process in nature, notably in the photosynthetic apparatus of plants, where the energy of light absorbed by accessory pigments is transferred to chlorophyll. It is also a valuable means of studying the structure of proteins and other biologically interesting systems.

Efficient energy transfer depends on three conditions being met (Förster, 1960). First, the rate of energy transfer is inversely proportional to the sixth power of the distance separating the donor and acceptor molecules; it is, therefore, only found at high concentrations when the two species are independently soluble. Second, the process is critically dependent on the relative orientation of the two transition dipoles, being maximally efficient when these are parallel and adjacent to one another and diminishing as they are slid apart or as an angle opens up between them. Finally, energy transfer increases in efficiency as the transition of the donor from its first excited state back to its ground state more closely matches the energy of the transition of the acceptor to its first excited state, i.e. as the overlap increases between the emission and absorption bands of the donor and acceptor, respectively. The last feature explains why this type of transfer is called resonance energy transfer, and suggests that apt analogies for the process can be found in the transfer of energy between tuning forks or pendula suspended on a common string.

Energy transfer may be inferred from the observation of sensitized fluorescence, i.e. the process naturally occurring in the chloroplast, where light absorbed by one molecular species reappears as the fluorescence of another. Such a process also occurs in proteins where light absorbed by tyrosine residues excites the fluorescence of tryptophan. Similarly, energy transfer allows haeme to quench the fluorescence of tryptophan residues in haemoglobin and explains the photodissociation of CO from CO haemoglobin by light absorbed by the aromatic amino acids of the protein moiety. The spatial requirements for energy transfer permit experiments such as these to be used to investigate the relative positions of donor and acceptor species in biological structures.

Energy transfer also takes place between molecules of the same species in proportion to the overlap of their absorption and emission bands. This transfer cannot be investigated by a change in the fluorescence spectrum, which remains unchanged, but can be studied by measurements of fluorescence depolarization. The latter technique is described below.

5.3.2.6. *Competition between decay processes*

As seen previously, internal conversion and vibrational energy loss return excited molecules to the lowest vibrational level of the first excited state. Here, intersystem crossing, fluorescence, quenching, photochemical sensitization and energy transfer all compete with one another for the remaining excitation energy. The competition between the different processes depopulating S_1 has practical use. For example, if energy absorbed by carotene promotes chlorophyll fluorescence in photosynthetic organisms, it may be argued that the same energy is promoting other processes, such as the photochemical evolution of O_2, that also depend on the excitation of chlorophyll.

5.3.3. LIFETIMES OF STATES AND EXCITATION COEFFICIENTS

The emphasis to this point has been on the energy states of molecules and the types of transition occurring between them. However, an absorption or emission spectrum not only reveals the energy of transitions but also the relative probability with which they occur. Transitions with high probability occur frequently and the permanence or "lifetime" of the state is correspondingly small; absorption is also great (Section 1.2.3). In contrast, improbable transitions are associated with states having long lifetimes and low absorption.

Several features of molecular states that determine transition probability have already been remarked upon. For example, transitions between states of the same symmetry are unlikely because they do not give rise to the linear translation of electron density that is necessary if interaction with the exciting light wave is to occur. For this reason, symmetrical molecules like benzene do not absorb strongly until substitution destroys their symmetry. Again, transitions involving two quantum-number changes are improbable; thus triplet states cannot be populated readily by absorption and their lifetime is very long. Finally, the promotion of an electron from a molecular orbital which overlaps only slightly with the orbital to which the electron is destined is improbable because the electron has to pass from one orbital to the other during the transition. As an example, $n-\pi^*$ transitions involve poorly overlapping orbitals: in consequence, absorption is low, lifetimes are long and the quantum yield of fluorescence is negligible (Kasha, 1961).

The terms "lifetime" and "absorption" now need to be defined more precisely. The transitions considered here are all first-order processes; the definition of lifetime is, therefore, the time taken for a given population of molecules within a state to decrease to $1/e$ of its initial value. Lifetime is represented by the symbol τ and it is easily shown that τ equals the reciprocal of the sum of the first-order rate constraints for the decay processes. The time molecular states "persist", quoted in earlier sections, in fact refer to typical values of their lifetime.

Absorption is measured in units of absorbance (Section 1.4.1), the latter term being synonymous with optical density, where (Section 2.2.2.2)

$$\text{Absorbance} = \log \frac{I_0}{I} = \varepsilon c e$$

I_0 is the intensity of light of wavelength λ passing through a reference cell containing solvent, while I is the corresponding intensity transmitted by a solution containing the substance under examination at concentration c. The depth of both cells is l. ε is the molar extinction coefficient; interestingly this is related to the cross-sectional area of the molecules that is effective in interrupting the light passing through the solution. The relation of ε to wavelength λ constitutes the absorption spectrum; thus, both the location and intensity of absorption are characteristic measures of molecular individuality. Measurements of absorbance are also extremely useful for analytical purposes; thus, absorbance is proportional to the concentration of a single substance and is additive for a mixture. The latter feature allows measurements made at several wavelengths to be used for analysing mixtures for their individual components (Beavan, 1961).

The probability of a downward transition accomplished by the emission of light, as already indicated, is proportional to the probability of the corresponding upward transition brought about by light absorption. Defining the radiative lifetime, τ_R, as the lifetime that would be observed if fluorescence were the only means of depopulating the excited state, its relation to the extinction coefficient at the maximum of the absorption band may be approximated by

$$\frac{1}{\tau_R} = 10^4 \, \varepsilon_{\max}$$

In practice, other decay processes compete with fluorescence, both diminishing its quantum yield and the lifetime of the state; indeed, it is only in the lowest singlet excited state that the competing processes are sufficiently slow for fluorescence to be measurable. Defining the quantum yield of fluorescence as q_f,

$$q_f = \frac{\text{Quanta emitted as fluorescence}}{\text{Quanta absorbed}}$$

$$\text{i.e. } q_f = \frac{k_f}{k_f + \sum_i k_i} = \frac{\tau}{\tau_R}$$

where k_f is the first-order constant for fluorescence, $\sum_i k_i$ is the sum of the rate constants for all other processes depopulating S_1 and τ is the measured lifetime of fluorescence. This definition is consistent with the observation that the quantum yield of fluorescence is independent of the wavelength of excitation, a result required to discuss the analytical application of fluorescence measurements (see below).

5.3.4. FLUORESCENCE EXCITATION SPECTRA AND FLUORESCENCE ASSAY

It is difficult to measure an absorption spectrum when the average absorbance is very small because the values of I_0 and I become almost indistinguishable. Should the substance concerned be fluorescent, the difficulty may be relieved by recording an excitation spectrum in which the fluorescence intensity is measured as a function of the exciting wavelength. For a wavelength λ

$$\varepsilon'_\lambda cl = 2\cdot 303 \log \left(\frac{I_0}{I}\right)_\lambda$$

The fluorescence observed at constant exciting-light intensity is equal to the light absorbed multiplied by a factor ϕ which comprises both the quantum yield of fluorescence (a constant independent of λ) and the response of the measuring instrument to unit fluorescence. Therefore, from the previous equation:

$$F_\lambda = \phi \left(\frac{I_0 - I}{I_0}\right)_\lambda = \phi(1 - e^{-\varepsilon'_\lambda cl})$$

Expanding $e^{-\varepsilon'_\lambda cl}$ as a series, we have for *small* absorbances

$$F_\lambda = \phi \varepsilon'_\lambda cl$$

F_λ is clearly proportional to the absorption spectrum of the substance whose fluorescence is observed. The presence of other substances that either do not fluoresce or fluoresce at an unobserved wavelength affects the excitation spectrum to the extent that their absorption diminishes I_0. Thus, impurities having an average absorbance similar to the substance under examination have a negligible effect on the excitation spectrum (Parker, 1968).

The last equation also provides the basis for the assay of concentration by measurements of fluorescence intensity. It will be noted that the factor ϕ requires the assay to be calibrated for each fluorimeter employed. Nonetheless, with reasonable quantum yields, the method has the advantage of providing a much more sensitive measure of concentration than absorptiometry, sometimes by several orders of magnitude. The greater sensitivity arises because the fluorescence observed is directly proportional to the exciting light intensity which may be increased, therefore, to any necessary value. In addition, because the relation of F_λ to ε'_λ is linear, excitation can encompass the entire absorption spectrum of the molecular species being assayed, making it even easier to increase the fluorescence to measurable values (Udenfriend, 1962).

5.4. Polarization of fluorescence

Measurements of the polarization of fluorescence are valuable because they provide directional information about molecules. It will be recalled that an electronic transition is equivalent, in classical terms, to the oscillation of an electric dipole and that, as such, it has a fixed direction within the molecular structure.

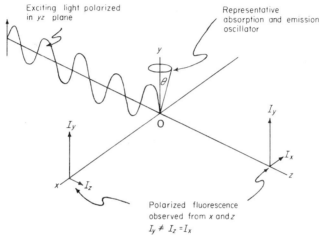

FIG. 5.6. The relative intensities of fluorescence polarized in directions parallel and perpendicular to the plane of polarization of the exciting light.

Figure 5.6 shows a light wave travelling in the z-direction, plane polarized in the yz plane, and impinging on a solution of absorbing molecules placed at the origin of the co-ordinate system. It can be shown that the chance of any molecule being excited by absorption of light increases with $\cos^2\theta$, that is, proportionately with decreasing values of the angle formed between its transition dipole and the direction Oy. In other words, absorption selects a group of molecules whose transition dipoles are preferentially in the Oy direction and symmetrically disposed about it.

It will now be assumed that the molecules are prevented from moving during the lifetime of the excited state, for example by cooling the solution to a rigid glass, and that the oscillator responsible for absorption is also responsible for emission. Under these conditions, fluorescence observed in the direction x has a greater component of intensity polarized in the direction y than it has in the z direction.

The degree of polarization, defined as

$$p = \frac{I_y - I_z}{I_y + I_z}$$

therefore takes its maximum value, 0·5. It should be emphasized that the maximum value of p is not unity, as might perhaps be thought, because absorption only selects preferentially, and not absolutely, in the direction Oy. There must, therefore, be a component of fluorescence intensity polarized in the xz plane. The minimum value of p is $-0·33$. This value is obtained when exciting molecules to higher singlet states whose absorption oscillators are at right angles to the emission oscillator of the fluorescent state. Figure 5.7 illustrates the process, taking naphthalene as an example.

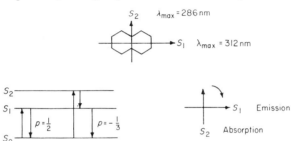

FIG. 5.7. Polarization of the fluorescence of naphthalene as a function of the wavelength of the exciting light and the relative orientation of the absorption and emission oscillators.

Figure 5.7 makes it clear that the measurements of p as a function of the exciting-light frequency may be used to determine the angles formed between the more energetic transition dipoles of a molecule and the lowest. In a similar fashion, if energy transfer is taking place (as in a protein, between aromatic amino-acid residues), the polarization of the sensitized fluorescence displays the relative orientation of the two oscillators involved.

Measurements of p may also be used to study the motion of large molecules in solution. Suppose, for example, that a given direction is strictly defined within a protein molecule; this direction may be that of the oscillator responsible for the $S_0 - S_1$ transition of a tryptophan residue or that of the corresponding transition of a fluorescent dye covalently bound to the protein. In either case, absorption of plane-polarized light photo-selects those molecules whose absorption oscillator is preferentially related to the plane of polarization. During the lifetime of the excited state, Brownian rotations randomize the orientation of the protein molecules so that, when emission occurs, the fluorescence is less polarized than it would have been had the solution been made sufficiently viscous to prevent movement. Measurements of p made in this way as a function of viscosity allow the determination of a characteristic time which defines the ease with which proteins rotate in solution. In such studies, the lifetime of fluorescence employed must not be so short that no depolarization occurs nor so long that complete randomization reduces p to zero. Dye molecules are, therefore, chosen for binding to the protein according to the rule that the larger the protein, the slower it rotates and the longer must the lifetime of fluorescence be (Weber, 1952).

Depolarization of fluorescence also occurs in rigid solutions when the absorbing molecules are present at a sufficient concentration to allow energy transfer between them. As already described, energy transfer is not exclusively limited to parallel and adjacent molecules; thus, each transfer progressively randomizes the original direction defined by absorption. Concentration depolarization is, therefore, a means of detecting energy transfer between like molecules, fulfilling the role played by sensitized fluorescence in detecting transfer between different species (Weber, 1954).

5.5. Environmental effects

Up to now, the electronic transitions of a molecule have been treated as if they were independent of its environment. This is not permissible, for the spectroscopic properties of dissolved molecules are sensitive to the solvent. Some part of these effects arises from short-range interactions, such as protonation, H-bonding or solvation. In such a category, solvation of the non-bonding electrons responsible for $n-\pi^*$ transitions may be included, the effect of which is so to stabilize the ground state that the energy of the transition is considerably raised. An effect of this sort is displayed by chlorophyll b where an $n-\pi^*$ transition, visible as the band of least energy in dry benzene, disappears under the penultimate $\pi-\pi^*$ band in the presence of traces of water, an observation of great biological interest.

Henceforth, however, these quasi-chemical interactions will be neglected and we shall concentrate on the effects that the physical properties of the solvent exert on the solute, the more so because the changes in spectroscopic properties of the solute that are observed may be used as a tool to investigate its micro-environment.

The principal causes of the changes observed in fluid solutions fall into two classes: dipole-dipole and dipole-induced dipole interactions. In the first class, a dipolar solute interacts with solvent dipoles, thus reducing the energy of the system and, in particular, stabilizing the molecular state of the solute, whether it be the ground or the excited state. In the second class, the solute dipole displaces the electrons and nuclei of the solvent molecules relative to one another so as to induce a solvent dipole; the interaction of the solute dipole with the induced dipole then reduces the energy of the system as before.

Particular instances of these interactions may now be considered. First, the energy of a molecular transition taking place in solution is always reduced by a dipole-induced dipole interaction caused by the instantaneous transition dipole displacing the electrons of the solvent molecules. The atomic nuclei of the solvent remain unaffected by the transition because of its rapidity. The extent of the energy reduction is governed by the polarizability of the solvent and is, therefore, a function of its refractive index.

More important effects than those due to the momentary transition dipole arise if the solute possesses a permanent dipole. For the changed electron distribution that follows excitation may result in a change in the dipole moment, which thereby affects the interaction of the solute with the solvent. Assuming, first, that the solvent is non-polar, the interaction is of the dipole-induced dipole type and changes immediately the transition is accomplished. Thus, if the dipole moment of the excited state, μ_e, is greater than the dipole moment of the ground state, μ_g, the excited state is stabilized more than the ground state and the energy of the transition is reduced. In contrast, if μ_g is greater than μ_e, the energy of the transition is increased. In both instances, the energy change depends on the polarizability of the solvent and, therefore, on its refractive index.

The effects already considered also operate when both the solute and the solvent possess a permanent dipole moment. However, in this situation, the overwhelming contribution to the stabilization of the ground and excited states is the interaction of the permanent dipoles of the solute and solvent, the one orientating the others around it. The mechanism involved here implies a relationship between the solvent-induced shift in transition energy and the dielectric constant of the solvent. An additional factor that has to be considered, however, in deriving the quantitative expression of the relationship, is the time it takes solvent molecules to change their position in response to the change in dipole moment of the solute that occurs on excitation. For the solute molecule in its ground state sets up a cage of oriented solvent dipoles round itself which can only be changed in about 10^{-11} s (a rotational period) whilst the transition itself is accomplished in 10^{-15} s. The initial state reached after excitation is, therefore, one in which the solute is under the strain of an inappropriate orientation of solvent molecules and has a correspondingly high energy. The initial state, called the Franck-Condon state, disappears as the solvent molecules readjust themselves to the new dipole moment of the solute, producing an equilibrium state of lower energy which persists for the lifetime of fluorescence. The same processes occur on emission of light, the Franck-Condon state produced initially disappearing as the solvent molecules relax. The mechanism is illustrated in Fig. 5.8, taking as an example a typical $\pi-\pi^*$ transition of a molecule where μ_e is greater than μ_g. It will be observed that solution of such a molecule in a polar solvent produces a small decrease in the energy of absorption when compared with that of the same transition observed in the gaseous phase, with a relatively large decrease (red shift) in the energy of emission (Mataga et al., 1956).

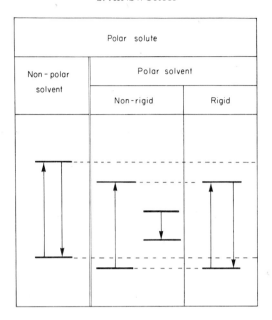

FIG. 5.8. The electronic-transition energies between the ground and first excited singlet states of a polar aromatic molecule in non-polar, non-rigid polar and rigid polar solvents. The ground-state dipole moment is lower than the excited-state dipole moment. (The transition energy in the non-polar solvent is somewhat smaller than the corresponding transition in the gaseous phase because of dispersive interaction with the solvent).

The effect of solvent viscosity on the transitions of a polar solute in a polar solvent is also represented in Fig. 5.8 for the extreme case where the solvent is a rigid glass. It will be noted that the energy of absorption depends on solvent polarity but, because there is no solvent rearrangement during absorption, not on the solvent rigidity. However, because the ground state solvent cage cannot relax, emission occurs from the Franck-Condon state and has an energy that in non-rigid media would be typical of a more non-polar environment. The distinction, observed here, between the effect of solvent rigidity on the absorption and emission frequencies of a solute, offers a means of deciding whether the changes in the fluorescence properties of the solute arise from changes in the polarity of the environment or from changes in its viscosity (Ainsworth and Flanagan, 1969).

In somewhat less rigid solutions, the Franck-Condon state relaxes to the equilibrium state at a rate comparable to fluorescence emission. Here, the average emission spectrum varies with viscosity, while time-resolved spectroscopy allows the decay of the Franck-Condon state to be studied by the change in emission spectrum following effectively instantaneous excitation with light flashes.

The effects of the environment on the spectroscopic properties of molecules have been widely employed in recent years to investigate biological systems. For example, the accessibility of aromatic amino-acid residues of proteins to solvent may be ascertained by noting the sensitivity of the residue to changes in solvent composition (Yanari and Bovey, 1960; Herskovits, 1965). Again, l-anilinonaphthalene-8-sulphonic acid (ANS) is one of a number of dyes which show a marked increase in their fluorescence intensities and a blue shift in their emission maxima on going from a polar to a non-polar environment. These dyes are bound by many proteins and, when bound, show changes in their emissions which suggest that the dye-binding sites are non-polar in character (Laurence, 1952; Stryer, 1965). The changing fluorescence properties of ANS have also been used to reveal conformational changes in the structure of enzymes, nerves and sub-cellular organelles (Lim and Botts, 1967).

References

Absorption and emission spectroscopy offers a powerful tool for the investigation of biological systems, the more so because it generally leaves the system unchanged. The following list of references, in conjunction with the *Science Citation Index* will direct the interested reader to a more detailed understanding of the method. In this connection, it should be noted that many of the references are much wider-ranging than implied in the text.

Ainsworth, S. and Flanagan, M. T. (1969). *Biochim. Biophys. Acta* **194**, 213–221.
Beavan, G. H. (1961). *Adv. Spectroscopy* **2**, 331–428.
Förster, Th. (1960). In "Radiation Research, Supplement 2" (L. G. Angenstine ed.), pp. 326–339. Academic Press, New York and London.
Herskovits, T. T. (1965). *J. Biol. Chem.* **240**, 628–638.
Kasha, M. (1961). *In* "Light and Life" (W. D. McElroy and Bentley Glass, eds), pp. 31–64. Johns Hopkins Press, Baltimore.
Laurence, D. J. R. (1952). *Biochem. J.* **51**, 68–180.
Lim, S. T. and Botts, J. (1967). *Arch. Biochim. Biophys.* **122**, 153–156.
Lower, S. K. and El-Sayed, M. A. (1966). *Chem. Rev.* **66**, 199–241.
Mason, S. F. (1961). *Quart. Rev.* **15**, 287–371.
Mataga, N., Kaifu, Y. and Koizumi, M. (1956). *Bull. Chem. Soc. Japan* **29**, 465–471.
Parker, C. A. (1968). "Photoluminescence of Solutions". Elsevier, Amsterdam.
Platt, J. R. (1956). *In* "Radiation Biology" (A. Hollaender ed.), Vol. 3, pp. 71–123. McGraw-Hill, New York.
Reid, C. (1957). "Excited States in Chemistry and Biology". Butterworths, London.
Stryer, L. (1965). *J. Mol. Biol.* **13**, 482–495.
Udenfriend, S. (1962). "Fluorescence Assay in Biology and Medicine". Academic Press, New York and London.
Weber, G. (1952). *Biochem. J.* **51**, 145–155.
Weber, G. (1954). *Trans. Faraday Soc.* **50**, 552–555.
Yanari, S. and Bovey, F. A. (1960). *J. Biol. Chem.* **235**, 2818–2826.

6. Optical Rotatory Dispersion and Circular Dichroism

D. G. Dalgleish

6.1. Introduction

Optical activity (p. 9) results from the different interactions of left- and right-circularly polarized light with asymmetric chromophoric groups in molecules, and is manifested by the two phenomena of circular birefringence and circular dichroism. The first of these defines the possession of different indices of refraction for the two forms of circularly polarized light, so that the plane of polarization of plane-polarized light passing through the solution is rotated, the well-known measurement of optical rotation. When measured as a function of wavelength, this constitutes an optical-rotatory dispersion (ORD) spectrum. The dispersive nature of this spectrum implies that an isolated chromophoric group gives rise to optical rotations at all wavelengths, not only in the region of its absorption band. Circular dichroism (CD) spectra arise from the difference in absorption coefficients for left- and right-circularly polarized light, and are therefore observed only at those wavelengths where absorption occurs. The ORD and CD spectra of an isolated chromophore are illustrated in Fig. 6.1.

These anomalous effects in the rotatory dispersion in the vicinity of an absorption band are known as Cotton effects, and the term is also applied to the extrema found in the CD spectra at these wavelengths. Cotton effects may be positive or negative and this may help to make individual components clearer than they are in absorption spectra, which are of only positive sign. The phenomena of ORD and CD are related to absorptions, since all three processes arise from the excitation of electrons from lower to higher energy states, and this allows information from both optical-rotatory and absorption spectra to be compared and analysed together. The assignment of bands in optical rotatory* spectra is then dependent on the analysis of the absorption spectra.

---

* Since the phenomena of ORD and CD are related, the terms "optical activity" and "optical rotatory properties" are used when the discussion involves use of either of the techniques. The terms ORD and CD are used in descriptions of experiments when only one of the techniques was used.

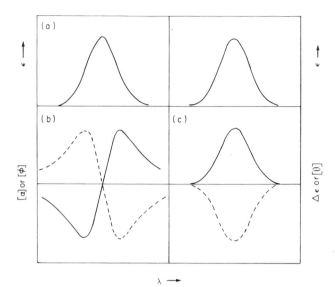

FIG. 6.1. Cotton effects caused in an isolated chromophore. (a) represents the absorption band, (b) positive and negative Cotton effects in the ORD and (c) positive and negative Cotton effects in the CD.

There are few definite rules governing the interpretation of optical rotatory spectra of biological polymers. It is generally impossible to predict whether, for example, a protein will show Cotton effects from aromatic side-chains, even if it is known that such side-chains are present in the molecule. Nor is it possible to lay down standard procedures for the analysis of the optical activity of macromolecules. The techniques are best used in conjunction with other forms of experiment, either physical or chemical, rather than in isolation, and, as with other forms of spectroscopy, it is generally easier to interpret changes in the optical rotatory properties, induced by various means, than to make a complete analysis of a single spectrum.

In this chapter, the measurement and origin of these optical properties will be considered, and then typical applications to studies of proteins and nucleic acids will be described.

6.2. Measurements

Details of instrumentation have been dealt with elsewhere (Jirgensons, 1973; Velluz *et al.*, 1965) and are not considered in the following discussion.

Plane-polarized light is the sum of two beams of left- and right-circularly polarized light of the same frequency. When this light is passed through an optically-active solution, one of the components is slowed down relative to the other, because of the circular birefringence. The result of this is that the two components emerge from the sample out-of-phase by a small amount, and when recombined give a beam of plane-polarized light whose plane of polarization is at an angle to the original plane (Fig. 6.2). If, at the same time, the wavelength of the incident light is suitable for exciting an electronic transition in the sample, the two components will be differently absorbed, so that they emerge not only out of phase but having different intensities (Fig. 6.2c), and the emergent beam becomes elliptically polarized, with its major axis tilted through the angle caused by the original rotation.

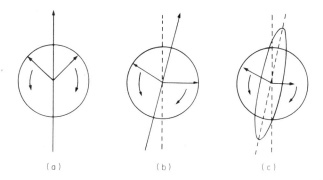

FIG. 6.2. The effect of an optically active medium on plane-polarized light. (a) Illustrates the original plane-polarized light as the sum of two circularly polarized beams, in (b) the optical activity α is caused by the different indices of refraction only, while in (c) both optical rotation and CD are active, giving elliptical polarization.

Measurement of the ORD spectrum is then simply a matter of measuring the optical rotation as a function of wavelength, the rotation at any wavelength being expressed in terms of the specific rotation ($[\alpha]_\lambda = 100 \, \alpha(c.l)$, the molecular rotation ($[\phi]_\lambda = [\alpha]_\lambda \times M/100$), or, in the case of proteins, by the mean residue rotation ($[m] = [\alpha]_\lambda \times M'/100$), where c is the concentration in gdl^{-1}, l the pathlength in dm, and M and M' are the molecular weight and mean-residue weight respectively. It is necessary to correct the observed values for the refractive index of the solvent, by multiplication by the factor $3/(n^2+2)$, where n is the refractive index of the solution at the appropriate wavelength. Values of these corrections have been tabulated by Fasman (1963).

Circular dichroism may be measured by determining the ellipticity of the emergent light, where the ellipticity is defined as that angle whose tangent is the ratio of the major and minor axes of the ellipse. Alternatively, CD may be determined by subjecting the sample to alternate pulses of left- and right-circularly polarized light, and determining the difference in absorption directly. In this case, the dichroism is expressed as the difference in molar extinction coefficients ($\Delta\varepsilon = \varepsilon_l - \varepsilon_r$), which is related to the ellipticity by the equation

$$[\theta] = 3300\Delta\varepsilon$$

A measure of the strength of an optically active transition is given by the area under the CD spectrum, and is known as the rotational strength. It is this, rather than the size of the maximum of the dichroism, which is a measure of the strength of the transition, since the rotational strength contains the contribution of the bandwidth, which is neglected if the maximum amplitude alone is used. The CD of a transition may be defined in terms of its rotational strength, its bandwidth and its wavelength of maximum amplitude.

The two phenomena of circular birefringence and dichroism are not independent, and may be related to one another by Kronig-Kramers transforms (Moscowitz, 1960, 1962);

$$[\theta(\lambda)] = -\frac{2}{\pi\lambda} \int_0^\infty [\phi(\lambda)] \frac{\lambda'^2}{(\lambda^2 - \lambda'^2)} d\lambda'$$

$$[\phi(\lambda)] = -\frac{2}{\pi} \int_0^\infty [\theta(\lambda)] \frac{\lambda'^2}{(\lambda^2 - \lambda'^2)} d\lambda'$$

(1)

Therefore, both types of measurement must yield the same information about the molecules under investigation. However, CD spectra generally offer the advantage that the different contributions to the spectrum may be resolved more easily than in ORD, in which the optical rotation at any wavelength is the sum of dispersive contributions from all of the chromophoric groups in the molecule. The dispersive nature of ORD offers the possibility of estimating the ORD spectrum at wavelengths shorter than those at which measurements can be made, since the shape of the ORD of a single chromophore remote from its absorption maximum is given by the Drude equation:

$$[\phi] = \frac{K}{\lambda^2 - \lambda_0^2}$$

(2)

Such extrapolations were the basis of the Moffitt-Yang (1956) relationship for estimating the α-helix content of proteins from measurements of their near u.v. and visible ORD, and more recently, the ORD spectrum below 185 nm has been inferred for proteins by an analysis of their far-u.v. ORD and CD spectra (Tsong and Sturtevant, 1969). This latter method depends on the analysis of the CD spectrum, transforming it into rotations by eq (1), and subtracting from the observed ORD spectrum, leaving the dispersive contributions from the bands below 185 nm.

The transforms (1) are also used in the analysis of ORD and CD spectra into individual components, since the correctness of an analysis of a CD spectrum can be checked by transforming the calculated CD components into rotations, summing, and comparing with the measured ORD spectrum. Only if the analysed contributions are correct will the calculated and observed ORD spectra coincide.

6.3. Sources of optical activity

Chromophoric groups which exhibit optical activity must be asymmetric, possessing neither a centre nor a plane of symmetry, and this may be caused by an asymmetry in the chromophoric group itself, or by an asymmetric perturbation of the chromophore by the surrounding groups in the molecule. Of the two causes, the latter is the more common, the most important representative of the former type being the disulphide group of cystine. Most of the chromophores which are considered here are symmetrical, but are optically active as the result of the asymmetry of their surroundings, provided by the α-carbon atoms of amino acids and the sugar units of nucleotides.

This asymmetry provides a basis for the interaction of electric and magnetic transition moments of the chromophore, essential for optical activity, as is seen in the equation for rotational strength,

$$R = Im\langle a|\boldsymbol{\mu}|b\rangle \cdot \langle a|\mathbf{m}'b\rangle$$

Here a and b represent the initial and final electronic states of the transition, and $\boldsymbol{\mu}$ and $\mathbf{m}$ are the electric and magnetic dipole operators for the transition. It follows that the optically-active transition must possess allowed components of both moments; this is not possible for symmetrical unperturbed chromophores, where only one moment is allowed. The relation with absorption spectra is clear from this, but absorption requires only one allowed transition moment (although it is weak for a magnetically allowed transition). Three interactions with the surroundings can permit the electric and magnetic dipole components to interact: the coupled oscillator, one-electron, and $\boldsymbol{\mu}$-**m** mechanisms.

The coupled-oscillator mechanism (Kirkwood, 1937; Moffitt, 1956) involves coupling between the transition moments of two adjacent chromophores; these must have a spatial relationship in which the interacting moments are non-parallel. (If the moments are parallel, only the absorption spectrum is affected; hyperchromism is shown if the chromophores are arranged head-to-tail, and hypochromism if they are stacked one above the other (Kasha, 1963).) If the coupled oscillators are the same or similar, as in DNA, then an exciton system is formed, in which the excitation is delocalized by interactions between neighbouring similar units. These exciton systems show splitting of the absorption and optical activity of the transitions about the monomer excitation wavelength, so that there are two components of opposite sign in their optical activity (see Fig. 6.5 for DNA).

The one-electron mechanism (Condon et al., 1937) arises from the electric and magnetic moments of a transition being perturbed by a static field generated by the surrounding groups in the molecule, and the **μ-m** mechanism (Bayley et al., 1969) enables the electrically allowed transition of one group to interact with the magnetically allowed transition of a neighbouring group, causing a compound excitation with the necessary properties for the generation of optical activity. This last mechanism is important in generating the optical activity of the amide groups of proteins, where the electrically allowed π-π* and magnetically allowed n-π* transitions of two adjacent peptides interact.

These mechanisms are clearly dependent on the geometry of the chromophore and its surroundings, and it is this which makes optical activity a sensitive measure of conformation. The dependence on conformational properties may be used either as a means of determining the conformation of a molecule, or it may be used to check the results of calculations of the preferred conformations of molecules.

From the knowledge of these mechanisms, it is possible to calculate rotational strengths directly, if the electronic ground and excited states can be formulated. These calculations must take into account all of the possible mechanisms for the production of optical activity, the contributions from which are not additive. Complete calculations have therefore been restricted either to simple systems, such as the amino acid tyrosine (Hooker and Schellman, 1970), or to regular structures, such as those of nucleic acids (Johnson and Tinoco, 1969), or the α-helix, β-sheet and polyproline I and II structures of proteins (Pysh, 1970). Although these calculations give results in agreement with observed spectra, they are not at present sufficiently accurate, because of necessary simplifications, to allow an explanation of the small changes in optical-rotatory properties which are seen during the reactions of macromolecules. While it is to be hoped that sufficiently accurate calculation methods will become available, interpretation of optical-rotatory properties is at present on a more empirical basis.

6.4. Polypeptides and the secondary structures of proteins

Although satisfactory calculations of the rotational strengths of polypeptides in the regular conformations of the α-helix and β-pleated sheet may be made, it is not easy to estimate the amounts of these regular structures present in proteins from studies of the optical activity and it is worth while considering critically the various methods which have been used to make these types of estimates.

Rotational strength is generated from the n-π* and π-π* transitions of the amide groups of the polypeptide, the transitions appearing in the range 190–230 nm. This rotational strength depends on the interactions between adjacent amide groups and, therefore, on the geometrical relationship between them. In regular structures such as the α-helix and β-pleated sheet, the relationship of adjacent amides is the same throughout the structure, so that both of these structures possess distinctive optical-rotatory properties (Fig. 6.3). Therefore, it should be possible to estimate the amount of either regular structure by detecting the appearance of its distinctive rotatory properties in the observed rotatory properties of the protein or polypeptide.

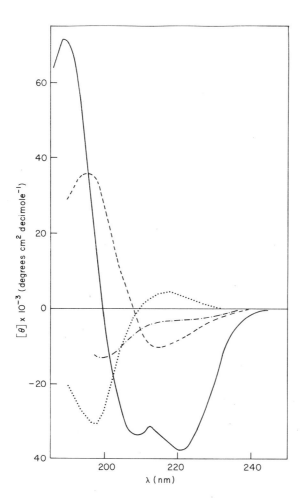

FIG. 6.3. The CD of poly-L-lysine in the α-helix (———), β-pleated sheet (-----), random solution form (· · · : ·) and the CD spectrum of the fraction F2b of chicken erythrocyte histone in its unordered conformation), which is similar to that of a film of poly-L-lysine (– · –).

The earliest method, which is still used occasionally, is the Moffitt-Yang equation (1956), which relates the plain rotatory dispersion curve of a protein above 350 nm to three constants, λ_0, a_0, and b_0 by a summation of Drude curves (eq. (2), p. 168) from the n-π^* and π-π^* absorptions, giving

$$[m']_\lambda = a_0 \frac{\lambda_0^2}{\lambda^2 - \lambda_0^2} + b_0 \frac{\lambda_0^4}{(\lambda^2 - \lambda_0^2)^2}$$

The value of λ_0 is generally taken to be 212 nm, and a plot of $[m']_\lambda (\lambda^2 - \lambda_0^2)/\lambda_0^2$ against $\lambda_0^2/(\lambda^2 - \lambda_0^2)$ gives a straight line from which values of a_0 and b_0 can be found. If the plot is not linear, the value of λ_0 is adjusted from 212 nm to the value necessary to straighten the line. From the value of b_0, an estimate of helix content may be calculated, knowing that completely α-helical and completely unordered homopolypeptides give b_0 values of -630 and 0 respectively. No estimates of β-structure content may be made by this method, although attempts have been made to do so by comparing the values of a_0 and b_0 from the same protein. The helix contents found by the use of the Moffitt-Yang method are of uncertain value, since the method relies on optical-rotation measurements at wavelengths far removed from the absorption bands and assumes that the visible rotatory dispersion is caused by contributions from the peptide bands only. No sources of optical rotation in the visible or near u.v. are considered, and it must be remembered that this indirect measurement was necessitated by the limitations of instrumentation, since it was not possible to make measurements into the far u.v. The method may, however, be used to determine whether or not synthetic polypeptides are in the α-helical conformation, since these compounds are more likely to be similar to the model compounds used to determine the standard values of b_0. Moreover, changes in Moffitt parameters of proteins caused by reaction or denaturation of the protein may be taken as indications of conformational changes in the molecule.

A similar method, by Schechter and Blout (1964), which also used the visible and near u.v. ORD spectrum of proteins, may also be employed, but suffers from much the same disadvantages as the Moffitt-Yang method in being indirect.

With the development of instruments capable of measuring into the far u.v., it became possible to measure the optical activity of the amide bands of proteins and polypeptides directly (Fig. 6.4). The methods now used to estimate the amounts of regular structure are based on comparisons of the observed optical activity of proteins with those of polypeptides known to exist in regular structures. For this purpose, the standards used are, for the α-helix, poly-L-lysine at pH 11·5, and for β-structure, poly-L-lysine at pH 11·5 after heating to 50°C or in 1% sodium dodecyl sulphate solution (Li and Spector, 1969). The remaining non-regular portions of proteins are assumed to have

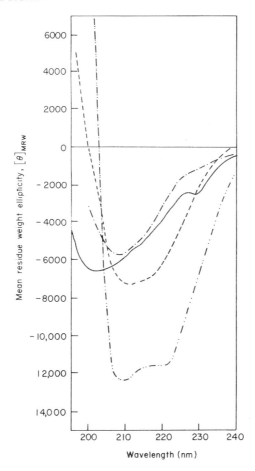

FIG. 6.4. Circular Dichroism spectra in the far u.v. of chymotrypsinogen (———), ribonuclease (– – – – –), bovine erythrocyte cupro-zinc protein (– · – · –) and β-lactamase 1 from Bacillus Cereus (– · · – · · –). These illustrate the type of variation found in proteins containing relatively small amounts of α-helix and β-sheet.

similar optical activity to unordered homopolypeptides, for example poly-L-lysine at low pH (Fig. 6.3).

The simplest and least accurate method using these model structures is to compare the protein optical activity at one wavelength in the far u.v. (generally 220 or 210 nm for the CD of the α-helix and 215 nm for β-structure or 233 nm for the ORD of the α-helix), where the contribution of the regular structure is a maximum (Fig. 6.3). If it is assumed that there are only two components present, then this single-wavelength analysis may give a more or less correct estimate of the regular-structure content, but it is not a method to be recommended, unless prior studies of the i.r. spectrum have shown that the protein lacks either one of the two regular structures.

A more satisfactory method is to analyse the entire far u.v. ORD or CD spectrum in terms of the three parameters, α, β and unordered, using as standards the optical activity of the model polypeptides described above (Greenfield and Fasman, 1969; Greenfield et al., 1967). These three-parameter fits have been shown to give results which are in reasonable agreement with those from X-ray crystallographic analysis of some proteins, but not of those proteins which contain little or no regular structure, such as chymotrypsinogen. This lack of reliability arises from the uncertainty as to the correct optical activity to use for the unordered portions of proteins. For, although the polypeptide models for α-helical and β-structures are good approximations to those found in proteins, the random polypeptide conformation in solution is not a good representation of the non-regular portions of proteins, which, although devoid of order, are nevertheless coiled in a very specific manner. Because of this, it has been suggested that a better model for non-regular protein should be unordered poly-L-lysine in a film, rather than in solution, since the film exercises constraint on the conformation of the polypeptide chain (Fasman et al., 1970), so that the optical activity of the film is different from that of the polypeptide in solution (Fig. 6.3) and appears to be similar to that of unordered protein molecules. No attempts to fit observed spectra with the three parameters of polypeptides in the α-helical, β-form and film state have yet been made.

With the development of knowledge of the tertiary structures of proteins from X-ray crystallography, extension of this type of analysis has been used. If the CD spectrum of the peptide region of a protein is represented by

$$X = f_\alpha X_\alpha + f_\beta X_\beta + (1 - f_\alpha - f_\beta) X_u$$

where the f's are fractions of helix and β-sheet and the X's are the ellipticities at any given wavelength, then from a knowledge of the ellipticities of any three or more proteins of known structure (i.e. f_α, f_β using the X-ray data), the contributions $X_{\alpha 1}$, X_β and X_u can be found (Chen et al., 1972). Five proteins were used in this calculation, giving a computed α-helix CD spectrum which was similar to that of poly-L-lysine, although about 12% less in magnitude. The β-sheet and unordered conformations gave CD spectra which were less similar to those of the reference polylysine, giving extrema which were less than half as large as expected. The calculated CD spectrum of the "unordered" conformation in proteins had, in addition, a negative extremum at 220 nm, in contrast to the positive extremum found for polylysine and similar to that found for the film state of the model and for unordered F2b histone (Fig. 6.3). Use of these calculated spectra gave reasonable fits to the CD spectra of five other proteins of known structure, although proteins with low helix content are still not easily fitted.

A further approach to the study of proteins is to try to estimate how many components are required to fit the CD spectra of proteins. Using the technique of matrix rank analysis (McMullen et al., 1967), the number of components required to fit a set of CD spectra from several different proteins was found to be greater than three (Dalgleish, 1972). Thus there may be some doubt as to whether three parameters can be considered adequate to fit the CD spectra of proteins. The nature of the other contributions remains obscure, but may arise from other backbone conformations, or from side-chain contributions.

Apart from the problem of which polypeptide structures should be used as reference structures, there are other problems attending on the analysis of the far u.v. optical activity of proteins. The type of curve-fitting mentioned above can allow only for three types of structure being present in proteins; if other regular structures are present, they may remain undetected. In addition, properties similar to those of the α-helix may be created by small groups of interacting peptides, as has been shown for small cyclic oligopeptides, which have CD spectra similar to those of α-helices although they cannot adopt such a conformation (Craig, 1968; Laiken and Craig, 1969). Thus, small loops in the polypeptide backbone of proteins may interfere with α-helix determinations.

A further complication in these analyses stems from the contributions of side-chain chromophores to the optical activity of proteins in the far u.v. It is certain that the side-chains of the amino acids tyrosine and cystine show optical activity in the far u.v., so that such contributions should occur in proteins (Hooker and Schellman, 1970; Coleman and Blout, 1968). It is not possible to isolate the effects caused by side-chains from the total optical activity in the far u.v., so that errors must be expected in the analysis of the total optical-rotatory properties in terms of regular structures.

It is, therefore, difficult at this stage to use optical activity to determine the absolute amounts of regular structure in proteins. The analyses of α-helix content are more likely to be correct for high helix contents, since the rotational strength of the α-helix is greater than that of the β-structure or unordered components. Since few proteins possess high helix contents, the factors mentioned above are likely to be important (see also Section 10.2.7).

However, as a measure of conformational change, these optical properties are useful, since it is not necessary in such cases to define the total conformations of the initial and final states of the protein. Therefore, so long as the optical properties of the initial and final states are known, it is possible to follow the course of reactions such as denaturation and renaturation by observations of the variation of the optical activity, often at only one wavelength (220 nm for CD and 233 nm for ORD) as the reaction proceeds. In experiments with the refolding of enzymes, the criterion of correct refolding can be taken as the possession of optical rotatory properties indistinguishable from those of the native protein before unfolding. In such a context, the CD or ORD spectrum of the protein should be studied not only in the far u.v. but also in the near u.v. and if necessary the visible, since in many cases the side-chain extrinsic optical activity is more sensitive to small conformational alterations of the protein than the backbone optical activity (Pflumm and Beychok, 1969).

A further use of the far u.v. optical activity is to ascertain if synthetic polypeptides are in the α-helical conformation, to confirm the results of calculations on their preferred conformations. In certain cases it has been calculated that a left-handed, rather than a right-handed, helical conformation is favoured and this may be checked by the use of ORD or CD (Yan et al., 1970). It should be noted that the CD spectrum of a left-handed helix of L-amino acid residues is not the mirror image of that produced by a right-handed α-helix. Poly-D-amino acids, however, show a mirror image of the right-handed α-helix CD spectrum.

The folding of short peptides derived from proteins has been studied by using optical rotatory properties and generally it has been found that peptides isolated from proteins by enzymatic or chemical cleavage do not show the measure of α-helix content expected from a knowledge of the structure of the native protein. This is especially clear in the case of peptides isolated from myoglobin (Atassi and Singhal, 1970; Singhal and Atassi, 1970), where the helix content of the original protein molecule is high, so that peptides isolated from it should show optical rotatory properties of high helix content. Experiments have shown that the α-helical character of the peptides, as estimated from their CD spectra, is much reduced, illustrating that protein folding does not depend solely on short-range interactions between amino acids.

6.5. Nucleic acids

The bases of DNA and RNA being planar, are not optically active, but their incorporation into nucleosides and nucleotides induces optical activity in their π-π* absorption bands. This optical activity depends on the nature of the base and on the glycosidic linkage with the ribose or deoxyribose. Thus the purine nucleotides show negative Cotton effects from the absorption at about 260 nm when linked β, but positive effects when linked α. Pyrimidines show similar behaviour, with the signs of the Cotton effects being reversed. The optical activity, in most cases weak, is caused by the perturbation of the base transitions by the sugar moiety; also, since some degree of free rotation is possible around the glycosidic bond, the presence of several different conformations reduces the optical activity.

When nucleotides are linked together, to give di- or trinucleotides, the rotational strength is increased by the coupling of the base transitions. The optical rotatory properties of these oligomers are sequence-dependent and, in principle, it is possible uniquely to determine the sequence of an unknown trinucleotide from its optical rotatory properties (Cantor and Tinoco, 1967).

In general, the rotatory strength of these oligomers is low, since the coupling of the base transitions is inhibited by the rotational freedom of the molecule. As the chain length of the oligomers is increased and there is a tendency to formation of more stable helical structures, the rotational strength increases dramatically, since the possibilities of coupled oscillator interaction between

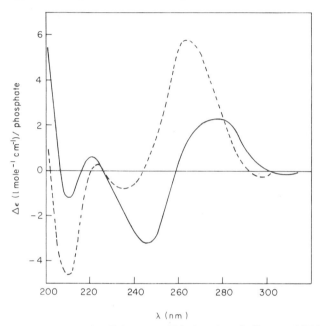

FIG. 6.5. The CD spectra of calf thymus DNA (———) and ribosomal RNA (-----). Solutions at pH 7·0.

the bases increase (Brahms *et al.*, 1966), and eventually a stable structure is produced which shows exciton-type interactions.

DNA and RNA show multiple Cotton effects in the near u.v., although the two types of nucleic acid show different rotatory properties (Fig. 6.5), the difference arising from their different structures. In solution, DNA exists in the B form, in which the bases are at right-angles to the helix axis. The consequence of this is that a conservative optical activity is produced, with two oppositely-signed Cotton effects about the maximum of absorption at 260 nm arising from the splitting of the base transitions caused by the exciton system of the nucleic acid. In RNA, the bases are tilted with respect to the helix axis, producing a non-conservative optical activity. The reason for the difference is thought to be that the interaction of the near u.v. transitions with the polarizability caused by transitions in the far u.v. tends to cancel out in DNA, but not in RNA. This is borne out by the non-conservative optical activity of DNA in 80% ethanol solution, where the DNA is believed to exist in the A

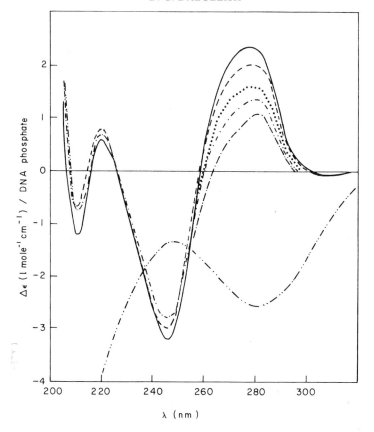

FIG. 6.6. The CD spectra of calf thymus DNA (2×10^{-4}M in base phosphate) with divalent metal ions (4×10^{-4}M), at pH 7·0. ——DNA; ––––with Mg^{2+}; ·····with Ca^{2+}; —·—·—with Fe^{2+}; —·—·—with Mn^{2+}; —··—··—with Hg^{2+}. Ghe Hg^{2+} spectrum is drawn half-scale.

form, which has tilted bases and an optical activity similar to that of RNA.

The optical activity of nucleic acids is dependent on the base composition; this effect is not such as to cause a variation in the shapes of the CD spectra, but induces alterations in the magnitudes of the extrema at 280, 250 and 220 nm (Gratzer et al., 1970). The optical activity is also altered by the binding of metal ions, the extent and nature of the variations induced being dependent on the particular metal ion bound. Thus, while Mg^{2+} affects the CD spectrum to only a small extent, Mn^{2+} causes a considerable alteration, and Hg^{2+} completely destroys the spectrum (Fig. 6.6). Similarly, ribosomal RNA shows changes in its CD spectrum when Mg^{2+} ions are bound. The alteration in the properties induced by metal ions may be caused by the separation of base-pairs by metal ions bound between them.

Thus, it can be seen that optical rotatory techniques are sensitive to conformational alterations in nucleic acids. Such experiments are of interest, especially when considering the complexing of nucleic acids with proteins in virus particles, ribosomes and nucleohistones. The large changes induced in the optical activity make it clear that the nucleic acid conformation does not remain as it is in free solution (Fasman et al., 1970a). Although in such cases the DNA adopts a conformation which is not known, since it is comparable to no other known nucleic acid spectrum, the use of the optical activity allows the elimination of several possible structures which might be proposed.

6.6. Extrinsic Cotton effects of proteins

An extrinsic Cotton effect may be defined as one which arises from a chromophoric group which is not part of the backbone structure of a protein or nucleic acid. This definition covers the Cotton effects of aromatic side-chains of amino acids and those of bound prosthetic groups such as haem or pyridoxal, as well as those induced in ligands binding to the macromolecule.

The optical rotatory properties of proteins may be divided into parts according to the region of the spectrum in which they occur. The region below 240 nm is principally caused by the amide chromophores of the backbone, that from 240 to 330 nm is caused by extrinsic effects from the aromatic and disulphide side-chains, and the region from 330 nm into the visible is due to the extrinsic effects of bound groups such as haem or metal ions, if these are present. In general, the optical activity of the second region is complex and contains contributions from some, if not all, of the tyrosine, tryptophan, phenylalanine and cystine side-chains, as well as from metal ions or prosthetic groups which also absorb in the visible. The positions and magnitudes of the Cotton effects of the amino acids depend on their local environment within the protein, so that the position of the spectra is dependent on such factors as the state of ionization of the side-chain, and whether it is situated in a hydrophilic or hydrophobic region of the molecule. These factors are also evident in absorption spectra. The local factors on which the sign and magnitude of the Cotton effects depend are the conformation of the surrounding peptides, the proximity of charged groups and the presence of other side-chains with transitions which may interact with the transition of the particular side-chain under consideration. In view of the complexity of these effects, it is found that some large proteins possess little optical activity in this region because the contributions from many side-chains cancel.

Figure 6.7 shows the near u.v. CD spectra of some simple proteins. The simplest of these to analyse is that of ribonuclease, since this contains no tryptophan. The analysis has been made (Horowitz et al., 1970) by comparing the absorption and CD spectra, at 298 and 77 K after repetitive scanning to eliminate noise. Analysis of the absorption spectrum showed that a number of bands were present, attributable to the vibrational structure of tyrosine transitions with their fundamental excitations in the range 282–289 nm; this showed that three different types of tyrosine residue were involved. From this, the CD spectrum was analysed in terms of these three types. Essentially the same shape of CD spectrum was found for each type, caused by the vibrational levels of one transition, as in the absorption spectrum. Three of the residues were found to be sensitive to solvent, confirming that they were on the surface, while the other three residues had absorption maxima shifted to the blue of these, and were not solvent-dependent. These residues, however, were of two types, indicating that not all of the three were in identical environments.

The side-chain optical activity of chymotrypsinogen appears to rise mainly from the tryptophan residues in the protein (Strickland *et al.*, 1969). This was illustrated by comparison of the optical properties with those of small model compounds and confirmed by the observation that modification of two of the tyrosyl residues by cyanuric fluoride did not alter the CD spectrum (Gorbunoff, 1969). Compared with the behaviour of the small model compounds, whose optical activity increased by some 20-fold on cooling from

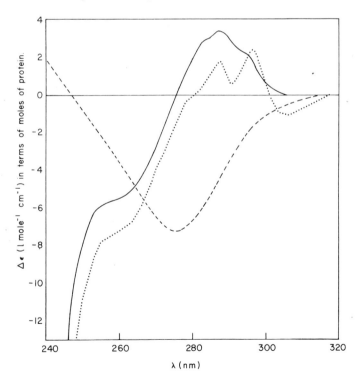

FIG. 6.7. The near u.v. CD spectra of lysozyme (———), ribonuclease (– – – –) and chymotrypsinogen (· · · · ·).

298 to 77 K, the magnitude of the optical activity of the chymotrypsinogen varied only slightly, illustrating that the residues responsible for the optical activity were buried in the protein, so that their freedom of rotation was severely curtailed. The fact that the residues are buried was confirmed by the position of the CD spectrum, which was red-shifted in comparison with that of model compounds.

The use of modification procedures is important in the analysis of near u.v. side-chain optical activity. These modifications may be as simple as the red-shift induced by the ionization of tyrosine, or may be a reactive modification or destruction of accessible residues. An attempt to analyse the CD spectrum of lysozyme has been made in this manner (Teichberg et al., 1970). By iodine oxidation of lysozyme, tryptophan 108 was destroyed; its contribution to the CD spectrum was therefore determined and found to be the major contribution to the near u.v. CD, although contributions from the other tryptophan and tyrosine side-chains were present. Binding of the tetramer of N-acetyl-D-glucosamine perturbed the near u.v. CD spectrum and the change was shown to be mainly caused by the change in orientation of tryptophan-108.

The optical activity of phenylalanine has been studied in small compounds, but there is difficulty in analysing its Cotton effects in proteins, since the transitions are very weak. Contributions have, however, been detected in insulin, and in denatured horseradish peroxidase C, although the native enzyme shows no contribution from its 23 phenylalanine residues, suggesting that their effects cancel to a great extent. Contributions from phenylalanine are also believed to occur in bovine erythrocyte cupro-zinc protein, since a band appears in the CD spectrum which cannot be attributed to any other cause (Wood et al., 1971).

The most difficult of the amino acids to discuss is cystine, whose chromophoric group is the intrinsically asymmetric disulphide. The position of its absorption and optical-rotation bands varies from 240 to 350 nm with the angle of the disulphide and, since the group is intrinsically asymmetric, the optical activity is determined both by the intrinsic effect and the effect of its environment; it is not certain which is the dominant factor. The fact that disulphide groups of opposite chiralities do not give equal and opposite CD spectra when they are incorporated into proteins is consistent with the analysis of the near u.v. CD spectrum of ribonuclease mentioned above; this suggested that a contribution from cystine was present, although there are two right-handed and two left-handed disulphides in the protein.

Most known haemproteins show extrinsic Cotton effects induced in the haem transitions. These Cotton effects are sensitive to the presence of ligands, the state of aggregation of the haemprotein, and the oxidation state of the haem group. As with the aromatic side-chains, the optical activity is induced by the asymmetry of the surrounding protein groups and the optical activity in myoglobin is believed to arise from the interaction with the π-π^* transitions of neighbouring aromatic side-chains, rather than with the peptide n-π^* or π-π^* or side-chain σ-σ^* transitions. It is, therefore, possible to use the extrinsic optical activity of the haem to detect conformational changes in the

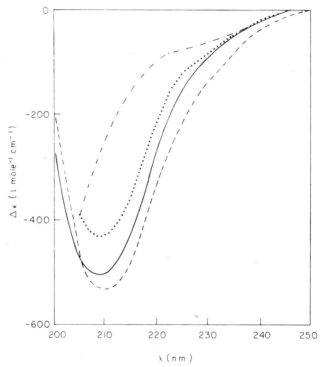

FIG. 6.8. Denaturation of erythrocyte cupro-zinc protein by urea. Far u.v. CD spectra: ———native protein; – – – – apoprotein with both Cu and Zn removed; · · · · · native protein in 8 M urea; – · – · – · – apoprotein in 8M urea.

haemprotein. Metalloproteins also show extrinsic optical activity in the absorption bands of the bound metal, which are destroyed when the metal is removed. These bands may be similar to those of haemproteins in that they are indicators of conformational change within the protein, and are sensitive to ligand binding. The importance of metal ions in maintaining protein structure in some cases has been illustrated by the attempts made to denature erythrocyte cupro-zinc protein with 8M urea. This protein contains both Cu and Zn ions bound to it and in the native state cannot be denatured by 8M urea, as shown by the relative constancy of the far u.v. CD spectrum. The removal of the metal ions, however, renders the protein susceptible to urea-denaturation, during which the far u.v. CD spectrum becomes very similar to that of unordered polypeptide in the film form (Fig. 6.8). Such experiments show the necessity of the metal ions in maintaining the protein conformation and also illustrate the "unordered" protein conformation as shown by different denatured proteins.

6.7. Extrinsic Cotton effects of ligands

A ligand which binds to a macromolecule may acquire extrinsic optical activity as a consequence of the asymmetric conformation around the binding site. In addition, the optical activity of the macromolecule may be altered as the result of a ligand-induced conformational change, or because of interaction between the ligand and groups on the surface of the macromolecule. In some cases, none of these effects occurs, and in others, all. The effect on the macromolecule may be detected by the alteration of its backbone optical activity, as described previously, and only the extrinsic Cotton effects generated in the absorption bands of the ligands will be considered here.

These Cotton effects may be simple or complex and, in view of this, it is preferable to use ligands whose absorption bands are not at the same wavelengths as the absorption bands of the macromolecule, if this is possible, so that the effect of the ligands on the side-chains of proteins may be analysed separately from the effects of the protein on the ligand. The extrinsic Cotton effects may then be used to determine binding isotherms, or to probe the stereochemistry of the ligand site; these two uses are to some extent dependent mutually, since it is not possible to determine the binding isotherm unless it is certain that no variation in the optical activity of bound ligand occurs (so that the stereochemistry of the binding site must remain constant, and the ligands once bound should not alter their optical activity when other ligands bind to the macromolecule). This variation can usually be detected at an early stage of the investigation, since plots of the magnitude of the extrinsic Cotton effect against ligand concentration are not hyperbolic. In such a situation, it is not possible to determine the binding isotherm by optical rotatory methods alone. On the other hand, if the binding isotherm has been previously determined by other methods, then the optical rotatory results may be used to discuss the stereochemistry of the binding site and the manner in which it alters as further ligands bind to the macromolecule.

6.7.1. PROTEINS

The simplest use of extrinsic Cotton effects is to diagnose whether binding occurs. Subsequently, the observations can be used in the manner described above.

A linear increase of the extrinsic Cotton effect of a ligand, denoting very strong binding, is illustrated by the binding of NADH to liver alcohol dehydrogenase (Ulmer and Vallee, 1965). The binding induces a very large Cotton effect, whose magnitude increases with the amount of ligand added to a given concentration of the enzyme. This increase is linear with ligand concentration up to a value of two moles of NADH/mole of enzyme, after which no further increase can be observed. Such an experiment illustrates that the binding is strong, since all available ligand appeared to bind to the enzyme; it also allows the number of binding sites in the enzyme to be measured.

A more complex Cotton-effect system is found when bilirubin binds to bovine or human serum albumin. Up to four moles of ligand bind/mole of protein and the optical rotatory properties of the complexes vary with the

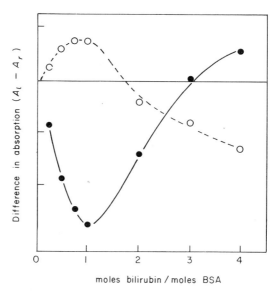

FIG. 6.9. Variation of the magnitude of the extrinsic Cotton effect of bilirubin bound to bovine serum albumin. Values calculated from Wooley and Hunter (1970). Since there is a double Cotton effect, contributions at 470 nm (●——●) and 400 nm (O---O) are shown.

amount of ligand bound (Blauer and King, 1970; Wooley and Hunter, 1970). Not only do the magnitudes of the extrinsic Cotton effects vary, but also the signs of the bands. The behaviour of the Cotton effects of the bilirubin-BSA complex, shown in Fig. 6.9, may result from changes in the interaction of the two dipyrrylmethene chromophores of the bilirubin as more dye binds. This alteration in conformation may be mediated by changes in the protein conformation, which alter the stereochemistry of the binding sites.

6.7.2. NUCLEIC ACIDS

It should not be assumed that ligands bound to DNA do not interact, or that conformational changes of the nucleic acid do not occur, since the structure of the nucleic acid is very susceptible to cooperative conformational changes, and offers a variety of similar, adjacent, binding sites. Therefore, before optical-rotatory measurements on extrinsic Cotton effects are made, it is necessary to have measurements of the binding isotherm.

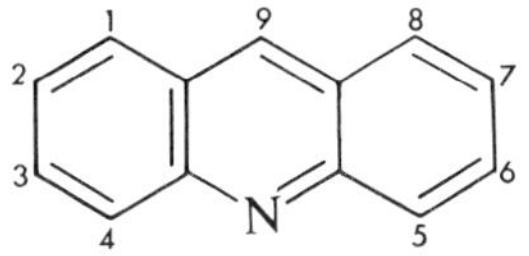

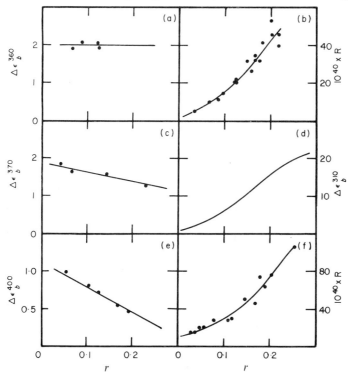

FIG. 6.10. Acridine

This is well illustrated by the study of the interaction of aminoacridines (Fig. 6.10) with DNA. In these investigations, it was found that the rotational strength of bound ligands was variable, the extent of the variation depending on the structure of the ligand (Dalgleish *et al.*, 1969, 1971, 1972, 1972a). The ligands which showed a strong cooperative increase of rotational strength (Fig.

FIG. 6.11. Variation of the extrinsic Cotton effects of several aminoacridines bound to DNA. (a) 1-aminoacridine; (b) 3,6-diaminoacridine; (c) 2-aminoacridine; (d) ethidium bromide; (e) 9-aminoacridine; (f) 3-aminoacridine. r is the ratio of bound ligand to DNA phosphate. The plots of (b) and (f) are of rotational strength, and the others of the magnitude of the extremum of the extrinsic Cotton effect.

6.11) all possessed a 3-amino group, since aminoacridines substituted in other positions gave linear plots of rotational strength against the amount of ligand bound. By comparing a wide range of substituted amino acridines, and building models, it was possible to make some estimate of the position in which the ligands bound to the nucleic acid. These positions are determined by two factors, first whether the ligand can intercalate between the base-pairs of the DNA, and second whether it possesses a 3-amino group which may H-bond to the ribose-phosphate backbone. The first of these factors cannot be determined by optical rotatory measurements alone, but can be found by either sedimentation or viscosity measurements. All of the aminoacridine molecules considered bind in slightly different positions, but the differences between them are small compared with the two determining factors described.

It has been shown that a group of "reporter" molecules (Fig. 6.12) have the interesting property of showing oppositely signed Cotton effects according to whether they are bound to DNA or to RNA (Gabbay, 1969). Moreover, they are not altered when the DNA is contained in nucleohistone, suggesting that the histone protein does not bind in the small groove of the DNA, in which it is thought that the reporter molecules bind.

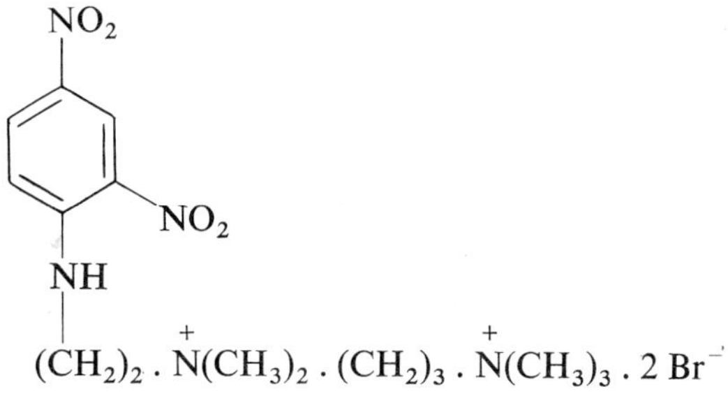

$$\text{Ar-NH-}(CH_2)_2 \cdot \overset{+}{N}(CH_3)_2 \cdot (CH_2)_3 \cdot \overset{+}{N}(CH_3)_3 \cdot 2\,Br^-$$

(where Ar = 2,4-dinitrophenyl)

FIG. 6.12

6.8. Summary

The foregoing discussion has made no mention of the optical rotatory properties of polysaccharides, since at this time only a few studies have been made on these molecules. The reason for this is that carbohydrates generally only absorb in the far u.v., which makes measurement difficult. It is only recently that suitable instruments have been developed which allow accurate measurement of optical rotatory properties in the far u.v., so that results on polysaccharides are lacking. It is known that the nature of the sugar units involved is important in determining the optical activity, as are the linkages between the monomers (Kabat et al., 1969). Polysaccharides also provide a suitable asymmetric environment for the production of extrinsic Cotton effects in ligands (Stone, 1969).

The uses of ORD and CD may, therefore, be summarized as being the determination of the conformations of proteins and nucleic acids, in both native and denatured states; the following of reactions or equilibria involving changes in conformation, especially the helix-coil transitions of polypeptides; and the determination of the steric influences of macromolecules on ligands which bind to them. While these investigations may also be made using other methods, the advantage of optical rotatory methods is the appearance of Cotton effects of both positive and negative signs and the sensitivity of these Cotton effects to conformational parameters. This sensitivity is lacking in absorption measurements, where little or no change may be seen in the electric dipole strength.

Although the potential of the optical-rotatory methods is great, it is not possible to exploit this to the full. This arises from the lack of knowledge of the behaviour of the chromophoric groups producing the optical activity, so that detailed analyses of the results are not always feasible. As more accurate descriptions of the sources of optical activity in macromolecules become available, it will become possible to make more detailed analyses of observed spectra. However, even at present, the methods have proved useful in many ways, even when interpretations of a semi-empirical nature have been made. Precision of instrumentation is still improving, allowing more accurate spectra (especially CD) to be recorded, and it is hoped that the provision of these spectra will render interpretations easier than at present.

References

Atassi, M. Z. and Singhal, R. P. (1970). *Biochemistry* **9**, 4252.
Bayley, P. M., Nielsen, E. B. and Schellman, J. A. (1969). *J. Phys. Chem.* **73**, 228.
Blauer, G. and King, T. E. (1970). *J. Biol. Chem.* **245**, 372.
Brahms, J., Michelson, A. M. and van Holde, K. E. (1966). *J. Mol. Biol.* **15**, 467.
Cantor, C. R. and Tinoco, I. (1967). *Biopolymers* **5**, 821.
Chen, Y. H., Yang, J. T. and Martinez, M. (1972). *Biochemistry* **11**, 4120.
Coleman, D. L. and Blout, E. R. (1968). *J. Amer. Chem. Soc.* **90**, 2405.
Condon, E. V., Altar, W. and Eyring, H. (1937). *J. Chem. Phys.* **5**, 753.
Craig, L. C. (1968). *Proc. Nat. Acad. Sci. U.S.* **61**, 152.
Dalgleish, D. G., Fujita, H. and Peacocke, A. R. (1969). *Biopolymers* **8**, 633.
Dalgleish, D. G., Feil, M. C. and Peacocke, A. R. (1972). *Biopolymers* **11**, 2415.
Dalgleish, D. G., Peacocke, A. R., Acheson, R. M. and Harvey, C. W. C. (1972a). *Biopolymers* **11**, 2389.
Fasman, G. D. (1963). In "Methods in Enzymology" (Colowick and Kaplan, eds), Vol. VI, p. 928, Academic Press, New York and Londin.
Fasman, G. D., Hoving, H. and Timasheff, S. N. (1970). *Biochemistry* **9**, 3316.
Fasman, G. D., Schaffhausen, B., Goldsmith, L. and Adler, A. (1970a). *Biochemistry* **9**, 2814.
Gabbay, E. J. (1969). *J. Amer. Chem. Soc.* **90**, 6574.
Gorbunoff, M. J. (1969). *Biochemistry* **8**, 2591.
Gratzer, W., Hill, L. R. and Owen, R. J. (1970). *Europ. J. Biochem.* **15**, 209.
Greenfield, N. J., Davidson, B. and Fasman, G. D. (1967). *Biochemistry* **6**, 1630.

Greenfield, N. J. and Fasman, G. D. (1969). *Biochemistry* **8**, 4108.
Hooker, T. M. and Schellman, J. A. (1970). *Biopolymers* **9**, 1319.
Horowitz, J., Strickland, E. H. and Billups, C. (1970). *J. Amer. Chem. Soc.* **92**, 2119.
Jirgensons, B. (1973). "ORD of Proteins and Other Macromolecules" (2nd edn). Springer-Verlag, Berlin and New York.
Johnson, W. C. and Tinoco, I. (1969). *Biopolymers* **7**, 627.
Kabat, E. A., Lloyd, K. O. and Beychok, S. (1969). *Biochemistry* **8**, 747.
Kasha, M. (1963). *Radiation Res.* **20**, 55.
Kirkwood, J. G. (1937). *J. Chem. Phys.* **5**, 479.
Laiken, S. and Craig, L. C. (1969). *J. Biol. Chem.* **244**, 4454.
Li, L.-K. and Spector, A. (1969). *J. Amer. Chem. Soc.* **91**, 220.
McMullen, D. W., Jaskunas, S. R., and Tinoco, I. (1967). *Biopolymers* **5**, 585.
Moffitt, W. (1956). *J. Chem. Phys.* **25**, 467.
Moffitt, W. and Yang, J. T. (1956). *Proc. Nat. Acad. Sci. U.S.* **42**, 596.
Moscowitz, A. (1960). In "Optical Rotatory Dispersion" (K. Djerassi, ed.), p. 150. McGraw-Hill, New York.
Moscowitz, A. (1962). *Advan. Chem. Phys.* **4**, 67.
Pflumm, M. N. and Beychok, S. (1969). *J. Biol. Chem.* **244**, 3982.
Pysh, E. S. (1970). *J. Chem. Phys.* **52**, 4723.
Schechter, E. and Blout, E. R. (1964). *Proc. Nat. Acad. Sci. U.S.* **51**, 695.
Singhal, R. P. and Atassi, M. Z. (1970). *Biochemistry* **9**, 4252.
Stone, A. L. (1969). *Biopolymers* **7**, 173.
Strickland, E. H., Horowitz, J. and Billups, C. (1969). *Biochemistry* **8**, 3205.
Teichberg, V. I., Kay, C. M. and Sharon, N. (1970). *Europ. J. Biochem.* **16**, 55.
Tsong, T. Y. and Sturtevant, J. M. (1969). *J. Am. Chem. Soc.* **91**, 2382.
Ulmer, D. D. and Vallee, B. L. (1965). *Adv. Enzymol.* **27**, 37.
Velluz, L., Legrand, M. and Grosjean, M. (1965). "Optical Circular Dichroism". Verlag Chemie, New York and London.
Wood, E., Dalgleish, D. G. and Bannister, W. H. (1971). *Europ. J. Biochem.* **18**, 187.
Wooley, P. V. and Hunter, M. J. (1970). *Arch. Biochem. Biophys.* **140**, 197.
Yan, J. F., Momany, F. A. and Scheraga, H. A. (1970). *J. Amer. Chem. Soc.* **92**, 1109.

Additional References

Several review articles should be consulted, for more thorough treatment of the topics discussed in this chapter.

Annual Reports of the Chemical Society: sections by Dalgleish, D. G. (1969–70) and Bayley, P. M. (1971–72); Specialist Periodical Reports "Amino-acids, Peptides and Proteins": section by P. M. Bayley in **5** (for 1972), p. 237.

For polynucleotides, see:
Yang, J. T. and Samejima, T. (1969). *Prog. Nuc. Acid Res.* **10**, 223.

A short review, containing many useful references is
Gratzer, W. and Cowburn, D. A. (1969). *Nature (London)* **222**, 426.

A detailed description of the optical activity of small peptides is given by
Urry, D. W. (1969). *Ann. Rev. Phys. Chem.* **19**, 477.

Finally, a review of protein studies, with emphasis on four specific proteins is
Timasheff, S. N. (1970). In "The Enzymes" (Boyer, ed.), 3rd edn, Vol. II, p. 381. Academic Press, New York and London.

7. Nuclear Magnetic Resonance

J. S. Leigh

7.1. Introduction

7.1.1. WHAT IS N.M.R.

Nuclear magnetic resonance (n.m.r.) spectroscopy is perhaps the most powerful technique available for study of detailed molecular phenomena in biopolymers. With its capability for observing *individual atoms* within a molecule, n.m.r. can yield information about both the "anatomy" and "physiology" of a molecule; i.e. both structural and dynamic information are available about the motion of molecules in solution. When used by a biological chemist, most other techniques of spectroscopy such as i.r. and u.v. must be interpreted only empirically; even qualitative interpretations are usually equivocal. N.m.r. data, on the other hand, are capable of yielding good qualitative, and sometimes quantitative, interpretations.

N.m.r. techniques have been used for several years in the study of biologically relevant molecules. Many rigorous texts are available and excellent reviews have been written by some of the pioneers of biological n.m.r.: Jardetzky and Jardetzky (1971), Kowalsky and Cohn (1964) and McDonald and Phillips (1970).

This chapter is not intended to be a rigorous treatment of the theory of magnetic resonance,* but to be a survey of the ways in which n.m.r. may be used to advantage in the study of biopolymers. Most of our discussions will centre on proton (^{1}H) n.m.r., although some space will also be devoted to other atoms of biological interest. Emphasis will be placed on those applications where "useful" information may be obtained, where the subjective term "useful" reflects the prejudices of the author.

7.1.2. A SHORT HISTORY OF N.M.R.

The phenomenon of nuclear magnetic resonance in bulk materials was first reported in 1946 by two groups in the U.S.A., headed by E. M. Purcell of Harvard and F. Bloch at Stanford; it remained strictly in the hands of physicists for about five years until chemical shifts were discovered in N compounds. In 1951, the first resolved proton n.m.r.—a three-line spectrum

---
* Some of the many excellent books dealing with n.m.r. theory are listed at the end of the chapter.

from the protons of ethyl alcohol—was published. With this observation, n.m.r. left the realm of physics and began to move into its role as the primary structure determining tool in organic chemistry.

Commercially built instrumentation first became available in the mid 1950's. The introduction of a 30 MHz proton spectrometer (0·7T magnet*) was followed quickly by improvements in magnet technology which allowed dramatic increases in resolution and sensitivity. Chemists were quick to realize the exquisite power of n.m.r. spectra to detect subtle structural changes in complicated molecules. The first n.m.r. spectrum of a protein (ribonuclease) was published in 1957. With the advent of practical superconducting magnet systems and their adaptation to the exacting requirements of n.m.r. spectrometers, n.m.r. really began to move into the realm of biochemistry, so that now it is possible in some cases to "see" individual protons in large biopolymers. The most extensively studied biopolymers are proteins. It was shown very early that the n.m.r. spectrum of a protein in its native form could not be simulated by a corresponding mixture of amino acids; the intricate folding at the polypeptide chain(s) makes major changes in both chemical shifts (line positions) and relaxation effects (line shapes). In principle, this should provide a wealth of structural and dynamic information about the protein. The trouble is that, in most cases, there is so much information that the complexity is

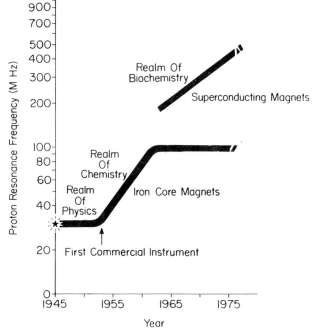

FIG. 7.1. The past, present and future (?) of n.m.r. spectrometers.

*0·7 tesla; one tesla(T) equals 10000 G = 10 kG.

overwhelming. Some attempts have been made to peck away at the edges of protein spectra. For instance, the C-2 proton of histidine residues may sometimes be individually resolved on the downfield end of the aromatic area. Studies of these residues have resulted in pK determinations for individual histidines in a few proteins. Recent improvements are now allowing more widespread applications (Fig. 7.1).

7.1.3. ELECTROMAGNETIC SPECTRUM

Electromagnetic radiation (Section 1.2.2) is characterized by a wavelength, λ, or frequency, ν, related by $\lambda \nu = c$, where c is the velocity of light. The spectrum of electromagnetic radiation as shown in Fig. 7.2, illustrates the range of the phenomenon; twenty orders of magnitude in frequency (or wavelength) are included. (For other units, see Table 1.1, p. 4.)

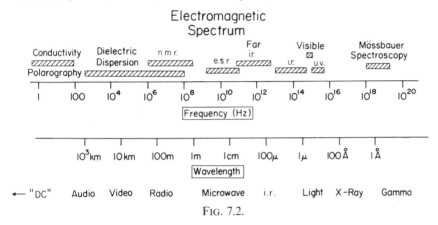

FIG. 7.2.

The quantum description of radiation is commonly used for wavelengths shorter than about $1\mu m$; quanta of radiation (photons) correspond to packets of electromagnetic energy. Classical (wave) and quantum (photon) descriptions are connected by the equation $\Delta E = h\nu$, where Planck's constant, h, is the conversion factor for expressing the (quantum) energy of a photon with (classical) frequency, ν. An energy-level diagram shows a photon $h\nu$ inducing a transition from one energy level to another, separated by an energy difference, ΔE.

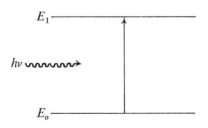

It is a general result (due to Einstein) that the probability of a photon inducing upward or downward transitions in a given system is identical (p. 5). In a large ensemble of systems, a net upward (absorption) or downward (emission) result depends only on the difference in the number of systems in E_0 and E_1. Thermal fluctuations are constantly inducing transitions among the available energy levels of any system. The condition of thermal equilibrium is defined by the Boltzmann equation, which relates the population ratio, n_1/n_0, of two energy levels by

$$n_1/n_0 = \exp\{-\Delta E/kT\}$$

For an energy separation corresponding to visible light (0·5μm), the Boltzmann factor at room temperature is so small that *all* systems exist in the ground (lowest-energy) state. On the other hand, for energy separations equivalent to n.m.r. frequencies (say 220 MHz), the ratio of spins in the excited (upper) state to those in the ground state differs from 1 by only 35 parts in a million. That is, in a collection of 2 000 035 nuclei at room-temperature thermal equilibrium, there would be, on the average, 1 000 000 spins in the excited state and 1 000 035 spins in the ground state. Net absorption of radiation, then, is observed for only those excess 35 out of 2 million, as compared with 100% of the molecules in optical spectroscopy.

7.1.4. SPINS IN A MAGNETIC FIELD

Nuclei which have an odd number of protons and/or an odd number of neutrons possess the property of spin, e.g. ^1H, ^2H, ^{13}C, ^{14}N, ^{17}O, etc. Nuclei with an even number of both protons and neutrons, such as ^4He, ^{12}C, ^{16}O, ^{20}Ne, ^{24}Mg, etc., have zero spin. As a consequence of having both charge and spin, nuclei possess magnetic moments μ_N which are proportional to the magnitude of the spin I, $\mu_N = \gamma_N \hbar I$. The constant of proportionality, γ_N, known as the gyromagnetic ratio, is characteristic of each nucleus. The interaction of the nuclear magnetic moment with an applied (static) magnetic field B leads to a (spin) Hamiltonian of the form $\mathscr{H} = -\mu_N \cdot B$. Quantum-theory solutions demand that the allowable nuclear-spin states be quantized. The component of I along the direction of the magnetic field can take only the discrete values I, $I-1, \ldots, -I$, as shown in Fig. 7.3. For a spin-1/2 system like a proton, there are only two levels, $+1/2$ and $-1/2$, the energy separation between which is $\gamma_N \hbar B$. An oscillating magnetic field with an angular frequency ω such that $\omega = \gamma_N B$ can induce transitions between these levels with an absorption of energy. Adjusting the magnetic field (sweeping) through the resonance condition generates an n.m.r. spectrum. Note that the resonant frequency for a given nucleus is directly proportional to the strength of the applied (static) field, B.

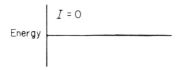
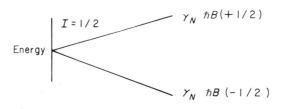
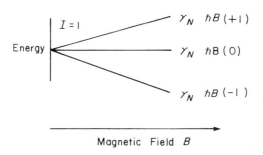

FIG. 7.3. Quantized energies of nuclei in a magnetic field; illustrated are cases of $I = 0$, 1/2 and 1.

7.2. Principles of n.m.r. spectra

7.2.1. GENERAL FEATURES

A simple high-resolution n.m.r. spectrum consists of number of (Lorentzian; see p. 41) absorption lines. For each, the observable quantities, schematically illustrated in Fig. 7.4, are:

1. position of the line or *chemical shift*;
2. intensity (integrated area);
3. line shape.

The integrated intensity of a line is proportional to the number of (roughly) equivalent nuclei which absorb there. Measurement of detailed line shape leads to the definition of the relaxation times, T_1 and T_2. The line shape may also be complicated by *spin-spin coupling* which leads to multiplet structure.

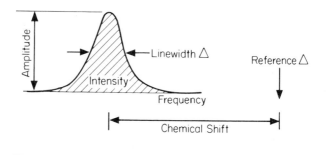

FIG. 7.4. The basic line, a hypothetical n.m.r. spectrum. Amplitude of radio-frequency absorption as a function of magnetic field (scan).

7.2.2. CHEMICAL SHIFTS

The presence of electrons around a nucleus causes a small shielding effect. A nucleus senses not just the external applied magnetic field but also a local field, which varies according to the chemical environment of the nucleus and arises from induced electronic currents. Thus, different protons in a molecule have different *chemical shifts*; the magnitude of the shielding effect is directly proportional to the applied magnetic field strength, B.

Figure 7.5 shows an ^{1}H n.m.r. spectrum of methanol; it is conventional to present highest fields at the right-hand side. The larger (upfield) line is from the three methyl protons and the smaller (low-field) line is from the hydroxyl proton. The areas under the lines are in the ratio of 3:1.

In a biopolymer such as a protein, each of the many kinds of proton in each amino acid has its own resonance. Figure 7.6 shows the n.m.r. spectra (taken at 220 MHz) of a few representative amino acids; the large line in the centre of each spectrum arises from the solvent water. D_2O (deuterium oxide) was used

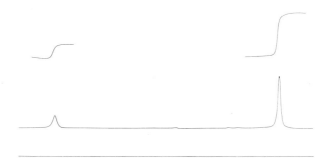

FIG. 7.5. N.m.r. spectrum of methanol. The separation between the OH peak on the left and the CH_3 peak on the right is ~1·6 ppm (see p. 197). Shown above each peak is the integral; the ratio of integrals is 1:3.

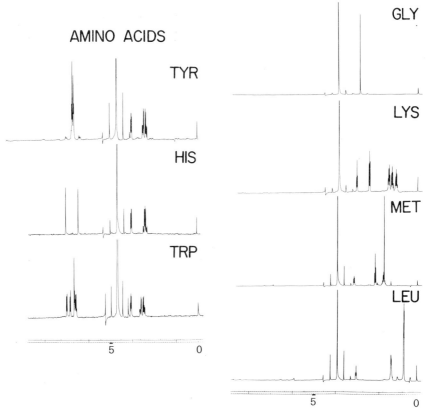

FIG. 7.6. 220 MHz spectra of some common amino acids at neutral pH in D_2O solution. Assignment of the peaks may be found in the review by Roberts and Jardetzky. Solutions were 25 mM in D_2O at neutral pH. Scan range ~8 ppm.

as the solvent in order to minimize the size of the line, but the few tenths of a percent residual protons in the heavy water still creates a rather large resonance. The spectrum is also complicated by spinning sidebands: the two pairs of peaks on either side of the water line are caused by spinning of the sample tube in order to reduce field inhomogeneities (see Section 7.3.1). The amino acids on the left side of the figure are aromatic and are the only ones with lines downfield from water. These downfield lines correspond to protons on the aromatic rings of the side chain, whereas aliphatic residues generally give rise to lines upfield from water. The small line on the right (highest field) in each spectrum is from tetramethylsilane (TMS), contained in a capillary tube within the sample, and commonly used as a reference for chemical-shift measurements. The multiplicity of the various lines is caused by spin-spin splittings discussed in Section 7.2.3.

Since the chemical shift depends on the strength of the applied magnetic field, it would be inconvenient to compare chemical shifts reported on a 100

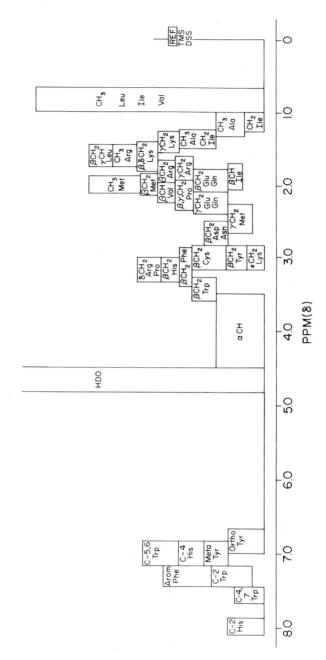

Fig. 7.7. Approximate chemical shifts of amino-acid protons in random-coil proteins.

MHz n.m.r. spectrometer with those from a 220 MHz machine, if the shifts were reported in frequency (or field) units. Thus the chemical shift is normalized to the frequency used and reported as shift in parts/million (ppm) from a reference line.

In proteins, the different amino-acid side chains do not have exactly the same chemical shifts as free amino acids, although they are usually not too far off. Local influences from nearby residues in the protein molecule cause additional shifts. In Fig. 7.7, the approximate positions of amino-acid protons in a protein are indicated. In the n.m.r. spectra for actual proteins (Fig. 7.8), the large peak (with spinning sidebands) in the centre is water. Although the general appearances of the protein spectra are similar, complex differences are

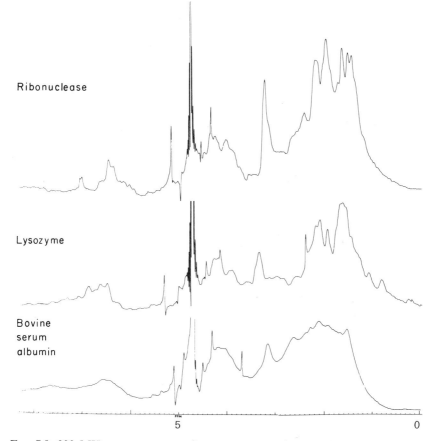

FIG. 7.8. 220 MHz n.m.r. spectra of some representative proteins in neutral D_2O solution. Ribonuclease and lysozyme are small proteins (MW $\sim$ 14 000); bovine serum albumin (BSA) is much larger (MW $\sim$ 66 000). Protein concentration of each sample was 60 mg cm^{-3}. The total scan range is 8 ppm; spectra were accumulated in about 3 min. by FT technique (Section 7.3.3).

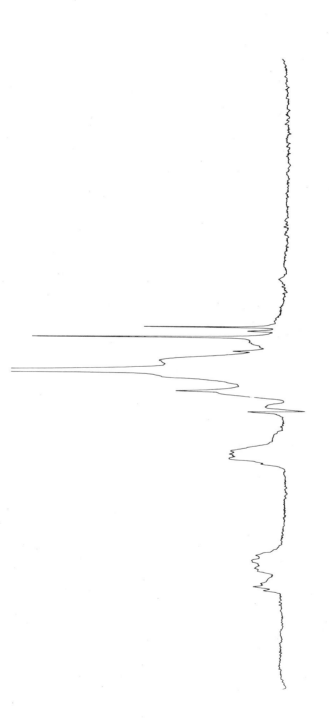

FIG. 7.9. 220 MHz n.m.r. spectrum of transfer ribonucleic acid (tRNA). Spectrum shown is of the partially "melted" form at 69°C, in 200 mM NaCl, D_2O, $pD \sim 7$, concentration 60 mg cm^{-3}. At lower temperatures, the rigidly folded form of tRNA predominates and much broader lines result. Total scan is ~ 12 ppm; as usual water is in the centre. (See also p. 299.)

also evident; protein spectra are not as well-resolved as amino-acid spectra. Lack of resolution stems from two distinct causes:

1. Individual proton lines in proteins (or any large molecule) are broadened because the molecules cannot tumble quickly enough to average out dipolar interactions.
2. Because of the large number of protons in a large molecule, the broadened lines tend to overlap one another rather extensively in most regions of the spectrum, so that only an envelope of lines is seen.

Both broadening and overlap are aggravated by size.

Nucleic-acid (RNA, DNA) spectra might be expected to be much simpler than those of proteins. Instead of more than 20 constituent parts (amino acids), only four common nucleotide bases appear in either RNA or DNA. In Fig. 7.9, which shows a proton n.m.r. spectrum of partially "melted" tRNA, the aromatic protons of the base rings stand out to low field (on the left). Individual nucleotide spectra are shown for comparison in Fig. 7.10. (Raman spectra of polynucleotides were discussed in Section 3.5.)

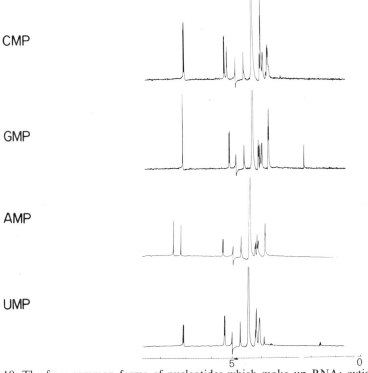

FIG. 7.10. The four common forms of nucleotides which make up RNA: cytidine monophosphate (CMP), guanosine monophosphate (GMP), adenosine monophosphate (AMP) and uridine monophosphate (UMP). Scan range ~ 8 ppm. Solutions were each 25 mM in D_2O at neutral pH.

7.2.3. SPIN-SPIN SPLITTING

The presence of spins on neighbouring nuclei can cause splitting of a resonance line. Since each spin has a magnetic moment, small magnetic fields exist near it. This local magnetic field is mainly of a dipolar nature; in a liquid where rotation of molecules is rapid, the dipolar field averages to zero. Dipolar fields are a major source of relaxation. There may, however, be a very small amount of scalar (non-angular-dependent) spin-spin coupling to other nuclei, transmitted through chemical bonds by small polarizations of electron orbitals. This coupling is independent of the strength of the applied magnetic field. The coupling constant, J, is specified in H_z.

In very simple situations, the splitting patterns are easily predictable. A molecule with two loosely coupled protons (a so-called AX spin system) produces an n.m.r. spectrum in which each proton line appears as a doublet, in which the individual lines within each doublet correspond to the two spin states of the other proton. The splitting J_{AX} of the resonance lines, identical in each doublet, results from the quantized states of the neighbouring spin. When the splitting is caused by more than one proton, then each line may be further split. A proton which is coupled to two non-equivalent protons (an AMX system) can show double-doublets. Coupling to two equivalent protons AX_2 (a methylene group) gives a characteristic 1:2:1 triplet due to overlap of two of the lines. Coupling to three equivalent protons AX_3 (a methyl group) yields a 1:3:3:1 quartet. Some simple cases are shown schematically in Fig. 7.11.

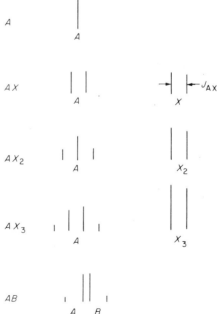

FIG. 7.11. First-order spin-spin splitting patterns.

If the chemical-shift difference between the different protons is of the same order of magnitude as the spin-spin splitting, then very complicated spectra arise. The simplest case, that of two spins, yields what is known as an *AB* pattern. Analysis of complex high-resolution spectra is discussed in some detail in the references listed in the bibliography.

Theoretical expressions for spin-spin coupling predict that a major determining factor is the dihedral angle between adjacent protons. While angles thus obtained may be used for structural purposes, it is unusual for a complete analysis to be possible with a sample approaching a biopolymer in complexity. Spin-spin splittings are not *usually* observed in larger molecules, due to increased linewidths, but they can sometimes be defected.

7.2.4. RELAXATION TIMES

Relaxation (p. 6) implies return to an equilibrium condition. The macroscopic description of spins in a magnetic field contains both a.c. oscillating components (transverse to the magnetic field) and d.c. components (parallel to the magnetic field). Relaxation is thus characterized by two distinct relaxation times. The so-called spin-lattice relaxation time, T_1, is the time constant for the spins to achieve an equilibrium population among the allowed energy levels in the applied magnetic field; it measures the time constant for the spins to exchange energy with their surroundings. Macroscopically, this corresponds to the static (d.c.) magnetization time. T_1 relaxation is mainly caused by the fluctuating local magnetic environment seen by each spin. The components at the resonance frequency of the randomly fluctuating field cause the transitions resulting in T_1 relaxation. Analogous processes are met in e.s.r. (Section 8.2.4).

The spin-spin relaxation time, T_2, is the time constant for spins to exchange energy amongst themselves (adiabatically). Macroscopically, it corresponds to the time constant for the "phase memory" of the transverse (a.c.) spin component. T_2 relaxation is caused by the components of the random local magnetic fields which are effectively non-fluctuating, that is random "static" fields which cause shifts in the resonance frequencies of individual spins. The T_2 relaxation time is related to the linewidth, Δ, by $T_2 = 1/\pi\Delta$.

The (magnetic) dipole-dipole interaction is the most potent cause of relaxation from a structure-determining standpoint. Since the field of a magnetic dipole falls off as the cube of the distance and relaxation rates are proportional to the mean-square fluctuating field, relaxation effects of dipole-dipole interactions fall off with the sixth power of the distance between the dipoles. The general form of the dipolar relaxation interactions may be written

$$1/T_{1,2} = D f_{1,2}(\tau_c)/r^6$$

D is a numerical coefficient containing physical constants and statistical factors, $f(\tau_c)$ is a function of the correlation time which modulates the interaction, and r is the distance between the interacting dipoles. The explicit forms of these functions, with typical numerical values, are given in Appendix 7.1, p. 216.

The correlation time, τ_c, may be identified with changes which interrupt (modulate) the dipolar coupling. It can be dominated by whichever is most rapid of (a) rotation of the dipoles relative to one another (as a molecular rotation), (b) a reversal of sign of one of the dipoles (a spin-relaxation process), or (c) a distance change (dissociation of a complex). This may be expressed mathematically as, $\tau_c^{-1} = \tau_r^{-1} + \tau_s^{-1} + \tau_M^{-1}$, where τ_r represents the rotational time, τ_s the dipole (spin) relaxation time and τ_M the chemical-exchange lifetime of the complex. The molecular reorientation time, or effective rotational correlation time, may be approximated for spherical molecules by:

$$\tau_r = M\bar{V}/RT$$

Here M is the molecular weight of the particle, $\bar{V}$ is the partial specific volume (equivalent to the reciprocal of the effective density of the molecule) and RT is the molar Boltzmann energy.

For proteins with $\bar{V} = 0.7$ at room temperature, we get $\tau_r \sim 3 \times 10^{-13} M$ s. Thus a protein with molecular weight of 100 000 has a rotational correlation time of about $\sim 3 \times 10^{-8}$ s.

7.3. Experimental methods

7.3.1. THE CONTINUOUS-WAVE HIGH-RESOLUTION N.M.R. SPECTROMETER

In order to detect the rather feeble n.m.r. absorption of the radio-frequency energy, a very sensitive radio-frequency receiver is connected to a coil wrapped around the sample. The sample, commonly contained in a 5 mm cylindrical tube, with a volume of approximately 0.5 cm³, is spun at a rate of 20 to 100 revolutions s⁻¹ by a small air turbine in order to average out static magnetic-field inhomogeneities in the plane perpendicular to the spinning axis. In the conventional mode of operation, the magnetic field is slowly swept through

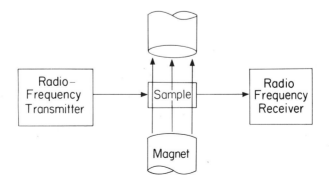

FIG. 7.12. Block diagram of n.m.r. spectrometer.

resonance and the r.f. signal from the receiver is detected and displayed on a recorder.

The main components of an n.m.r. spectrometer are (a) the magnet, (b) the radio-frequency receiver and (c) a radio-frequency transmitter. A block diagram of an n.m.r. spectrometer looks much the same as for any other type of spectrometer (see Fig. 7.12), except that the sample must be positioned in a strong steady magnetic field.

It is important that all parts of the sample be bathed in precisely the same intensity of magnetic field. Otherwise they will resonate at different frequencies and broaden the observed line. Conventional iron-core electromagnets have been steadily improved throughout the years so that it is now possible to achieve magnetic field strengths of more than 23 000 G (2·3 T) which are stable to within 1 part in 10^9 for reasonable periods of time with a homogeneity of the same order of magnitude.

Spectrometers utilizing superconducting magnets, however, promise to take over most of the biological applications of n.m.r. A comparison of the geometry of iron core and superconducting magnets is shown in Fig. 7.13.

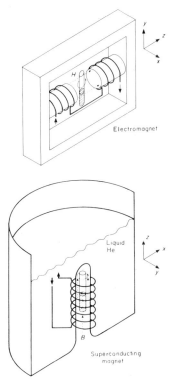

FIG. 7.13. Geometry of (upper) (Fe) electromagnet and (lower) superconducting magnet (solenoid).

Recent developments, which make available solenoids that can produce stable homogeneous fields of more than 8 T, will greatly increase the effective resolution that can be achieved in complex biological samples. Future developments will undoubtedly see further increases in the field intensity and useable volume of superconducting magnets.

Two rather different types of detection schemes were used by the original discoverers of n.m.r. and both these methods, with minor modifications, continue to be used. Bloch described "nuclear induction" signals; i.e., the nuclear spins induced a radio-frequency voltage in a receiver coil when irradiated with an r.f. magnetic field from nearby transmitter coils. The transmitter and receiver coils are aligned orthogonally so as to eliminate any direct coupling, an arrangement known as a double or crossed-coil probe.

Purcell's group on the other hand, used only a single coil, which served both as receiver and transmitter. Resonance was detected by monitoring changes in the impedance of the coil. Cancellation of the large transmitter signal was achieved by incorporating the coil into one arm of a bridge circuit. An excellent description of the details concerning various elaborations of these schemes may be found in Andrew (1969).

7.3.2. COMPUTER AVERAGING

In order to increase the signal-to-noise ratio (Section 1.5.2) of n.m.r. spectra, a technique known as multi-scan averaging may be used. A small digital computer is used to store a given spectrum. A second spectrum of the same sample may then be scanned and added (point by point) to the first spectrum stored in the computer memory. In the resultant (added) spectrum, while the n.m.r. peaks will be twice as big, the noise will be different on the second scan. The addition of two (incoherent) noise scans will again produce noise, but the amplitude will be only $\sqrt{2}$ times larger than the noise on each individual scan. Generalization to N scans will give N times as much signal and $\sqrt{N}$ times as much noise. The net gain in signal-to-noise ratio is thus equal to $\sqrt{N}$, so that 100 scans will produce 10 times better signal-to-noise ratio.

7.3.3. FOURIER TRANSFORM N.M.R.

Fourier-transform (FT) n.m.r. (Section 1.5.3) is a powerful new technique. While its advantages in high-resolution n.m.r. were originally pointed out by Ernst in 1966, the technique has only been fully exploited more recently.

If a spin system is displaced from equilibrium (by hitting it with an r.f. pulse, say), a transient signal called a free-induction decay (f.i.d.) is obtained. This transient is formally related to the conventional (frequency) spectrum by the mathematical operation known as Fourier transform. The transient response contains the same information as the more conventional frequency (or field) spectrum; either may be obtained from the other by a Fourier transform. If we denote the transient signal by $S(t)$ and the frequency spectrum by $S(\omega)$, then (ignoring normalization constants)

$$S(t) = \int_{-\infty}^{\infty} (\cos \omega t - i \sin \omega t) S(\omega) d\omega$$

$$S(\omega) = \int_{-\infty}^{\infty} (\cos \omega t - i \sin \omega t) S(t) \, dt$$

For a simple spectrum, the transformations are reasonably straight-forward; Fig. 7.14 shows the transient responses and equivalent spectra for three simple cases. Some points are worth noting.

1 The transient response of a Lorentzian line is an exponential decay.
2. A broad line corresponds to a fast decay and a narrow line to a slow decay.
3 Multiple spectral lines give multiple frequencies which show up as beats in the transient signal.

A major advantage of FT-n.m.r. is that, for a given signal-to-noise ratio, the amount of time necessary to acquire a spectrum decreases with the ratio of line-width to sweep-width. For polymer spectra with line-widths of a few Hz, with signals which stretch over a few thousand Hz, the time savings may be enormous. Furthermore it is reasonably easy to obtain relaxation times T_1 and T_2, even in very complicated spectra, by suitable pulsing schemes.

The reason for the increase in sensitivity is that all lines in a spectrum are observed at the same time. In n.m.r. the major source of noise in a well-designed spectrometer is the thermal noise from the receiver coil.* A conventional swept spectrum looks at only one line at a time. If a way could be found to look at more than one line at a time (this would not necessarily cause any extra noise since the receiver coil is generating most of the noise), then the amount of time necessary to accumulate a full spectrum would be reduced. In principle, this could be achieved by using many separate receivers and transmitters, all operating at slightly different frequencies. One could conceivably space the different frequencies about one line width apart and thus a complete spectrum could be accumulated in the time normally taken to scan just one line of the spectrum, that is a few T_2's. While this technique is clearly impracticable, a much simpler way exists: a short intense r.f. pulse can excite all the lines in a spectrum simultaneously.† All lines in a spectrum will then respond with a free-induction decay at their own particular resonant frequency. The receiver sees all of the spectral lines simultaneously. Each component in the f.i.d. decays

* An r.f. amplifier with a 1·5 dB noise figure contributes only about 20% of the total spectrometer noise.
† In order to do this, it must be shorter than the reciprocal of the spectral width, that is of the order of a few tens of μs.

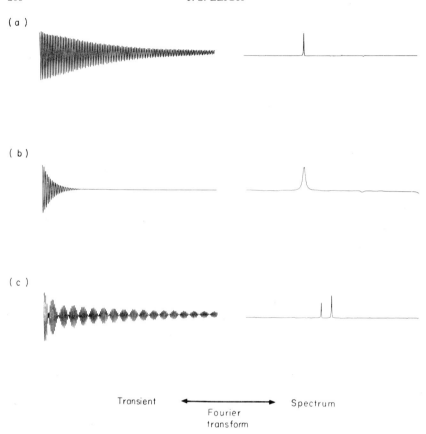

FIG. 7.14. Fourier-Transform n.m.r. transients (as a function of time) and corresponding spectra (as a function of frequency) for some simple cases.

(exponentially) with a characteristic time constant corresponding to its relaxation time, T_2. In order to get a reasonably faithful representation of a line (after we take the FT), the f.i.d. must be accumulated for a period of time equal to a few T_2's. This method is thus equivalent to having many transmitters and receivers. We achieve a saving in time roughly equal to the ratio of the total spectral width (SW) to the individual linewidths (Δ). This technique is quite practicable; the only complications introduced are the necessity for a high power, a pulsed r.f. transmitter and a computer to perform the FT's.

How the accumulation of N spectra can increase the signal-to-noise ratio by a factor of $(N)^{1/2}$ has already been discussed (Section 7.3.2). Thus a time saving by the FT technique may be turned into a sensitivity increase by accumulation of multiple spectra. In a given amount of time, then, the sensitivity increase over a conventionally scanned spectrum amounts to $(SW/\Delta)^{1/2}$. This represents a rather considerable factor which becomes larger and more important with higher-frequency spectrometers.

In Fig. 7.15, a comparison of a scanned spectrum and an FT spectrum of the same sample with the same spectrometer is shown. In 300 s, the FT technique achieves results vastly superior to the 1000 s conventional scan. An accumulation of additional FT spectra for 1000 s results in roughly a 30-fold increase in signal-to-noise ratio over the scanned spectrum taken in the same amount of time.

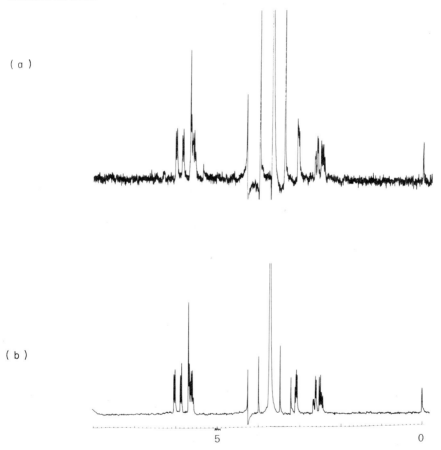

FIG. 7.15. N.m.r. spectra of 25 mM tryptophan in D_2O at neutral pH; scan range is 8 ppm (TMS (external) reference on right): (a) Scanned spectrum under optimal conditions, total scan time 1000 s; (b) Fourier-transform spectrum accumulated in 300 s.

7.4. Paramagnetic interactions

7.4.1. UNPAIRED ELECTRONS

The magnetic moment of an unpaired electron (responsible for e.s.r., dealt with in Chapter 8) is about three orders of magnitude stronger than the magnetic moment of most nuclei. Dipolar fields from an electron spin are of the order of 10000 G (1 T) at a distance of 1Å (and fall off as the inverse cube of distance). Dipolar fields, which are angular-dependent (anisotropic), can cause both relaxation and shift effects in the n.m.r. behaviour of nearby nuclei.

7.4.2. PARAMAGNETIC RELAXATION EFFECTS

Since the magnetic moment of an electron spin is 657 times that of a proton, the inherent relaxing ability of an unpaired electron is almost 500000 times that of a proton. Thus the presence of paramagnetic species in an n.m.r. sample can easily dominate the relaxation of nuclear spins. In effect, this establishes the paramagnetic centre as a reference point in a complex from which the predominant relaxation effects come. Distances to the paramagnetic centre can be estimated from the magnitude of the dipolar interaction.

Some biological molecules have natural, built-in, paramagnetic reference points. In many other cases, it is possible to add a paramagnetic centre by covalently binding one to a protein (e.g. a spin label; see also Section 10.2.5) or by using a paramagnetic substrate. An example of the types of results which may be obtained by these methods is illustrated by some studies on creatinse kinase carried out in the laboratory of Professor M. Cohn.

The enzyme creatine kinase (CK) catalyses phosphoryl transfer between adenosine triphosphate (ATP) and creatine:

$$\text{ATP} + \text{Creatine} \underset{M^{2+}}{\overset{CK}{\rightleftharpoons}} \text{ADP} + \text{Phosphocreatine}$$

A divalent metal ion (Mg^{2+}, Ca^{2+}, Mn^{2+}, etc.) is needed for activity and a variety of enzyme substrate complexes may be formed. Paramagnetic reference points can be introduced into the enzyme substrate complex by: (a) spin-labelling an enzyme sulfhydryl group with a stable nitroxide free radical (see Section 10.2.5) and (b) use of manganous ion as activator. Proton relaxation rates of substrate groups were measured in the presence of the paramagnetic reference points (individually) and distances between substrate protons and the reference points were calculated. Electron paramagnetic resonance (e.p.r. or e.s.r.) studies were used to determine the distance between reference points; use of the calculated distances enabled a molecular model to be constructed (Fig. 7.16).

Another type of interaction which can be a significant relaxation mechanism is the hyperfine or contact coupling. This scalar (non-angular-dependent)

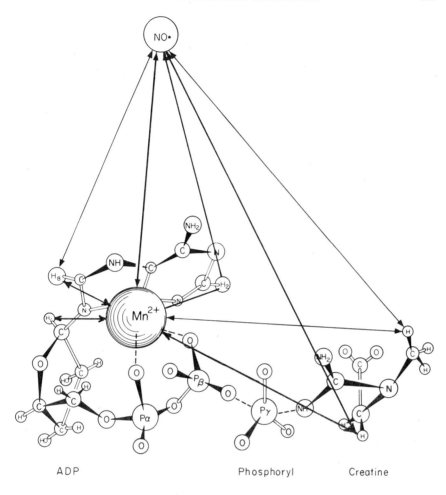

FIG. 7.16. Molecular model of MnADP—(phospho)creatine—creatine-kinase complex.

interaction is transmitted through chemical bonds by polarization of s orbitals. Nuclear-nuclear couplings of this kind give rise to the spin-spin splittings observed in n.m.r. spectra. Electron-nuclear coupling causes the hyperfine structure observed in e.s.r. spectra (Section 8.2.6) and in n.m.r. spectra causes both relaxation effects and "contact" shifts. Estimates of molecular-orbital densities can be obtained from the contact interaction.

7.4.3. PARAMAGNETIC SHIFTS

Paramagnetic centres can cause large shifts in the resonance frequency of nearby nuclei. Two types of paramagnetic shifts may be distinguished, contact

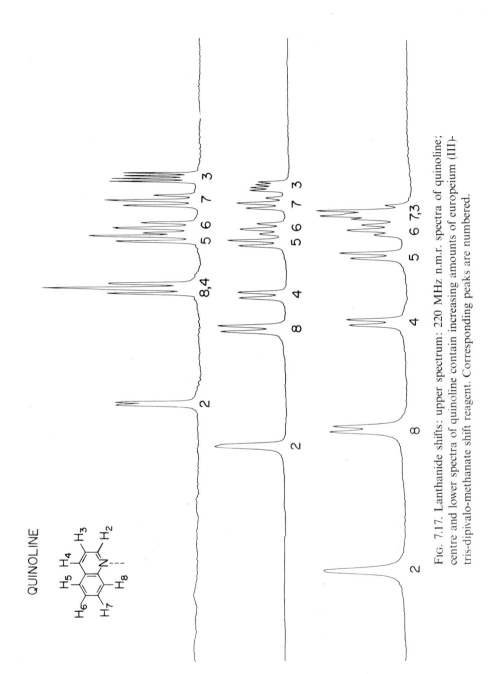

Fig. 7.17. Lanthanide shifts: upper spectrum: 220 MHz n.m.r. spectra of quinoline; centre and lower spectra of quinoline contain increasing amounts of europeium (III)-tris-dipivalo-methanate shift reagent. Corresponding peaks are numbered.

and pseudo-contact. The contact shift arises from delocalization of unpaired electron spin into *s* orbitals at a nucleus and is a measure of electron density at the position of the shifted nucleus. Contact effects are transmitted through chemical bonds. The pseudo-contact shift arises from dipolar interactions and may be used as measure of average distance between the unpaired electron and the nucleus. Pseudo-contact effects are transmitted through space. Both contact and pseudo-contact shifts are inversely proportional to absolute temperature. So far, it has proved difficult experimentally to ascertain the relative contribution of these two different shift mechanisms except in special cases.

In order that the shifts, contact or pseudo-contact, are not obscured by excessive line broadening (relaxation), the electron-spin relaxation time of the paramagnetic centre must be very short. The pseudo-contact mechanism may be expected to be very useful in obtaining structural information about biopolymers, since the shifts are proportional to the inverse cube of the electron-nuclear distance.

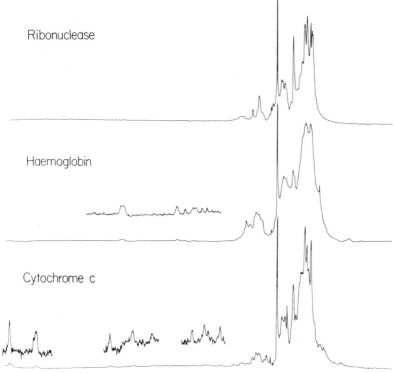

FIG. 7.18. Paramagnetic shifts in proteins. The two paramagnetic proteins, haemoglobin and cytochrome c, show lines far outside the normal 8 ppm range of diamagnetic proteins (ribonuclease, for example). Total scan range for all spectra ~45 ppm. (Parts of the spectra are shown at higher gain.)

The use of pseudo-contact shifts to measure distances in small organic complexes is simplified by the use of *shift reagents*. Certain chelates of rare-earth (lanthanide) ions with very short electron relaxation times (and thus ineffective as "relaxers") cause large pseudo-contact shifts. An example of the use of shift reagents is shown in Fig. 7.17. Increasing concentrations of shift reagent (a europium complex, in this case) cause progressively larger shifts in the spectrum of quinoline. Those protons which are closest to the metal ion (which is loosely bound to N) experience the greatest shifts.

Figure 7.18 shows two examples of paramagnetic shifts in haemoproteins, with the ribonuclease spectrum for comparison. In contrast to ribonuclease, both haemoglobin and cytochrome c have observable peaks far downfield from the rest of the "normal" protein spectrum. These peaks are associated with groups on the haem ring and protein residues liganded to the haeme-Fe. The use of paramagnetic shifts in protein n.m.r. has been reviewed by MacDonald and Phillips (1970) and Wüthrich (1970).

The possibility of combining the techniques of paramagnetic *relaxation* rates with paramagnetic *shift* studies offers outstanding possibilities in structural chemistry. The pseudo-contact shift is proportional to $(3\cos^2\theta - 1)/r^3$ and relaxation rates are proportional to $1/r^6$. Measurements of both shifts and relaxation rates may then allow the determination of both distance and direction from the reference point (see Section 10.2.2).

7.5. Chemical exchange

Nuclei in different chemical environments usually have different chemical shifts. What happens to an n.m.r. spectrum if a nucleus exchanges chemical environments at some finite rate? If the rate of exchange is very slow, then nothing much happens; a separate resonance is observed for each different chemical environment. At the other extreme, if the chemical exchange rate is very fast, only a single resonance is observed at the (weighted) average between the two environments. At intermediate rates of exchange, very interesting effects may occur (see also p. 302).

Development of theories to describe the observed effects was first attempted by modifying the Bloch equations* to include the effects of chemical exchange. The conceptual framework of McConnell, which extended the efforts of earlier workers, provides a very useful intuitive approach. Whereas the modified Bloch equations are unable to handle spin-spin interactions exactly, the density-matrix method, described by Kaplan and Alexander, is much more general and can, in principle, handle any system. If the exchange rate can be controlled experimentally (by temperature adjustment or addition of

---

* Details of the so-called Bloch equations, a set of simple differential equations proposed in 1946 by Professor F. Bloch to describe magnetic resonance phenomenologically, may be found in one of the texts listed in the bibliography.

catalysts), then it can usually be determined. Temperature variation also allows evaluation of the thermodynamic parameters of exchange.

Some interesting results of chemical exchange may be seen in Fig. 7.19 which shows the proton spectrum of an aqueous solution of the ammonium ion. The exchange of protons between ammonium ion and water is base-catalyzed. At low pH, the symmetrical NH_4^+ ion exchanges very slowly and the spectrum appears as a triplet, due to spin-spin coupling to the ($I = 1$) ^{14}N nucleus. As the pH is raised, the relaxation rate of the ^{14}N spin is increased by the increasing proportion of asymmetric NH_4OH molecules in solution and the triplet structure broadens. Further increases in pH first collapse the triplet to a single line and, at still higher pH, it broadens again as chemical exchange with water becomes more rapid. By pH 6, the NH_4^+ line and the H_2O line have merged into a single narrow line.

The uses of paramagnetic ions have been discussed as reference points for determining structures. Frequently, the relaxation (or shift) effects are too strong; chemical exchange then allows the effects to be diluted. If the nucleus

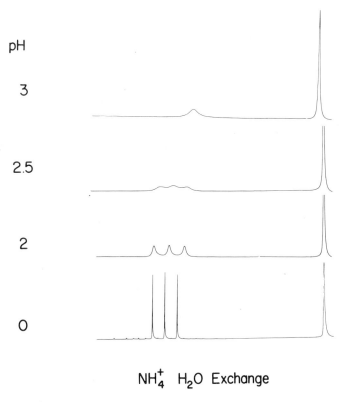

FIG. 7.19. Chemical exchange: 220 MHz n.m.r. spectra of ammonium ion at various pH's.

being observed is in great excess over the number of paramagnetic centres, and exchange is rapid, then an average effect will be seen, the weighted average of the two different environments, A and B (the "free" species and the "bound" species). That is,

$$\text{Observed Effect} = P_A \text{Effect}(A) + P_B \text{Effect}(B)$$

where the effect is a relaxation rate ($1/T_1$ or $1/T_2$) or a shift $\Delta\omega$. Mole fractions P_A and P_B denote the proportion of free and bounded nuclei. Thus, in situations where the exchange rate is more rapid than the (bound) relaxation rates and the (bound) shifts,* the observed effects may be adjusted to a conveniently measurable amount by judicious choice of concentrations (p. 302).

7.6. Multiple resonance experiments

7.6.1. DOUBLE RESONANCE

The term double resonance is an overall one for the effects of applying more than one radiofrequency to a sample. These may be grouped roughly into the categories (a) spin decoupling, (b) nuclear Overhauser effect and INDOR and (c) saturation transfer. We will attempt a short description of each technique, with an indication of its applicability (if any) to practical n.m.r. of biopolymers.

7.6.2. SPIN DECOUPLING

As the name implies, spin decoupling is erradication by irradiation—i.e. removal of spin-spin coupling by shining on a second radiofrequency. When two non-equivalent nuclei, say A and X, are coupled by (scalar) spin-spin interactions, the spectrum of each nucleus is split into a characteristic multiplet structure (p. 200). In order that the splitting be well resolved, the relaxation rate $(T_1)^{-1}$ of the coupled nucleus must be smaller than the magnitude of the splitting, J. That is, if the nucleus A senses the spin state of nucleus X (by a spin-spin interaction) and X changes its spin state (relaxes), then the resonance frequency of A will jump to the new resonant frequency corresponding to that state. If X is a simple spin-1/2 nucleus (say, a proton), then X has two possible spin states, $+1/2$ and $-1/2$. A will be split into two lines, one resonant frequency for each X spin state. If X changes its spin state (relaxes) only very slowly, then the observed spectrum of A is a well-resolved multiplet, whereas if X relaxes very quickly, then A finds its resonant frequency jumping back and forth. This is reminiscent of chemical exchange, where a given nucleus alternates between two chemical environments which have different resonant frequencies (chemical shifts). Thus, with X relaxing quickly, A senses only the average condition, so that the splitting will collapse and only a single line will be seen.

---

*This is an n.m.r. definition of "rapid exchange".

If (while observing the spectrum of nucleus A) a second radiofrequency* is applied at the resonant frequency of X, the r.f. will induce transitions among the spin states of X. If these transitions occur rapidly enough (they will if the r.f. is sufficiently intense), then the multiplet structure will collapse. The condition for collapse is that $B_1 \gg \pi J$. This effect can be very useful in a complex spectrum; it provides unequivocal proof that two sets of spins are coupled. Collapse of multiplets greatly simplifies complex spectra and, since with collapse all lines of a multiplet are bunched into a single line, signal-to-noise ratio may be increased several-fold.

If the second radio-frequency is generated by a noisy source, then a whole range of frequencies may be irradiated or decoupled at the same time. Noise decoupling is valuable in ^{13}C resonance; resonances are split by protons directly attached and also by those on neighbouring atoms. These splittings may be collapsed by irradiation of the protons. The spectral simplification and accompanying increases in intensity are almost mandatory for successful ^{13}C spectra in complicated molecules. An additional increase in intensity of the ^{13}C resonances is caused by the nuclear Overhauser effect (NOE).

7.6.3. NUCLEAR OVERHAUSER EFFECT

Irradiation at the resonant frequency of a given spin leads to deviations from the thermal distribution of population of the spin states. Saturation of a resonance line is caused by the near equalization of populations. If two spins A and X are coupled, then saturation of one spin will lead to changes in the level distributions of the other spin. If the spin being irradiated and the one being observed are both nuclei, this readjustment in populations is called the nuclear Overhauser effect (NOE).

The redistribution of population among the spin levels causes the intensities of observed resonances to change by a so-called fractional enhancement, defined as

$$\left(\frac{\text{Area of } A \text{ when } X \text{ is saturated}}{\text{Area of } A \text{ normally}} \right) - 1$$

The fractional enhancement may approach $\gamma_X/2\gamma_A$ for dipolar coupling and $-\gamma_X/\gamma_A$ for scalar coupling. These limiting values of enhancement are only achievable when the major relaxation mechanism of spin A (but not necessarily X) is through the coupling with X. Any additional relaxation mechanisms will hold the observed enhancement well below these maxima. As mentioned previously, this enhancement phenomenon is experimentally useful with nuclei such as ^{13}C, for example, where increases in sensitivity by a factor of four are possible. The Overhauser effect is not limited to nuclei but has been observed in both electron-nuclear couplings and in electron-electron interactions. Since dipolar coupling is distance-dependent, the NOE may, in principle, be used to determine structures. This technique has not been heavily exploited in the past

* An audio-frequency side-band may be used in homo-nuclear decoupling.

but in the future promises to be an exceedingly useful tool in determining biopolymer structures.

If the intensity of a given transition is observed (by sitting on top of a particular line) and the "Overhauser frequency" is *swept*, then a spectrum results which shows only those transitions coupled to the one being observed. With complicated spectra having many overlapping peaks, this technique, known as INDOR (internuclear double resonance), could be very useful.

7.6.4. SATURATION TRANSFER

Chemical exchange transfers a nucleus between two different environments. If exchange is so slow that separate resonances may (in principle) be observed, saturation of one resonance will be carried over to the other, an effect which may be utilized to measure exchange rates. Furthermore, it may be useful in eliminating (by saturation) the ever-present and usually annoying residual HDO solvent line. Strong irradiation of a position in a spectrum which corresponds to the resonance of an exchangeable (say N-H) group may effect saturation of the water line.

Saturation transfer has been used in studies of electron exchange between cytochrome c molecules in aqueous solution. Extensions of this technique may prove useful in other studies of biological electron transfer.

APPENDIX 7.1
Relaxation equations

This appendix lists the appropriate forms used in the general dipolar relaxation equation (p. 201),

$$1/T_i = D_i f_i(\tau_c)/r^6 \tag{1}$$

Nuclear relaxation by other nuclei

For pairs of like spins, I, on the same molecule (e.g. H_2O)

$$D_1 = 2\gamma_I^4 \hbar^2 (I(I+1)/5 = 2D_2 \tag{2}$$

$$f_1 = \tau_c/(1+\omega_I^2\tau_c^2) + 4\tau_c/(1+4\omega_I^2\tau_c^2) \tag{3}$$

$$f_2 = 3\tau_c + 5\tau_c/(1+\omega_I^2\tau_c^2) + 2\tau_c/(1+4\omega_I^2\tau_c^2) \tag{4}$$

with

$$1/\tau_c = 1/\tau_r.$$

For pairs of unlike spins, I and S, on the same molecule (e.g. HDO), the results for the $f(\tau)$'s are identical to those above and the D's are given by

$$D_1 = 4\gamma_I^2\gamma_S^2\hbar^2\,S(S+1)/15 \tag{5}$$

$$D_2 = D_1/2 \tag{6}$$

with

$$1/\tau_c = 1/\tau_r.$$

Nuclear relaxation by unpaired electrons (see p. 202)
The dipolar terms for a nuclear spin, I, interacting with an electron spin, S

$$D_1 = 2\gamma_I^2\gamma_S^2 h^2 \, S(S+1)/15 = 2D_2 \qquad (7)$$

$$f_1 = 3\tau_c/(1+\omega_I^2\tau_c^2) + 7\tau_c/(1+\omega_s^2\tau_c^2) \qquad (8)$$

$$f_2 = 4\tau_c + 3\tau_c/(1+\omega_I^2\tau_c^2) + 13\tau_c/(1+\omega_s^2\tau_c^2) \qquad (9)$$

where $\gamma_s = g\mu/h$ and $1/\tau_c = 1/\tau_r + 1/\tau_s + 1/\tau_M$.

For calculation of distances (to paramagnetic spins), equations 1, 7 and 8 may be solved for r. Numerical values for ^{1}H nuclei interacting with $S = 1/2$ and $S = 5/2$ paramagnetic spins give,

$$r(\text{Å}) = R_i[T_i f_i(\tau_c)]^{1/6}$$

^{1}H – ($S = 5/2$) : $R_1 = 819$
$R_2 = 731$
^{1}H – ($S = 1/2$) : $R_1 = 544$
$R_2 = 484$

TABLE 7.1

Symbol	Definition	Units (S.I.)	Numerical Values
γ_1	Proton gyromagnetic ratio	rad·s^{-1}·T^{-1}	2.675×10^8
γ_s	Electron gyromagnetic ratio	rad·s^{-1}·T^{-1}	1.759×10^{11}
h	Planck's constant	J·s·rad^{-1}	1.054×10^{-34}
g	Electronic "g" factor	---	2.00 (for Mn^{2+})
μ_s	Bohr magneton	J·T^{-1}	9.27×10^{-24}
S	Electronic spin quantum number	---	5/2 (for Mn^{2+}); 1/2 (for NO.)
r	Average electron-nuclear distance	m	
τ_c	Dipolar correlation time	s	3.0×10^{-11} (Mn-H$_2$O)
τ_e	Hyperfine correlation time	s	7.6×10^{-9} (Mn-H$_2$O)
τ_r	Rotational correlation time	s	3.0×10^{-11} (Mn-H$_2$O)
τ_s	Electron spin relaxation time	s	1.0×10^{-8} (Mn-H$_2$O)
τ_M	Chemical lifetime (of complex)	s	3.2×10^{-8} (Mn-H$_2$^{17}O)
ω_1	Nuclear resonance frequency ($2\pi\nu$)	rad·s^{-1}	1.38×10^9 (for ^{1}H at 5.167 T, 220 MHz)
ω_s	Electron resonance frequency	rad·s^{-1}	9.10×10^{11} (for $g = 2.00$ electron at 5.167 T)
A	Hyperfine coupling constant	rad·s^{-1}	3.9×10^6 (for Mn-^{1}H$_2$O)
k	Boltzmann constant	J·K^{-1}	1.38×10^{-23}

Nuclear relaxation by hyperfine coupling

If hyperfine coupling is present between nucleus and electron, the following terms should be added to the dipolar relaxation rates:

$$1/T_{1H} = \tfrac{2}{3} S(S+1)A^2 \left(\frac{\tau_e}{1+\omega_I^2 \tau_e^2} \right) \tag{10}$$

and

$$1/T_{2H} = \tfrac{1}{3} S(S+1)A^2 \left(\tau_e + \frac{\tau_e}{1+\omega_s^2 \tau_e^2} \right) \tag{11}$$

APPENDIX 7.2

Paramagnetic shifts

Contact shift ΔB_c in n.m.r. spectra caused by hyperfine coupling A, to an unpaired electron(s) S, is given by:

$$\Delta B_c/B_0 = -A h \gamma_s S(S+1)/\gamma_I 3kT.$$

Pseudo-contact shift, due to dipolar coupling to an unpaired electron a distance r away, is given by

$$\Delta B_p/B_0 = -(3\cos^2\theta - 1)(g_\parallel^2 - g_\perp^2)\mu_s^2 S(S+1)/9kT r^3$$

Here $g_\parallel$ and $g_\perp$ refer to the values of the (anisotropic) electron spin parallel and perpendicular to the electron spin symmetry axis, and θ is the angle between the direction vector from the electron spin to the nucleus and the symmetry axis. Both the pseudo-contact-shift equation and the one above for contact shifts are given in a form appropriate for a slowly tumbling biological macromolecule.

References

Abragam, A. (1961). "The Principles of Nuclear Magnetism". Oxford University Press, London.

Andrew, E. R. (1969). "Nuclear Magnetic Resonance" *in* Cambridge Monographs on Physics (Feather, N. and Shoenberg, D., eds).

Bovey, F. A. (1972). "High Resolution N.M.R. of Macromolecules". Academic Press, New York and London.

Carrington, A. and McLachlan, A. D. (1967). "Introduction to Magnetic Resonance". Harper and Row, New York.

Cohn, M., Leigh, J. S., Jr., and Reed, G. H. (1971). Mapping active sites of phosphoryl-transferring enzymes by magnetic resonance methods. *Cold Spring Harbor Symposium on Quantitative Biology* **36**, 153 (1971).

Dwek, R. A. (1973). "Nuclear Magnetic Resonance in Biochemistry." Clarendon Press, Oxford.

Emsley, J. W., Feeney, J. and Sutcliffe, L. H. (1965). "High Resolution Nuclear Magnetic Resonance Spectroscopy", in two volumes. Pergamon Press, Oxford.

Hecht, H. G. (1967). "Magnetic Resonance Spectroscopy". Wiley Interscience, New York.

James, J. L. (1975). "Nuclear Magnetic Resonance in Biochemistry." Academic Press, London and New York.

Jardetzky, O. and Wade-Jardetzky, N. G. (1971). Applications of nuclear magnetic resonance spectroscopy to the study of macromolecules. *Ann. Rev. Biochem.* **40,** 605.

Kowalsky, A. and Cohn, M. (1964). Application of n.m.r. in biochemistry. *Ann. Rev. Biochem.* **33,** 481.

McDonald, C. C. and Phillips, W. D. (1970). "Proton Magnetic Resonance Spectroscopy of Proteins" *in* Biological Macromolecules **4,** 1 (Fasman, G. D. and Timasheff, S. N., eds). Marcel Decker, New York.

Mildvan, A. S. and Cohn, M. (1970). *Advan. Enzymol.* **33,** 1.

Phillips, W. D. (1973). "N.m.r. of Paramagnetic Molecules" (G. N. La Mar, W. de W. Horrocks, Jnr and R. H. Holm, eds), Chapter 11. Academic Press, New York and London.

Pople, J. A., Schneider, W. G. and Bernstein, H. J. (1959). "High Resolution Nuclear Magnetic Resonance". McGraw-Hill, New York.

Roberts, G. C. K. and Jardetzky, O. (1970). Nuclear magnetic resonance spectroscopy of amino acids, peptides and proteins. *Advan. Protein Chem.* **24,** 447.

Slichter, C. P. (1963). "Principles of Magnetic Resonance". Harper Row, New York.

Ts'o, P. O. P., Schweizer, M. P. and Hollis, D. P. (1969). Contribution of nuclear magnetic resonance to the study of the structure and electronic aspects of nucleic acids. *Annals N.Y. Acad. Sci.* **158,** 256.

Wüthrich, K. (1970). "Structural Studies of Hemes and Hemoproteins by n.m.r. Spectroscopy", *in* Structure and Bonding, Vol. 8, p. 53 (P. Hemmerich *et al.*, eds). Springer-Verlag, New York.

8. Electron Spin Resonance

J. H. Keighley

8.1. Introduction

Whereas early applications of electron spin resonance spectroscopy was limited to irradiated materials and certain organo-metallic compounds, the development of experimental techniques has facilitated the examination of materials from which little information was obtainable previously. During the last few years, biological systems have become an important and growing example of this class of materials.

Electron spin resonance is a spectroscopic technique which detects the presence of unpaired electrons in a sample. These may occur as a consequence of incomplete chemical bonding, as in free radicals, or they may be present unpaired in atomic orbitals, in certain transition metals, such as Fe, Co and Ni. Both situations lead to the property of paramagnetism which is detected by the e.s.r. spectrometer.

Although there is some analogy with n.m.r. (Chapter 7), the complete mathematical theory of e.s.r. is distinct and, before we consider any results obtained from biologically important materials, it is necessary to outline the fundamentals of this theory.

8.1.1. THE BASIC PRINCIPLES OF ELECTRON SPIN RESONANCE

A spinning electron possesses two properties which characterize it as an elementary particle: a spin angular momentum and an associated spin magnetic moment. In a magnetic field B_0, the electron experiences a torque which tends to align the magnetic moment in the direction of B_0; because of the spinning motion of the electron, the moment precesses about the field describing a cone (Fig. 8.1) with a frequency proportional to B_0. If a compass needle is placed between the poles of a magnet, the system has minimum energy when the north pole of the needle is directed towards the magnet south pole. In principle, a higher energy state also exists when the needle is turned, with the absorption of energy, through 180° to be aligned anti-parallel to the applied field. This is a macroscopic analogy with the tendency of the electronic moment to be aligned either parallel or anti-parallel to the magnetic field in which the electron is placed. If the energies of the two states are E_2 and E_1,

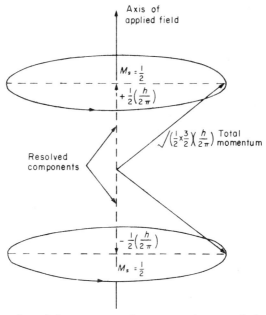

FIG. 8.1. Precession of electron magnetic moment about applied magnetic field.

where $E_2 > E_1$, a transition can be induced from E_1 to E_2 by absorption of incident radiation of frequency v:

$$hv = E_2 - E_1 = \Delta E$$

where h is Planck's constant.

Spin angular momentum and spin magnetic moment are vector properties which are aligned mutually antiparallel and which precess about the applied magnetic field direction. In general, the spin angular momentum can assume only discrete values,

$$\hbar[S(S+1)]^{\frac{1}{2}}$$

where the spin quantum number $S = \frac{1}{2}$ for a single electron, and $\hbar$ is the fundamental unit of angular momentum, $\dfrac{h}{2\pi}$. Along any given direction of an applied field, only two components of the spin angular momentum can be resolved, which have values of $+\frac{1}{2}\hbar$ and $-\frac{1}{2}\hbar$, where $\pm\frac{1}{2}$ are the values of the magnetic quantum number m_s, and the associated magnetic moments are $-\beta$ and $+\beta$, respectively. β, the Bohr magneton, the fundamental unit of magnetic moment, is defined by the relation

$$\beta = \frac{eh}{2mc}$$

where e is the electronic charge, m the electronic mass and c the velocity of light.

When the external field and the magnetic moment are parallel, $m_s = -\frac{1}{2}$ represents a condition of minimum energy; when $m_s = +\frac{1}{2}$, the applied field and magnetic moment are antiparallel and hence the energy is higher.

The energy of interaction between a magnetic dipole with a magnetic field B is $-\mu_B B$, where μ_B is the component of the magnetic moment along B. Since, in this case, the moment is restricted to two values $-\beta$ and $+\beta$, corresponding to $m_s = +\frac{1}{2}$ and $-\frac{1}{2}$, respectively, the energy of interaction between electron spin and magnetic field can have only two values, $+\beta B$ and $-\beta B$. Substitution into the Bohr equation

$$E_1 = E_0 - \beta B$$

and
$$E_2 = E_0 + \beta B$$

where E_0 is the initial energy in zero magnetic field so that

$$v = \frac{2\beta B}{h}$$

The orbital motion of the electron, however, gives rise to an orbital magnetic moment (which links with B), βM_L, where M_L is the magnetic quantum number and can take values in the range $L, (L-1), \ldots 0 \ldots -(L-1), -L$, where L is the total orbital angular momentum of the state. The energy of interaction, E_{ML}, of this magnetic moment with the field is

$$E_{ML} = \beta M_L B$$

The spin quantum number for an electron, $S = \frac{1}{2}$, when L and S are strongly coupled, gives a resultant total angular momentum J, the energy levels are given by

$$E_{MJ} = g_J \beta M_J B$$

Here g_J is the Landé splitting factor, a rational number of value dependent on L, S and J. M_J is an additional magnetic quantum number with values in the range $J, (J-1), \ldots 0 \ldots -(J-1), -J$. The Landé factor g_J is given by

$$g_J = 1 + \frac{S(S+1) + J(J+1) - L(L+1)}{2J(J+1)}$$

It can be calculated precisely for atomic systems with a strong Russell-Saunders $(L+S)$ coupling between the total orbital and total spin angular momenta; for transitions of free radicals, however, the condition is written

$$hv = g\beta B$$

where g is an experimental g-value or spectroscopic splitting factor; this cannot be calculated and allows for a mixture or coupling of orbital and spin momentum. It is a measure of the coupling between L and S and so is proportional to the splitting of the original level. For a free electron, $g = 2 \cdot 00232$, which represents the condition when the spin orients in the applied magnetic field, unimpeded by any other interactions. (The deviation from the calculated value of $2 \cdot 00000$ is a relativistic correction which allows for the orbital velocity of the electron.) Hence, for a completely free electron, the required transition frequency in Hz is

$$v = 2 \cdot 80 \times 10^6 B$$

where B is the strength of the applied field in kilogauss (kG). In terms of the S.I. unit Tesla (1T = 10 kG),

$$v = 2 \cdot 80 \times 10^5 B_T \text{Hz}$$

where B_T is the strength of the applied field in T.

In order to obtain a large energy difference, radiation in the microwave region is used, so that, for example, with $v = 9 \cdot 4$ GHz (where 1 GHz = 1 kMHz), the applied field is about 3·36 kG or 0·336 T.

For a continuous detectable absorption, energy must be lost by the high energy electrons so that they return to the lower energy state. This relaxation process (Sections 1.2.3 and 7.2.4) can occur by radiative emission whereby a photon of energy of frequency v is emitted from each electron. An equilibrium is thus established whereby electrons are absorbing energy from the incident radiation and an equal number are losing energy to return to the ground state; the net absorption of energy can be detected. Other relaxation processes that occur (Section 8.2.4) enable energy to be lost by the high-energy electrons.

In order to obtain a net absorption of energy, more electrons must be in the lower energy state. The distribution of electrons between the two levels is given, as for n.m.r. (p. 192), by the Maxwell-Boltzmann equation

$$n_2/n_1 = \exp\{-hv/kT\} = \exp\{-\Delta E/kT\}$$

where n_1 and n_2 are the numbers of electrons in the lower and upper energy states, respectively; k is the Boltzmann constant, T is the absolute temperature and ΔE is the energy separation between the two levels. Although the situation is less unfavourable than in n.m.r. (Section 7.1.3), high sensitivity requires $(n_1 - n_2)$ to be large. It is often easier to achieve this by lowering the temperature (using a dewar filled with liquid N_2) than by raising the radiation frequency, although the latter can sometimes be useful.

8.2 Interpretation of spectra

Information from e.s.r. spectra about the nature of free radicals can be classified under five headings (compare Section 1.4).

8.2.1. INTENSITY

The intensity of the spectrum is a measure of the number of free radicals present. Strictly, the integrated area under the absorption curve is proportional to the number of unpaired electrons in the sample, but this may often be approximated as the peak height of the absorption curve or as the peak-to-peak amplitude of the first-derivative curve measured under specific conditions.

8.2.2. g-VALUES

For free radicals, deviations of g-values from the free-spin value seldom exceed $\pm 0.5\%$; coupling between the spin and orbital motion of the unpaired electron is quenched as a consequence of the hybridization of the bonding electrons of the carbon atom in organic compounds. For non-organic radicals, g-values can vary widely. Transition-group metals are distinguished from free radicals in the location of unpaired electrons in inner atomic orbitals characteristic of the element; some of these are considered in Section 8.2.3.

8.2.3. ELECTRONIC SPLITTING

Whereas it has been assumed so far that each species under study is associated with only one unpaired electron, two unpaired electrons can co-exist on a single atom; such entities play an important role in studies of biological molecules and are often associated with enzymatic activity.

Two unpaired electrons associated with a molecular obital, as in the triplet-state system, can be regarded as possessing spin angular momentum only, so that the two momenta can be added vectorially to give total spin $S = 1$. Such a combined momentum and magnetic moment may act as an independent unit in a magnetic field; the three possible consequential orientations correspond to the resolved quantum numbers $M_S = +1$, 0 or -1, respectively. When a magnetic field B is applied (Fig. 8.2), the energy levels diverge with increasing B,

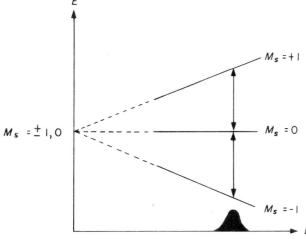

FIG. 8.2. Electronic splitting of magnetic energy levels.

while the $M_S = 0$ level remains unchanged in energy. Normal transitions occur at resonance between the $+1$ and 0 or 0 and -1 levels, i.e. $\Delta M_S = 1$, but, since transitions occur at the same value of B in each case, a single absorption line is observed.

Because of the surrounding molecular structure, such unpaired electrons are unlikely to be situated in a region of zero electric field, so that the energy of the $M_S = 0$ level is altered (Fig. 8.3a). While the selection rule $\Delta M_S = \pm 1$ still

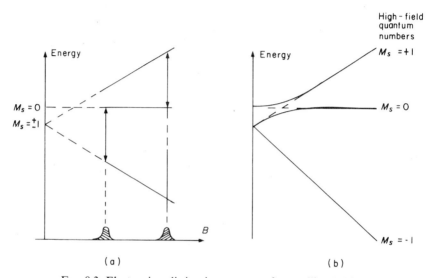

FIG. 8.3. Electronic splitting in presence of crystalline field.

applies, the $+1, -1$ transitions occur at differing values of B to give two absorptions. Competition between the applied field (which is dominant when large) and the electric field (which is dominant at low B) generated with the structure of the compound prevents the energy levels from intersecting (Fig. 8.3b). It also produces anisotropy measured as an angular variation when the direction of B is varied with respect to the crystal axis. Thus the electronic splitting depends on electric field, Δ, induced by the relative orientation, θ (Fig. 8.4), of B with respect to the crystal axis.

A transition-metal atom may have up to five unpaired electrons in the $3d$ shell, e.g. while electrons are also present in the $4s$ level. All such $3d$ electrons can align to give a total electron spin S up to 5/2. Metal atoms with similar characteristics occur in organo-metallic compounds such as haemoglobin and myoglobin which contain Fe (Section 8.6.1.). Manganous ions with $S = 5/2$ and in the ^6S spectroscopic state behave similarly; covalently bound species do not, however. Since the ^6S spectroscopic state is an orbital singlet state with no angular momentum, the magnetic moments arise solely from electronic spin angular momentum and the g-value is close to free spin. If no electrostatic

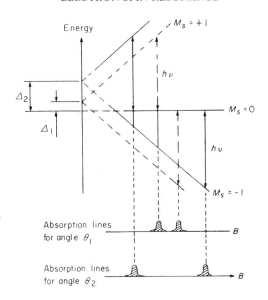

FIG. 8.4. Anisotropic and isotropic characteristics of electronic splitting.

(electric) fields are present, then all possible orientations of the $S = 5/2$ total spin will have the same energy (Fig. 8.5); when an external magnetic field is applied, the electrons will take up different orientations with respect to B, depending on the value of M_S. Since transitions between different energy states must obey the quantum rule $\Delta M_S = \pm 1$, only one absorption is detected (Fig. 8.5a).

In the presence of internal electric fields, however, as in a normal crystal, the energy levels will be separated as the triplet state (Fig. 8.5b). Hence allowed transitions will induce five separate absorptions.

8.2.4. LINE WIDTH

The width of an absorption line is a measure of the interaction between an electron and its environment. Of the relaxation processes additional to radiative emission, whereby the energy of electrons in the higher energy state is dissipated to allow them to return to the ground state, the most important is the spin-lattice interaction (cf. Section 7.2.4. on n.m.r.). This occurs when the spin energy is shared with the thermal vibrations of surrounding molecules and is characterized by a relaxation time, T_1, the time for the excess spin energy to

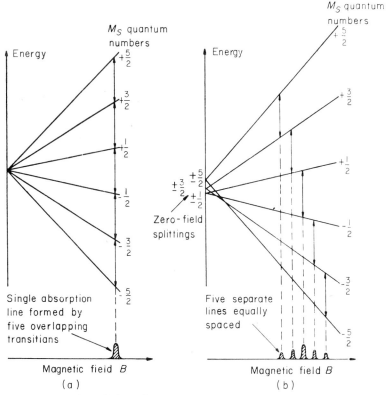

FIG. 8.5. Electronic splitting of manganous ions (a) in absence, (b) in presence of crystal-field splitting.

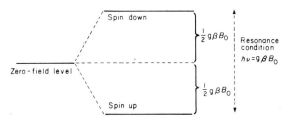

FIG. 8.6. Electron resonance condition $h\nu = g\beta B$ to give subject.

fall to $1/e$ of its original value. As a consequence of relaxation, the line is broadened, as defined by the Heisenberg uncertainty principle, i.e.

$$\Delta t . \Delta E = \hbar$$

where Δt is the lifetime of the state and ΔE is the uncertainty in the energy content. Lowering the temperature decreases the thermal motion and hence increases T_1.

Spin energy can be converted into thermal energy only through spin-orbit coupling, since one electron can interact with thermal vibrations only through orbital motion of the electron. Crystalline structure produces fields which tend to decouple the spin of an electron from its orbital motion, so that the spin-lattice interaction is reduced, as is the width of the absorption line. Thus, spin-lattice relaxation time and line-width can give an indication of the crystallinity of a sample.

Spin-spin relaxation (p. 201) is a second process which causes broadening; it involves a dipolar interaction between an electron, on the one hand, and surrounding electrons and nuclei on the other, to produce a field which the unpaired electron experiences in addition to the applied field B. Thus, when radiation of frequency v is incident on a sample, transitions will occur over a range of B to an extent dependent on the magnitude and orientation of the local field produced by the spin-spin interaction. This relaxation process, of characteristic time T_2, tends to broaden the original absorption line into a Gaussian curve, whereas spin-lattice relaxation produces a Lorentzian shape. The broadening, Δ, induced by these two interactions is defined by

$$\Delta = \frac{1}{T_1} + \frac{1}{T_2}$$

Two other phenomena influence absorption band-width significantly. Exchange narrowing occurs when the spin-spin interaction is reduced as a consequence of rapid electron exchange between the orbitals of different molecules. Saturation broadening occurs, together with a reduction in the peak height, as a result of weak relaxation processes when more radiation is absorbed than can be dissipated by exchange processes.

8.2.5. STABILITY

Whereas paramagnetic characteristics of transition metals and rare earths are quite stable, the e.s.r. spectra of free radicals observed over a period of time depend on the experimental conditions of temperature, humidity and atmosphere. While this is a consequence of the high reactivity of free radicals in general, in addition electron-migration processes in the sample can be particularly important in proteins (see Section 8.5.3.).

8.2.6. HYPERFINE SPLITTING

So far, attention has been confined to the effects of energy-level distribution on the electronic transitions of atomic systems; now the influence of neighbouring atoms, with which an unpaired electron in a free-radical system may interact, will be discussed. Figure 8.6 illustrates resonance of an isolated unpaired electron.

If the nucleus has a spin and magnetic moment (i.e. it is one of the hundred or so nuclear isotopes with spin $I > 0$ which can engage in n.m.r. (Section 7.1.4.)), its interaction with the orbital of an unpaired electron causes electronic energy levels to split slightly. Thus the magnetic moment of a single proton can be aligned either parallel or antiparallel to the applied field, as defined by $M_I = \pm\frac{1}{2}$, where I is the nuclear spin quantum number. Hence, in an applied field B_0, the electron actually experiences a magnetic field of $B_0 \pm \Delta B$, depending on the direction of the proton's magnetic moment. Since at resonance the electronic magnetic moments change direction while the nuclear spin remains unaffected, transitions occur only between levels with the same quantized components of nuclear spin.

At constant radiation frequency, absorptions occur at two magnetic field strengths (Fig. 8.7),

$$B_1 = B_0 - \Delta B$$

and

$$B_2 = B_0 + \Delta B$$

where

$$B_0 = \frac{h\nu}{g\beta} \quad \text{and} \quad B_2 - B_1 = 2\Delta B.$$

In general, for a nucleus with spin I, there are $(2I+1)$ equally spaced absorption lines; the spacing depends on the value of the *spin density* at the nucleus and is a measure of the time spent by the electron in the vicinity of the nucleus. Since the number of lines produced by hyperfine splitting depends on the spin quantum number of the enclosed nucleus, unpaired electrons associated with C or O atoms, for example (where $I = 0$), will give a single absorption line: for ^{1}H and ^{19}F nuclei ($I = \frac{1}{2}$), two absorptions are detected, while for ^{14}N and ^{2}D nuclei ($I = 1$), three absorptions of equal intensity arise.

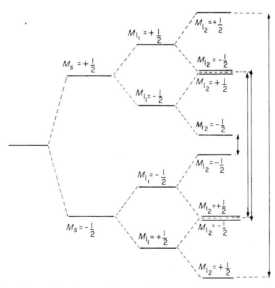

FIG. 8.7. Nuclear hyperfine interactions with successively one and two equivalent protons; arrows indicate triplet transitions for case of two protons.

In many cases, the orbital of the electron can interact with several nuclei which produce splitting; for n identical nuclei with spin I, there will be $(2nI + 1)$ symmetrically distributed absorptions. Thus, as Fig. 8.7 shows, two equally coupled protons produce three absorptions with an intensity distribution of 1:2:1, and three protons form four lines with intensity ratios 1:3:3:1. The doubly degenerate energy levels (Fig. 8.7) give a more intense line in accord with the number of possible arrangements.

When an electron is coupled with two protons to different extents, the spectrum may be considered to arise from the splitting into components by the stronger interaction followed by a further split into subcomponents by the weaker one (Fig. 8.8). Similar effects occur with multi-nuclear interactions. Other nuclei may induce smaller splittings than protons and may give characteristic multiplets (e.g. triplet peaks of equal intensity from a nitrogen radical); thus, as a probe of the environment of the unpaired electron, the shape of the e.s.r. spectrum can yield considerable structural information.

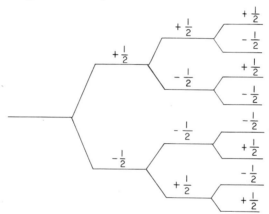

FIG. 8.8. Nuclear hyperfine interactions with non-equivalent protons (cf. Fig. 8.7).

8.2.7. ANISOTROPIC SPLITTING

So far, the above discussion of the interaction of an unpaired electron with a proton has neglected the influence of the orientation of the applied magnetic field with respect to the dipole axis; this leads to anisotropic splitting of the spectrum, allowed for in the general Hamiltonian equation which represents the total energy of the system. Energy change in an electronic transition is an average of the perturbing potential over the whole of the unperturbed system. Of the terms in the general Hamiltonian

$$\mathcal{H} = \mathcal{H}_E + \mathcal{H}_N + \mathcal{H}_C + \mathcal{H}_{LS} + \mathcal{H}_P$$

$\mathcal{H}_E$ depends on the kinetic energy of the electron, the attraction between electrons and nuclei, and the repulsion between the electrons; $\mathcal{H}_N$ represents the interaction energy of the electron spin with nuclear spin; $\mathcal{H}_C$ describes the effects of an internal crystalline field, as discussed on pp. 225 and 262; $\mathcal{H}_{LS}$ depends on spin orbit coupling; and $\mathcal{H}_P$, the term mainly responsible for paramagnetism, accounts for the Zeeman splitting. Anisotropic effects arise mainly from $\mathcal{H}_N$, which involves both a dipole-dipole interaction and a classical interaction between nuclear and electronic magnetic moments:

$$\mathcal{H}_N = 2\gamma\beta\beta_N \sum_i \left\{ \frac{(l_i - s_i)}{r_i^3} I + \frac{3(r_i s_i)(r_i I)}{r_i^5} + \frac{8\pi}{3} \delta(r_i)(s_i I) \right\}$$

In this, γ is the nuclear gyromagnetic ratio; β is the Bohr magneton; β_N is the nuclear magneton; l_i is the orbital angular momentum for the ith electron; s_i is the spin angular momentum for the ith electron; I is the nuclear spin vector; and $\delta(r)$ is a delta function (close to zero unless the electron density has finite value at the nucleus). Since the spin-orbit interaction l_i is very small in free radicals, this equation approximates to

$$\mathcal{H}_N = 2\gamma\beta\beta_N \sum_i \left\{ \frac{s_i I}{r_i^3} + \frac{3(r_i s_i)(r_i I)}{r_i^5} + \frac{8\pi}{3} \delta(r_i)(s_i I) \right\}$$

The first two vector terms are responsible for anisotropic effects but the delta function in the third term is scalar. Free radicals in non-crystalline phases do not show anisotropy as a consequence of random structural orientation. Furthermore, hyperfine splitting should occur only for electrons in s atomic orbitals or σ molecular orbitals. However splitting will also occur for π electrons since some σ orbital character of an excited state may be combined so that the hybridized orbital has a finite value at the nucleus.

Although often averaged to zero in liquid and powder specimens, the anisotropy can cause broadening. The anisotropic terms can be written

$$\mathcal{H}_{N_A} = C \sum_i \frac{3\cos^2\theta_i - 1}{r_i^3}$$

where C is a constant and θ_i is the angle between the direction of the applied field and r_i, $\mathcal{H}_{N_A}$ has a maximum value of $\left|\frac{2C}{r^3}\right|$ for $\theta = 0$ and a minimum value of $\left|\frac{2C}{r^3}\right|$ for $\theta = \pi/2$. Since $(3\cos^2\theta - 1) = 2$ for $\theta = 0$ and $0 = \pi$, it is clear that only two dipolar orientations correspond to this condition (i.e. parallel and anti-parallel) whilst an infinite number of possible orientations arise for $\theta = \pi/2$. Hence the doublet spacing which corresponds to $\frac{2C}{r^3}$ is very weak compared with that of $\frac{C}{r^3}$ which is observed experimentally.

Anisotropy can be regarded as originating in terms of a classical interaction between two dipoles. If two dipoles, of magnetic moments μ_1 and

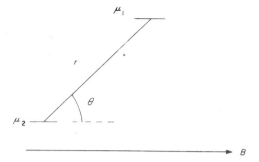

FIG. 8.9. Interaction between two magnetic dipoles, μ_1 and μ_2.

μ_2, are separated by a distance r at an angle θ to the direction of the applied magnetic field B (Fig. 8.9), then the classical interaction energy is

$$V = \frac{\mu_1 \mu_2}{r^3}(3\cos^2\theta - 1)$$

In a ĊH radical with planar sp^2 hybridization, the unpaired electron is located in a p orbital perpendicular to the bonding plane (Fig. 8.10). The spatial distribution of the orbital, together with the varying electron density, will cause the classical interaction V between the magnetic moments of the spinning proton and electron to vary with the position of the electron in its orbital, since r and θ will not both be constant. Consequently, the interaction energy will exhibit an average value

i.e.
$$V_{av} = \mu_1\mu_2 \left(\frac{3\cos^2\theta - 1}{r^3}\right)_{av}$$

The maximum and minimum values of $(3\cos^2\theta - 1)$ will be 2 and -1 when $\theta = 0$ or $\pi/2$, respectively. Similarly, the interaction will be zero when $3\cos^2\theta$

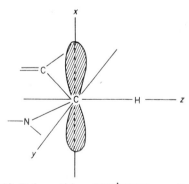

FIG. 8.10. Polypeptide —NHĊHCO— radical.

= 1 or $\theta = 54\cdot7°$. With the spinning proton as the centre of axes (Fig. 8.11), the values can be further subdivided by axes a and b, which are inclined at $54\cdot7°$ to the y axis, so that the graph is divided into zones of positive and negative interaction to give an overall splitting which depends on the zones occupied by the p orbitals.

The ĊH radical situated on three-dimensional axes in which the p orbital is aligned along the x axis will now be considered; the C—H bond is along the y axis, and the z axis is perpendicular to both these directions in the plane of the bonding orbitals. When the applied magnetic field B is aligned along the z axis, the p orbital (Fig. 8.13) is almost totally within the negative region of interaction, so that V_{μ_z} is negative. With B along the x axis (Fig. 8.14), the p orbital is located in regions of both positive and negative interaction, so that the overall interaction tends to zero; and with B along the y axis (Fig. 8.12), the p orbital is situated in a region of positive interaction. Measurements have shown that the isotropic coupling A_{iso} is about 20 G whilst A_{μ_x}, A_{μ_y} and A_{μ_z} are $0, -10$ G and $+10$ G respectively. Since the measured coupling $A = A_{iso} + A_\mu$

$$A_x = 20 \text{ G}, \quad A_y = 30 \text{ G}, \quad A_z = 10 \text{ G}$$

Hence, with B applied in turn along the x, y and z axes, the doublet splitting is 20 G, 30 G and 10 G respectively. Imperfect ordering in a crystalline specimen, and especially in polyatomic molecules of biological origin, can cause observed splittings to arise from a combination of these couplings.

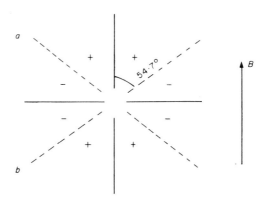

FIG. 8.11. Regions of positive and negative dipolar interactions.

8.3. Experimental methods

8.3.1. THE E.S.R. SPECTROMETER

For the observation of electron spin resonance, the requirements are a source of radiation, a sample cell, a magnetic field and a detection system. Since ΔE

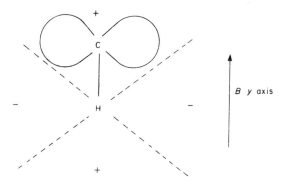

FIG. 8.12. —ĊH radical orbital interactions with B along y axis.

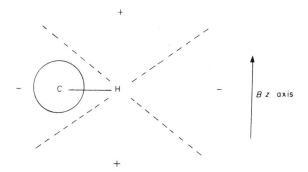

FIG. 8.13—ĊH radical orbital interactions with z axis.

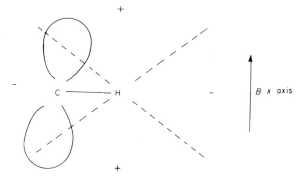

FIG. 8.14. —ĊH radical orbital interactions with B along x axis.

$= hv = g\beta B$, the ratio of the populations in the two levels is given (Section 8.1.1) by

$$n_2/n_1 = \exp\{-\Delta E/kT\} = \exp\{-g\beta B/kT\}$$

Evidently the population difference, and hence spectrometer sensitivity (Section 8.1.1), is enhanced by increase of B. However, with increasing frequency of the corresponding microwave radiation, the physical size of the waveguides for radiation transmission decreases so that, as a compromise, a magnetic field of a few kG is usually used with radiation of approximate frequency 10 KMHz (10 GHz).

In a typical arrangement (Fig. 8.15), microwaves from a klyston oscillator are transmitted along waveguides to a resonance cavity, within which the sample absorbs energy at resonance, located between the poles of an electromagnet. The amplified output of the crystal detector is observed on an oscilloscope or chart recorder so that any change with B of energy transmitted through the cavity can be measured. Fuller experimental details are given by Ingram (1969).

Within the cavity, electrical and magnetic components of the electromagnetic standing waves cause specific spatial distributions for the electric and magnetic fields (Fig. 8.16). Maximum microwave magnetic field and minimum electric field, necessary in e.s.r., occur throughout a plane for a rectangular cavity (a), or along an axis for a cylindrical cavity (b); the latter design, with its higher capacity for concentrating the microwave magnetic field, gives a higher sensitivity.

Samples (especially large ones in homogeneous magnetic fields) which contain highly polar groups, such as water, may interact strongly with the

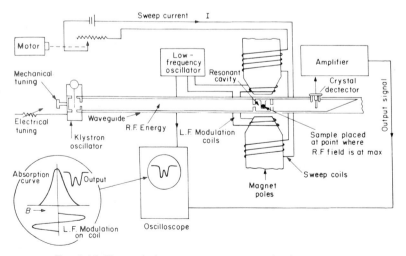

FIG. 8.15. Transmission e.s.r. spectrometer (Squires, 1963).

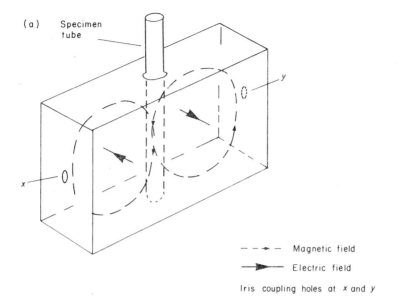

FIG. 8.16. Sample cavities (a) rectangular; (b) cylindrical (Ingram, 1958).

electric field, and so obscure the magnetic resonance. Aqueous systems are thus difficult; for such samples, operation at a low frequency or the use of a thin liquid cell in a rectangular cavity are recommended (see pp. 299 and 304).

8.3.2. THE DETECTION SYSTEM

With the detection of microwave radiation by semiconducting crystal-diode rectifiers, as in all modern instruments (Fig. 8.15), energy absorbed by the sample at resonance as B is slowly varied causes only a small change in the d.c. crystal current (Fig. 8.17), which is difficult to detect at the high noise levels in d.c. amplifier circuits. A low-frequency modulation (e.g. 50 Hz current applied to the electromagnet so that the resonance point is swept through 100 times s^{-1}) enables the crystal-current variations to be amplified and displayed on an oscilloscope; the sensitivity is sufficient only to enable the detection of a minimum free-radical concentration of 10^{19} spins. For greater sensitivity, phase-sensitive detection and a low-intensity magnetic-field modulation (Fig. 8.18) are customary; B is slowly swept through the resonance condition and, while the signal-to-noise ratio for the detector increases with the frequency of modulation, the simultaneous distortion of hyperfine lines occurs. Typically, 100 kHz modulation of the applied magnetic field is effected by passing the high-frequency current through a wire loop encircling the sample.

Since the a.c. amplitude of the crystal output measures the rate of change of the absorption as B is slowly swept through the resonance condition, the first derivative of the absorption curve is measured; distortion is minimized by limiting the modulation to 1/10th of the line width (Fig. 8.18a). After selective amplification, this 100 kHz signal is mixed with a 100 kHz reference signal of constant amplitude but variable phase to produce a d.c. signal fed to a pen recorder. In the transmission e.s.r. spectrometer (Fig. 8.15), energy passing through the cavity to reach the detector crystal decreases at resonance.

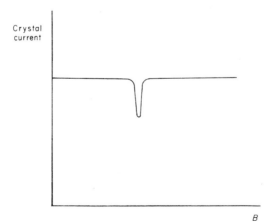

FIG. 8.17. Graph of crystal current *versus* magnetic field through an e.s.r. resonance.

ELECTRON SPIN RESONANCE

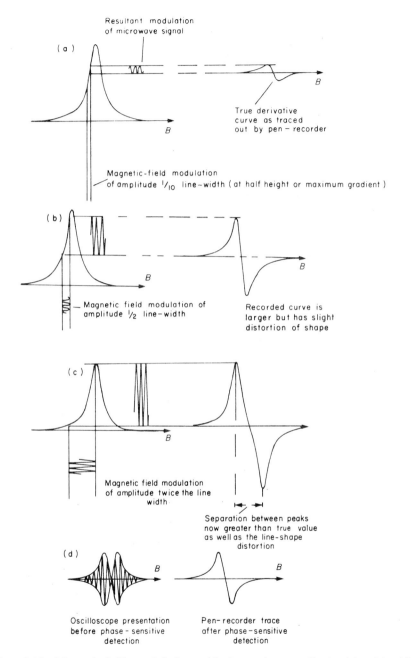

FIG. 8.18. Magnetic-field modulation with increasing amplitude (a) → (c); (d) represents output before and after phase-sensitive detection for low modulation amplitude (minimal distortion) (Ingram, 1958).

Sensitivity is improved with a balanced bridge system (Fig. 8.19) in which energy fed along arm 1 is balanced between arm 2 (which contains the cavity) and reference-arm 3, which has adjustable attenuater and phase controls, so that no energy passes into arm 4 in the off-resonance condition. At resonance, however, waves from arms 2 and 3 are no longer equal in amplitude and opposite in phase so that, when energy is absorbed in the cavity, an output signal is incident on the detector crystal.

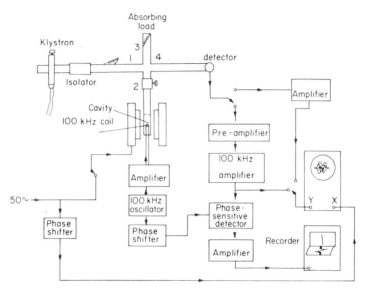

FIG. 8.19. Balanced-bridge spectrometer (Ingram, 1958).

8.3.3. THE MAGNETIC FIELD

Around a nominal 3000 G, the magnetic field may be swept through a range from 500 to 5000 G when biological materials are examined. Comparison of the spectra from different samples requires calibration of the spectrometer for g-value and magnetic-field strength; at constant frequency, the field strength alone may be calibrated throughout its range and hence any g-value calculated. Examples of samples which are used for calibration purposes include the stable free radicals diphenyl picryl hydrazyl (DPPH), which has a g-value of 2·0036, and ultramarine; these two give singlets with a field separation of 43 G at 9·4 kMHz. Alternatively, a dilute solid solution of Mn^{2+} in MgO (which gives a six-line spectrum of about 90 G splittings) is used.

8.3.4. SAMPLE PREPARATION

The large and expensive homogeneous magnets, together with small samples, which are needed to detect narrow, closely positioned, e.s.r. lines, are rarely necessary for biological materials. For paramagnetic materials, the previously dried sample is enclosed in a thin-walled, low-loss quartz (Spectrosil) tube, which is evacuated to ensure the absence of moisture. Samples which must be irradiated to induce free-radical formation are placed in such a thin-walled, high-purity quartz tube and dried thoroughly before final evacuation, since water vapour or oxygen can react with radicals produced. These tubes give no detectable signal when irradiated with X- or γ-rays or u.v. light for periods up to eight hours.

8.4. Amino acids, peptides and polypeptides

Since the characteristic properties of proteins depend on the molecular structures and sequence of the constituent amino acids, together with the configuration of the resultant polypeptide chain, discussion of amino acids must precede that of biological proteins.

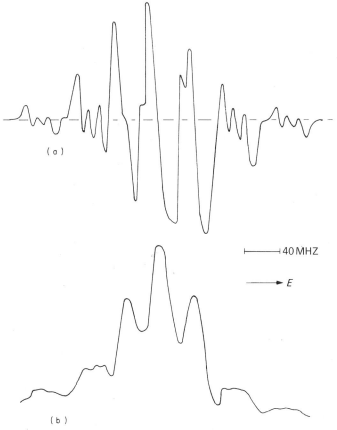

FIG. 8.20. Derivative (a) and absorption (b) spectra of glycine crystal with B along [100] axis (Ghosh and Whiffen, 1959).

8.4.1. AMINO ACIDS

Free radicals are induced when amino-acid samples are exposed to ionizing radiation. γ-rays from a ^{60}Co source and X-rays from a 50 kV generator produce identical species, but u.v. light of wavelengths ≤ 0.33 μm produces little radiation damage unless, as in some peptides and proteins, there is absorption by specific chromophores or specific conjugated side chains.

Among many e.s.r. studies on amino acids which do not contain S, those on glycine (Ghosh and Whiffen, 1959) illustrate differences between the spectra of single crystals in which measurements of the anisotropic characteristics of the unpaired spin help to establish the nature of the species present (Figs. 8.20 and 8.21) the individual spectra of which are still distinctive. Whereas from measurements with B along different crystal axes, Ghosh and Whiffen had interpreted variations in the low-field positions of the spectra as indicative of $^{+}$NH$_{3}$—ĊH—COO^{-} radicals, Weiner and Koski (1963) showed with deuteriated crystals that two radical species ṄH$_{4}$ and ĊH$_{2}$COO^{-}, were present. From the first-derivative line spectra of glycine and deuteriated glycine (glycine-d_{2} or $^{+}$NH$_{3}$–CD$_{2}$–COO^{-}) (Fig. 8.21), for B aligned along the $[1, 0, 0]$ axis, differences due to the presence of CD$_{2}$ groups are evident.

The deamination of amino acids when exposed to ionizing radiation was

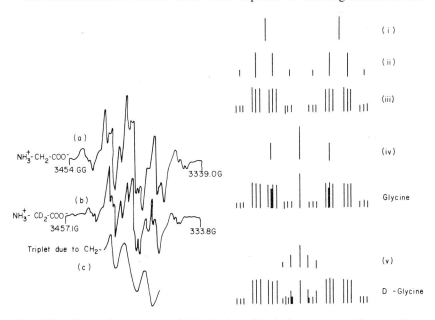

FIG. 8.21. Absorption spectra of (a) glycine; (b) glycine-d_{2}; and (c) normalized difference between (a) and (b). Stick diagrams represent (i) doublet due to B; (ii) quartet due to $B_{1} = B_{2} = B_{3}$; (iii) additional triplet splitting by ^{14}N; (iv) triplet due to CH$_{2}$COO^{-}; and (v) quintet due to CD$_{2}$COO^{-}. Glycine is sum (i)+(ii)+(iii)+(iv); glycine-d_{2} is sum of glycine+ (v). (Ghosh and Whiffen, 1959.)

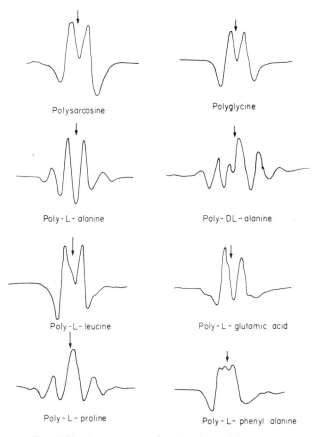

Fig. 8.22. E.s.r. spectra of polyamino acids.

shown by analysis of single-crystal spectra of L-alanine (Miyagawa and Gordy, 1960), recorded at several field orientations and at several microwave frequencies, which revealed the $CH_3 \dot{C}H\ COO^-$ radical. Later, Sinclair and Hanna (1967) confirmed that, although the unpaired electron was formed on the carbonyl group at 80 K, warming to 150 K induced reversion to the room-temperature species, albeit at a different orientation.

Sulphur-containing amino acids have a special interest in that disulphide crosslinks between adjacent polypeptide chains in proteins can form radicals. In contrast to other amino acids, polycrystalline cystine gives a closely similar e.s.r. spectrum to that obtained from a parent biological material. In the spectra of irradiated cystine, cysteine, and their hydrochlorides, small differences in peak separation and in relative signal heights of fine-structure peaks (Fig. 8.23) are attributed to variations in internal crystalline fields and in the rotational freedom of constituent molecular groups. Small but real

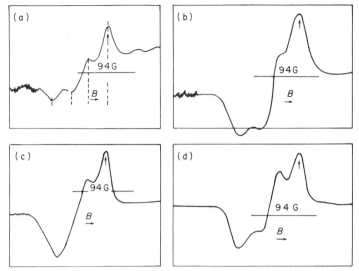

FIG. 8.23. E.s.r. spectra of (a) cystine; (b) cystine 2HCl; (c) cysteine; and (d) cysteine. HCl. Irradiation at 0·2537 μm. 94 G is width calibration; vertical arrow indicates DPPH g-calibration: horizontal arrow indicates direction of field sweeps (Bogle 1962).

differences in the spectra of γ-irradiated cystine from that obtained by u.v. irradiation (Fig. 8.24) are discussed in Section 8.5.

The resonances connected with cystine and cysteine are both thought to be associated with the formation of S radicals:

i.e. $\quad$ R—CH$_2$—S—S—CH$_2$—R $\rightarrow$ R—CH$_2$—Ṡ

and $\quad$ R—CH$_2$—S—H $\longrightarrow$ R—CH$_2$—Ṡ

However, several sets of measurements (Kurita and Gordy, 1961; Box and Freund, 1964: Akasaka, 1964) on single crystals of cystine hydrochloride (Fig. 8.25) with differing crystal orientation about B show changes in spectral patterns as the sample is warmed from 77 K to room temperature. Figure 8.26 represents a sample (a) immediately after irradiation, (b) after standing for nine days at 201 K and (c) after a further 10 h at room temperature. The spectra of the radical species proposed are designated α, β according to the

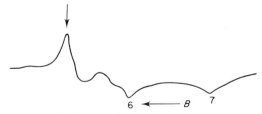

FIG. 8.24. E.s.r. spectrum of γ-irradiated cystine (not to same scale as Fig. 8.23; see also Table 8.1, p. 261).

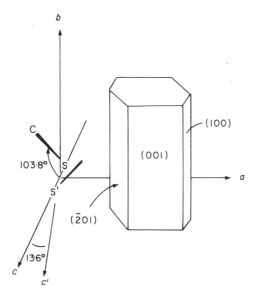

FIG. 8.25. Orientation of disulphide bonds in L-cystine dihydrochloride crystal (Akasaka, 1964).

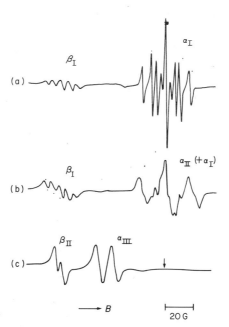

FIG. 8.26. L-cystine. 2HCl irradiated at 77 K (a) as irradiated; (b) after 9 days at 201 K, and (c) after further standing for 10 h at room temperature. (Akasaka, 1964.)

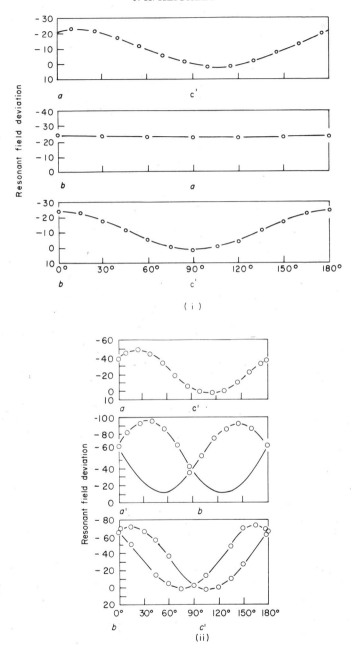

Fig. 8.27. Deviations (G) of resonant field for L-cystine. 2HCl radical from that of DPPH for (i) α_I; (ii) α_{II}. Thus top graphs in (i) and (ii) are for B about axis perpendicular to plane containing a and c' axes. (Akasaka, 1964.)

position of the unpaired electron. Above 201 K, a_I changes to a_{II} with a halving of radical concentration, whereas β_I radicals change to β_{II} with negligible loss in radical concentration. All the spectra show remarkable variation (anisotropy) in g-value with B. Figures 8.27, 8.28 and 8.29 show the deviation, ΔB, with resonant field at the centre of each absorption from that of DPPH versus orientation B in the direction of planes ac', ab and bc'. ΔB is related to deviation in the spectroscopic splitting factor:

$$\Delta g = -2\Delta B/B$$

Compared with the variation in g, the hyperfine structure showed negligible anisotropy in a_I, a_{II} and a_{III}, results which suggested interaction of the unpaired spin with protons of the adjacent CH_2 group (i.e. β-protons). In the ac' plane, β_{II} was an isotropic doublet and β_I showed five lines of intensity ratio 1:2:2:2:1 (a triplet-triplet structure) with poor resolution in the other plane directions.

The large g-factors suggested that the a_I spectra were associated with the ionized disulphide bond (Fig. 8.30) rather than free radicals involving C atoms. The isotropy of the β_I triplet-triplet structure also implied coupling with protons of the β-methylene group. a_{II}, however, was considered to arise from a semi-stable radical, R—CH_2—$\dot{S}$ (Fig. 8.31), produced by disulphide rupture in radical a_I, with no rotation of the C—S bond allowed, whilst the a_{III} room-temperature spectrum resulted from the allowed rotation about this bond. However, as there is some uncertainty about the exact nature of the cystine-type radicals, this will be discussed later.

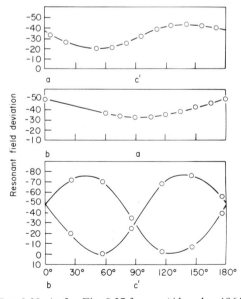

FIG. 8.28. As for Fig. 8.27 for a_{III} (Akasaka, 1964).

8.4.2. PEPTIDES AND POLYPEPTIDES

In contrast to spectra of amino acids (except cystine), the e.s.r. spectra of peptides and polypeptides are, in general, independent of the number of peptide groups involved. Thus cystine and poly-L-cystine give an asymmetric triplet; glycyl-glycine and (glycyl)$_n$ glycine give doublet spectra. While the

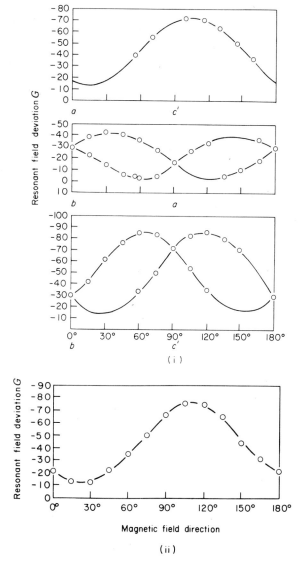

FIG. 8.29. As for Fig. 8.27: (i) for β_I; (ii) for β_{II} (Akasaka, 1964).

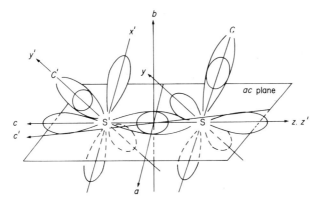

FIG. 8.30. Orbital structure of —C—S—S—C— group (Akasaka, 1964).

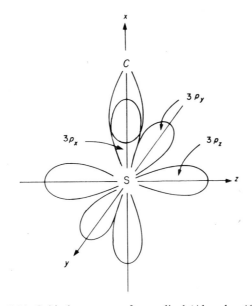

FIG. 8.31. Orbital structure of a_{II} radical (Akasaka, 1964).

basic spectrum from a peptide or polypeptide depends on the side chain at the a-C atom, i.e.

$$\begin{array}{c} \text{H}_a \\ | \\ —\text{C}—\text{C}^a—\text{N}— \\ \| \quad | \quad | \\ \text{O} \quad \text{R} \quad \text{H} \end{array}$$

spectra of mixed peptides are often complex and difficult to interpret (Meybeck and Windle, 1969a and b; McCormick and Gordy, 1958).

Drew and Gordy (1963) explained the observed spectra of polyamino acids in terms of the radical

$$-\underset{\underset{H}{|}}{N}-\underset{\underset{R}{|}}{\overset{\cdot}{C}}-\underset{\underset{O}{\|}}{C}-$$

in which a proton from the α-C was lost during the irradiation. Thus coupling of the unpaired spin on the α-C atom with β-protons produces a polyglycine ($R = H$) doublet, a poly-L-alanine ($R = CH_3$) quartet, and a poly-L-valine ($R = CH(CH_3)CH_3$) triplet. Figure 8.32 shows a selection of peptide and polypeptide spectra, measured at room temperature.

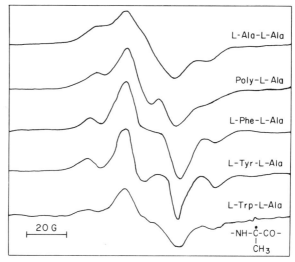

FIG. 8.32. E.s.r. spectra of peptides and polypeptides ascribed to radical shown (Meybeck and Windle, 1969, b).

8.5. Proteins

8.5.1. STRUCTURE AND RADICAL FORMATION IN PROTEINS

Radical formation in proteins is closer to that in polyamino acids than to radical formation in a mixed peptide (Section 8.4.2). Fibrous proteins, particularly wool in the k.m.e.f. (keratin, myosin, epidermin, fibrinogen) group and silk, will be examined in detail, since they exemplify many of the properties of biological materials.

8.5.2. RESONANCES FROM UNIRRADIATED MATERIALS

In their untreated states, wool and silk contain unpaired electrons stable to atmospheric oxygen and water vapour. The exact nature of the active sites and the mechanism of their formation are uncertain but it is likely that they are associated with either (a) the crystalline or ordered regions of the fibres (where oxygen and water penetration is minimal), or (b) aromatic side chains. Evidence from studies of wool irradiated with u.v. light of wavelength greater than 0·33 μm supports (b), since radicals formed behaved similarly to those detected in natural wool. While first it was suggested that radicals formed by sunlight irradiation of wool on the sheep's back, closer studies revealed (Eaton and Keighley, 1969) that the resonance is produced by two radical species with absorption bands of differing widths centred at the same g-value (Fig. 8.33).

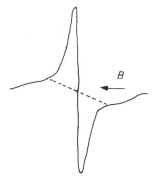

FIG. 8.33. Resonance of natural wool (Eaton and Keighley, 1969).

This is consistent with the simultaneous presence of radicals from both mechanisms (a) and (b), since the unpaired electrons in the ordered regions, with a smaller T_1, produce a narrower absorption than free electrons associated with aromatic side chains (as in poly-L-phenyl alanine, poly-L-tyrosine and poly-L-tryptophan). This interpretation is supported by raising the sample temperature, which facilitates electron migration through increased amplitude of molecular motion to sites of lower energy, so that the sharper component of the natural resonance is diminished.

In addition to this natural resonance, wool exhibits a free-radical resonance which increases in intensity with fibre pigmentation. Such unpaired electrons are associated with the pigment granules, composed of a natural product melanin, prevalent throughout nature; although the specific structure varies with the source, melanin is a polymerization product of enzyme-generated quinones. When polymerized by reductive condensation, indole-5,6-quinone gives the structure which is present in certain melanins, while hair melanin is a derivative of dihydroxyphenyl alanine. The unpaired spin in such melanin radicals is believed to be formed on an oxygen atom of one hydroxyl group with the loss of a proton (Mason, 1960) and the high stability of these radicals arises from resonance through the highly conjugated polymer.

The number of radicals in the granules increases both on irradiation (Allen and Ingram, 1961) with u.v. light and also by heating the sample above 333 K (Edwards, 1961). Irradiation of black Welsh-Mountain wool with X- and γ-rays give similar results (Fig. 8.34) with the singlet of the melanin radicals superimposed on the normal wool spectrum (Keighley, 1971). Although the "radiation-induced" melanin radicals are detected at $g = 2 \cdot 004$, as for the natural melanin radicals, the absorption induced by irradiation is broader.

Boiling pigmented wool in water for 1 h causes the resonance to increase, but continued boiling for 5 h gives a doublet absorption caused by an additional narrow absorption at $g = 2 \cdot 003$. Since the new radical has a g-value closer to free spin than that of the original species, the unpaired electron is more nearly free and less strongly coupled to the molecular structure. However, changes in g may be induced by changes in coupling between the unpaired electron on its incipient site and the conjugated structure of the melanin, so that the environment of the induced radical must differ from that of the original species and is more strongly coupled to the conjugated system. This view is supported by the observation that free radicals associated with O atoms characteristically have g values near $2 \cdot 004$.

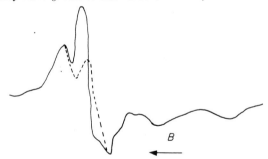

FIG. 8.34. E.s.r. spectrum of pigmented (full line) and unpigmented (broken line) wool (Keighley, 1971).

The changes induced by hot-water treatment probably arise by hydrolysis which, when prolonged, changes the colour of the granules from black to brown; this is consistent with conjugation decreasing during the treatment. Thus melanin exhibits three varieties of radical, each with a distinctive g-value and band-width. Since band-width is related to the degree of order of the molecular environment of the unpaired electron, radiation-induced radicals must be formed in regions of greater molecular disorder than those radicals initially present; differences in stability confirmed this, since radiation-induced radicals exhibited a spectrum which decayed with time. Thus radicals induced by water hydrolysis differ from the natural radicals in environment and in the extent of their coupling to the conjugated system; such differences are consistent with the presence of two or more phases within the melanin granules and of regions with a range of molecular order with corresponding extents of conjugation.

8.5.3. FREE RADICALS INDUCED BY IONIZING RADIATION

Following Gordy's e.s.r. examination of radiation-induced radicals in proteins in 1955, it has been established that the spectrum arises from three radical species, characterized as (a) a doublet, (b) a quartet and (c) an asymmetric triplet (Fig. 8.35). Originally the asymmetric triplet was assigned to the cystine residue in the protein structure but recently (Keighley and McKinley; Barkakaty and Keighley) understanding of the nature of the radical species induced in proteins has been transformed. As with the e.s.r. spectra of amino acids, so the spectra of proteins with and without S can be considered separately. The difference between the two types is associated with the distinct nature of a cystine resonance compared with those of other protein resonances. Although fibrous proteins are made up of large numbers (18 or 19 in wool) of different amino acids combined together in various sequences, all (apart from the cystine radical) give almost identical e.s.r. spectra when irradiated. If, as might be expected, irradiation induces changes randomly throughout the sample, at least 18 different radical species would be formed; this would give a broad singlet spectrum yielding negligible information. In practice, however, apart from the cystine absorption, only two radical species are detected, characterized by a doublet and a quartet. The explanation is that,

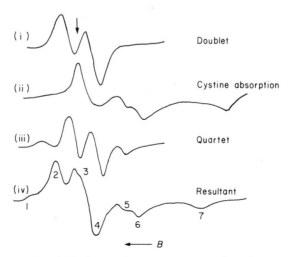

FIG. 8.35. Composite e.s.r. spectrum of wool

after initial random radiation damage, electron migration throughout the protein leaves the unpaired electrons of lowest energy. Thus a protein is energetically most stable and the two radical species represent the condition of highest stability. Such species have been shown to be associated with the chemical reactivity of proteins; chemical modification of reactive end groups, for example, modifies the radicals formed (Keighley, 1968).

The extent to which charge migration is conditioned by molecular structure was shown in Truby's (1962) study of energy-transfer processes in irradiated alkyl disulphides. Samples of several chain lengths (methyl to octadecyl) on either side of the disulphide bond, with differing extents of crystallinity, indicated that amorphous or disordered samples produced more free radicals on the S atoms than on C atoms; the octadecyl disulphide contained the highest proportion (30 %) of radicals on the aliphatic chain. As n, the number of C atoms on either side of the disulphide bond, exceeded 4, crystalline samples contained an increasingly higher proportion of C than S radicals. It was concluded that localization of energy depended on both chain length and crystallographic configuration; hence, in general, charge migration is limited in regions of high molecular order and is extensive in regions of molecular disorder.

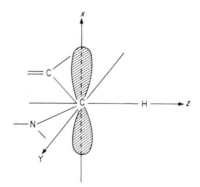

FIG. 8.36. —CO—ĊH—NH— radical.

In non-S-containing proteins, protein doublet and quartet absorption spectra were originally associated with an unpaired electron on the α-C atom of the peptide group in glycine (Gordy, 1959) and alanine (Keighley, 1968) residues, respectively; as in the case of synthetic polypeptides, the doublet was attributed to the —CO—ĊH—NH— radical (Fig. 8.36) and the quartet to the

$$-\text{CO}-\underset{\underset{\text{CH}_3}{|}}{\text{Ċ}}-\text{NH}-$$

structure. Recent evidence has undermined this interpretation, since the quartet was detected in samples containing negligible alanine while the doublet was detected in samples in which only very small quantities of glycine were present. Orientation studies on silk suggested that the ĊH radical could account for the anisotropic doublet splitting of 26 G with fibre axis parallel and 13 G when perpendicular to B. The corresponding calculated or theoretical splittings are 22·8 and 14·1 G, if the orientation of the supposed ĊH radical in silk (Barkakaty and Keighley) is considered in terms of the α-helical structure of wool, in which the x, y and z axes of Fig. 8.36 are aligned at 55°, 37° and 80° to the helix axis, respectively (approximately to the fibre axis). Since the doublet and quartet absorptions are associated with the more highly ordered regions of the protein structure, the experimental splittings of 17 and 15 G for wool are incompatible with calculations based on the ĊH radical. X-ray diffraction shows that the wool protein takes up an α-helix while the silk protein structure is of extended chains in a pleated-sheet configuration. The structure is stabilized by —CO . . . HN— hydrogen bonds: intrachain in the α-helix and shorter interchain ones in the extended structure. The difference between the isotropic splittings of 15·6 and 17·0 G for randomly oriented samples of wool and silk implies a different electron-proton interaction and is sufficient to rule out the ĊH radical, since isotropic splittings do not change with radical orientation.

Recent experiments (Keighley and McKinley) show that extended deuteriation of proteins *will* induce changes in the e.s.r. spectrum. Since the doublet and quartet absorptions are associated with highly ordered structural regions, deuteriation of the stable polyglycine structure can only be effected after extended treatment. The spectral pattern, also determined for silk and wool after extended deuteriation, changed from the doublet to that shown in Fig. 8.37; it arises from a doublet superimposed on an apparent singlet which, in fact, is an unresolved triplet absorption. Since $I = 1$ for D, a triplet is expected but the splitting is so small (~ 7 G) that resolution of the components is difficult. Since the deuteriation influences the nature of the splitting, only replaceable protons can be involved; O—H, N—H and S—H protons will exchange, but not C—H. If it is taken that the variation of doublet isotropic splitting occurs with protein structure, it is clear that the N—H groups deuteriate to form N—D. (OH groups are present in proteins only at the ends of main chains and side chains and are not preferentially oriented.) Thus the free radical is formed on the carbonyl group and interacts with the proton across the H bond, C—O . . . . H—N or, in deuteriated form, C—O . . . . D—N.

From the length of the H bond in proteins, a splitting of approximately 12 G would be predicted compared with 20 G for a ĊH radical; the measured value lies between these. By comparison with proteins in the natural state, irradiation induces radical formation on the O atom with consequent decrease

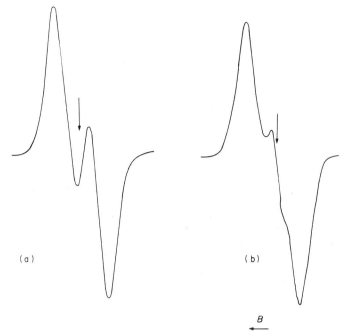

Fig. 8.37. E.s.r. spectrum of (a) polyglycine and (b) deuteriated polyglycine at 77 and 295 K.

in strength and an increase in length of the C—O bond; as a result of shortening of the H bond, coupling between the electron and the proton increases.

In a fibrous protein, the α-helix can be transformed into the extended pleated-sheet β-configuration by a fibre extension of 100%. Since the intermolecular H bond in the β-structure is shorter than the intramolecular bond in the α-structure, the doublet splitting observed in proteins is in agreement with the formation of a free radical on the carbonyl group. The "isotropic" splitting of a randomly oriented silk sample (β-structure with smaller O—H distance) will thus be greater than the "isotropic" splitting from randomly oriented wool (α-structure). Measurements of the secondary free-radical populations of proteins by tritiated free-radical scavenger techniques (Reisz and White, 1966–1970) detected no radical formation at glycine and alanine residues in the protein structure and did not support the concept of the "alanine" radical (Keighley, 1968). Further, the e.s.r. spectra of feather keratin and pig bristle exhibit the shoulders supposedly due to the "alanine" radical, although alanine is a constituent of neither keratin. The quartet spectrum of poly-L-alanine, with intensity ratios 1:3:3:1, was attributed to an unpaired electron on the α-C atom which interacts equally with three β-C protons to give a splitting of 17·5 G. The observation of similar splittings of about 23 G in

$(CH_3)_2 \dot{C}$ COOH and $CH_3 CH_2 \dot{C} CH_3$ casts doubt on the validity of the proposals, particularly since the spin densities (see p. 230) on alanine residues in proteins and poly-L-alanine are unlikely to differ much. Whilst absence of glycine activity (Reisz and White, 1966–1970) can be explained in terms of the carbonyl radical and, since tritiation of N—H . . . O═C groups will exchange with protons in aqueous solution, no such explanation can be offered for the "alanine" quartet. However, activity was found on proline, lysine, arginine and methionine residues (I to IV), all three of which possess a γ-methylene ($-CH_2-$) group in a position remote from the influence of any directing groups.

```
            α                          α                           α
    —N—CH—CO               —NH—CH—CO—                 —NH—CH—CO—
    /      \                       |                           |
   CH₂    βCH₂                   βCH₂                        βCH₂
     \   /                         |                           |
      γ                           γCH₂                        γCH₂
     CH₂                           |                           |
                                  CH₂                         CH₂
      proline  I                   |                           |
                                  CH₂                          NH
                                   |                           |
                                  NH₂                         C═NH    III
                                           II                  |
                                        lysine                NH₂
                                                            arginine
```

```
          α
   —NH—CH—CO—
        |
       CH₂
        |
       CH₂    IV
        |
        S
        |      methionine
       CH₃
```

Structures I to III can form hydroxy derivatives at the γ-position (γ-hydroxy methionine is inherently unstable and is not found in nature) and so the possibility of a link between the sites of radical formation and of chemical reactivity will be examined. For a radical formed on the γ-methylene group of lysine,

$$\begin{array}{c} \diagdown \quad \diagup H \\ N \qquad (H_\beta)\ (H_\alpha)\ (H_\beta) \\ \diagdown \qquad \downarrow \quad\ \downarrow \quad\ \downarrow \\ H-C-CH_2-\dot{C}H-CH_2-CH_2-NH_2 \\ \diagup \\ C{=}O \\ \diagup \end{array}$$

likely splittings are about 20 G for the proton bonded to the carbon atom carrying the unpaired spin (α-splitting) and 25 G for protons bonded to adjacent C atoms (β-splitting). The primary quintet due to four β-protons will be split to an unknown extent by the α-proton to give a six-line spectrum with relative heights approximately 1:5:10:10:5:1 (Fig. 8.38). Since this represents a small portion of the total protein absorption, the outer peaks will not be detected and the resonance will appear to have intensity ratios 1:2:2:1. The cyclic configuration of proline prevents continuation of an α-helical structure in the polypeptide chain, so that proline often terminates helical segments of the protein structure. Since electron migration is limited in the ordered or helical regions of the protein structure, the "doublet" and quartet components of the spectrum, which change similarly with time and temperature, can be associated with the ordered regions. Of the three amino-acid residues I to III in a protein which show activity after tritiation, the proline residues are almost certainly responsible for the free radicals formed that give the protein spectral quartet. The smaller recorded splitting of 17·5 G than that expected from α- and β-proton couplings is explained in terms of spatial distribution of the unpaired-electron orbital relative to the orientations of the C—H bonds of β-protons in the distorted ring structure and of spin density (lower in a cyclic ring system). Spin density in electron spin resonance, a measure of the time spent by an electron on a given nucleus, in proteins is likely to differ from that of other compounds because of charge migration.

Although limited in the ordered structural regions, electrons migrate over much greater molecular distances in the disordered regions, so that electronic vacancies occur on the cystine residues. Not all of these can participate in radical formation, and the number of cystine radicals formed is independent (Keighley, 1968) of the total cystine content. Among tentative explanations for the variation in the reactivity of cystine in proteins with a number of chemical reagents has been the establishment of an equilibrium which depends on the reaction medium. The influence of structural factors on "reactivity" of residues is shown by the effect of specific chemical pretreatment; the most reactive cystine groups also form radicals on exposure to ionizing radiation (Keighley, 1968).

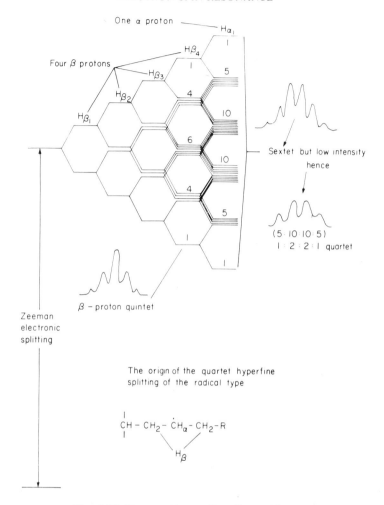

FIG. 8.38. E.s.r. spectrum of proline residue.

Previous proposals about the nature of the cystine radical in the amino-acid base the detection of a triplet (Fig. 8.35) on the association of an unpaired electron on the S atom with two β-protons on the adjacent methylene group. If such β-protons are replaced by other groups, e.g. methyl groups as in penicillamine (Henriksen, 1969), the spectral pattern is almost unchanged (Fig. 8.39). The same conclusion about the nature of the triplet absorption also applies to cysteine. It is clear, therefore, that a new explanation for the observation of the cystine resonance is required.

Now the relative magnetic field splittings and g-values of the low-field peaks in the e.s.r. spectra of fibrous proteins and polycrystalline cystine vary with the sample source (Table 8.1). Although the splitting has been ascribed to the

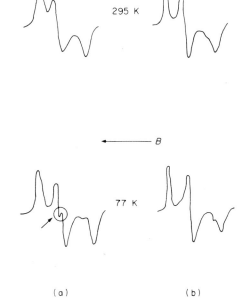

Fig. 8.39. E.s.r. spectra of (a) cysteine and (b) penicillamine.

crystalline field, this does not explain the difference between cystine (Fig. 8.35) and cysteine spectra (Fig. 8.39).

Change-transfer in proteins allows electrons to be dislodged from the structure during irradiation by Compton-effect secondary-electron irradiation; thus unpaired electrons can form on the cystine residues by electron donation from the disulphide (—S—S—) bridge (favoured energetically by the two lone pairs on each S atom). Other hypotheses have involved the formation of three-electron bonds, positive ions and negative ions, depending on the condition of measurement. Since similar radicals can be formed in proteins by mechanical degradation, then, in the absence of secondary electron migration, the concept of electron donation rather than electron acceptance by the disulphide group must be accepted. This is supported by dielectric-constant measurements of wool cloth (Keighley and McKinley) which show that, when irradiation ceases, positive and negative charges combine to leave only unpaired electrons associated with the system.

The system

$$R-CH_2-\ddot{S}-\ddot{S}-CH_2-R' \xrightarrow{\text{Irradiation}} R-CH_2-\ddot{S}-\ddot{S}-CH_2-R'$$
$$\downarrow$$
$$R-CH_2-\ddot{S}^{\cdot+} \quad \cdot\ddot{S}-CH_2-R'$$

TABLE 8.1. Spectroscopic splitting factors measured at peak 6 and peak 7 (Fig. 8.24), and mean separation between peaks 6 and 7 for different samples of wool proteins.

Samples	g (peak 6)	g (peak 7)	Separation/G (peak 6–7)
1. South American merino	2·035	2·068	55·84
2. Scottish blackface	2·034	2·067	55·56
3. New Zealand Romney	2·033	2·062	54·86
4. Herdwick	2·034	2·065	53·02
5. Shropshire	2·035	2·069	52·11
6. Lincoln	2·037	2·064	51·82
7. Devon long	2·033	2·064	49·85
8. Alpaca	2·037	2·069	49·28
9. Welsh greasy	2·035	2·071	49·24
10. Cheviot	2·033	2·064	49·02
11. Wensleydale	2·033	2·061	48·98
12. Southdown	2·036	2·071	48·26
13. Welsh pick	2·036	2·070	47·94
14. Lonk	2·032	2·059	47·31
15. Welsh mountain black	2·037	2·071	45·47
16. Swaledale	2·035	2·060	45·17
17. Cotswold	2·033	2·061	44·13
18. Dorset horn	2·034	2·064	40·89

should yield two products, a biradical, R—CH$_2$—$\ddot{\text{S}}\cdot^+$, and a monoradical, R'—CH$_2$—$\ddot{\text{S}}$: The exact site of extraction of the electron is unknown but the most stable electronic state of the ion has two unfilled orbitals.

If splitting is discussed in terms of S radicals only, before the influence of any neighbouring groups is considered, the S monoradical will produce a singlet absorption ($I = 0$ for S). Since the biradical ion will exhibit electronic splitting (Section 8.2.3.), doublet absorption will ensue, owing to electric fields in the molecule enhanced by the ion present. Thus (Fig. 8.40), the S monoradical and biradical can combine to give a total absorption almost identical with that observed. Small changes in protein structure will split the doublet components differently and will change g-values of the component peaks. The greater doublet splitting of polycrystalline cystine than for cystine in fibrous proteins is a consequence of the higher electric field in the more highly ordered crystalline structure. The variations with B of this doublet splitting for a single crystal can also be accounted for in terms of electronic splitting and the presence of electric fields (see Section 8.2.3).

Since substitution of the methylene protons adjacent to the disulphide S atoms by methyl has little influence on the (final) "cystine" spectrum, the unpaired electrons in the two radical species are probably in orbitals oriented along the C—S bond.

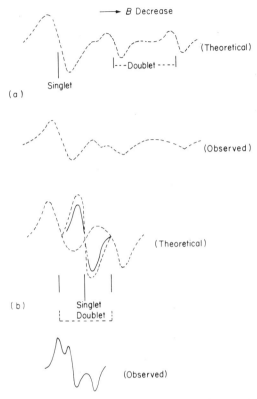

FIG. 8.40. Theoretical and observed spectra of (a) cystine; and (b) cysteine.

As shown in Fig. 8.40(b), the spectrum of cysteine also arises from a superimposed singlet and doublet. Two different mechanisms generate radical species, i.e.

$$R-CH_2-S-H \longrightarrow R-CH_2-\ddot{S} + H^+$$
$$R-CH_2-S-H \longrightarrow R-CH_2-\overset{+}{\ddot{S}\cdot} + \dot{H}$$

Since spectra are examined after irradiation, highly reactive $\dot{H}$ radicals are unlikely to be detected owing to combination with identical species or other molecules present. Because of the different structures, crystal-field splitting and g-values of the three components will differ in Fig. 8.40(b) from those in cystine.

8.6. Applications of e.s.r. to the action of blood

8.6.1. STUDIES OF HAEMOGLOBIN

The planar haem structure (Ingram, 1969) (Fig. 8.41), an essential component in many important biological compounds, constitutes the heart of the

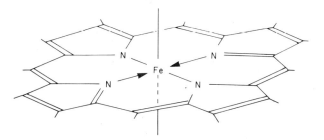

FIG. 8.41. Planar haem structure (see also p. 303; Ingram, 1969)

haemoglobin molecule. A central Fe atom is coordinated to four N atoms, with a fifth Fe coordination bonding to the protein; the sixth coordination is with O_2 or CO_2, according to whether the molecule is at that instant transporting O_2 or CO_2 waste product in the body's respiratory process. In haemoglobin (early studies were carried out on myoglobin with a single haem plane), e.s.r. studies can help to determine the bonding between the Fe atom and the ligands, including the group at the sixth coordination site.

While ferrous iron (Fig. 8.42) is not paramagnetic when covalently bonded, ferric compounds possess unpaired electrons in both ionic and covalent states. In such a system (Sections 8.2.2 and 8.2.3), five such electronic transitions can be obtained from six energy levels only if the energy of the microwave quantum is significantly greater than the splitting induced by the internal crystal field (zero-field splitting). The only transition observed is between the two components at the $S = \pm\frac{1}{2}$ level (Fig. 8.43), with a g-value which varies between 2 and 6, depending on the orientation of the applied field relative to the axis of the crystalline electric field. Since it can be shown theoretically that the total spin

		Electron configuration			Total spin
		3d — 4s — 4p			
Ionic bonds	Fe^{2+}	(⇅)(↑)(↑)(↑)(↑) ○ ○○○			2
	Fe^{3+}	(↑)(↑)(↑)(↑)(↑) ○ ○○○			$\frac{5}{2}$
Octahedral covalent d^2sp^3 complexes	Fe^{2+}	(⇅)(⇅)(⇅)○○ ○ ○○○			0
	Fe^{3+}	(⇅)(⇅)(↑)○○ ○ ○○○			$\frac{1}{2}$

FIG. 8.42. Spin states of Fe atoms in haemoglobin (Ingram, 1969).

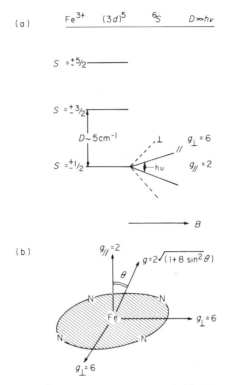

FIG. 8.43. Crystal-field splitting, D, of haemoglobin (Ingram, 1969).

vector lies along the haem plane, measurements for g with myoglobin crystals in the cavity have established this direction (Fig. 8.44) with respect to the three crystal axes. Similar results from haemoglobin (Fig. 8.45) showed that there are four separate g-value variations (A, B, C and D), each of which corresponds to a single haem plane. Haem proteins are discussed further on pp. 289 and 303.

8.6.2. STUDIES OF FIBRINOGEN

Fibrinogen is the protein in blood involved in clotting when a blood vessel is ruptured. The long and complex series of biochemical reactions involved may be summarized:

Fibrinogen → fibrin monomer + peptides;

Fibrin monomer → fibrin polymer (urea-insoluble);

Fibrin polymer → physiological clot (urea-insoluble).

Between the tyrosine and histidine residues in the fibrinogen molecule, nine intermolecular H bonds may form; this happens when the fibrin monomer polymerizes. Conversion to an insoluble clot involves the formation of isopeptide links between the γ-carbonyl groups of glutamine residues and the ε-amino groups of lysine residues in adjacent molecules. The initial

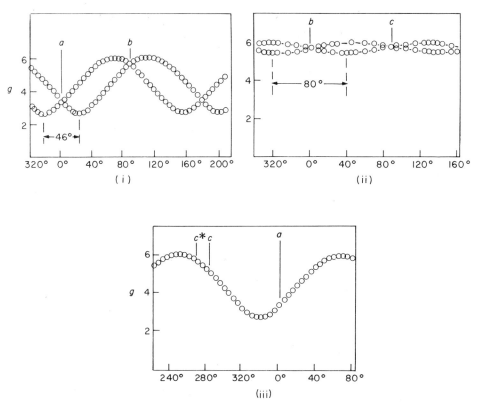

FIG. 8.44. Myoglobin: g-value variations with angle for rotations approximately about three crystallographic axes: (i) c, (ii) a, (iii) b (Ingram, 1969).

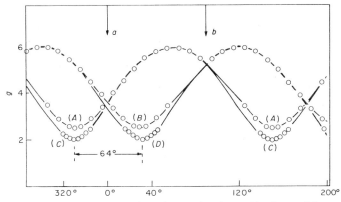

FIG. 8.45. Variations with orientations in g-value in (100) plane of haemoglobin (Ingram, 1969).

stabilization of the system involves the formation of crosslinks which involve Ca ions as essential components and which also act as a catalyst. Clotting depends on the concentration of Ca ions present. In the e.s.r. spectra (Earland et al., 1972) of samples treated with Ca ions (Table 8.2) in differing concentrations (Fig. 8.46), studies of other proteins show that Figs. 8.46(a, b, c and d) are superimpositions of a singlet, doublet and quarter, whereas Fig. 8.46(e) is a narrow singlet absorption superimposed on a triplet (shown diagrammatically in Fig. 8.47). The remarkable spectral change from Fig. 8.46(d) to 8.46(e) occurs near a concentration of 0·3· g of $CaCl_2 6H_2O$ per 5 ml of solution; the 1:2:1 spectral pattern splitting of 20 G indicates that the radical is associated with two a-C-atom protons, i.e. of the type —$\dot{C}H_2$.

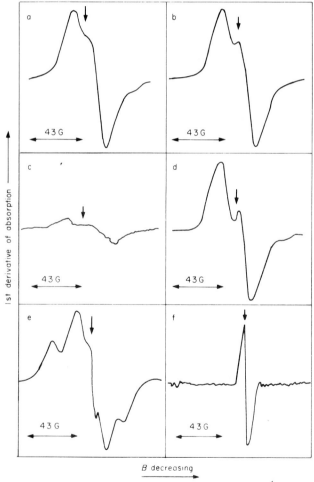

FIG. 8.46. E.s.r. spectra of Ca-treated fibrin samples as listed in Table 8.2. Vertical arrows denote DPPH reference. (Earland et al., 1972.)

Further, since the singlet is much narrower than those in Figs 8.46(a–d), it is from a radical present in a structurally well-ordered region. Thus the structures associated with Fig. 8.46(e) clearly differs from those associated with Figs. 8.46(a–c); the presence of Ca ions at the higher concentration accelerates clot formation. Since nine histidine and tyrosine residues are involved in H-bonding during clotting, and since the Ca ion concentration is low in (c) and (d) but sufficient in (e) to allow the presence of Ca^{2+} ions at each H-bonding and cross-linking site, it was concluded that the Ca^{2+} ions were bonded into the structure. X-ray diffraction data support this conclusion.

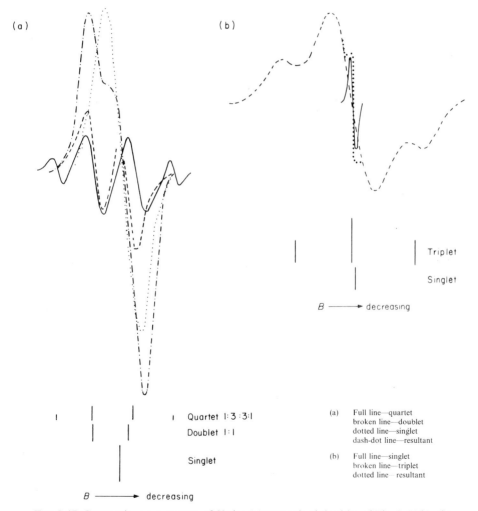

FIG. 8.47. Composite e.s.r. spectra of fibrin: (a) unresolved doublet of Fig. 8.46(b); (b) triplet structure of Fig. 8.46(e) (Earland et al., 1972).

TABLE 8.2. Properties of fibrinogen and fibrin clots formed in the presence of increasing amounts of calcium; spectra are shown in Figs. 8.46 and 8.47.

Sample	Calcium present ($CaCl_2 \cdot 6H_2O$) (g 5 ml)	State of clot	E.s.r. spectrum
(a) Fibrinogen	—	Unclotted	Unresolved doublet Quartet Singlet
(b) Non-calcified fibrin	—	Opaque gel soluble in 5M urea	Unresolved doublet Quartet Singlet
(c) Calcified fibrin	0·0002	Opaque gel insoluble in 5M urea	Unresolved doublet (low radical yield)
(d) Calcified fibrin	0·080	Opaque fibrous clot insoluble in 5M urea	Unresolved doublet
(e) Calcified fibrin	0·30	Clear gel insoluble in 5M urea	Triplet Singlet
(f) Calcified fibrin	3·0	No clot formed	Singlet, very low intensity

Since tyrosine and histidine residues are joined by H-bonding, the results were regarded as consistent with the structure in which only the phenolic O atom of the tyrosine and tertiary atom of the histidine coordinate to the Ca ion. In view of the likely instability of such a structure, and the known tendency of Ca^{2+} to octahedral coordination, the electrically neutral structure (Fig. 8.48) was proposed; this involved either aspartic acid or glutamic-acid residues, and the extra H bond imparts additional stabilization to the histidine $\rightarrow Ca^{2+}$ coordination. From models, the O—N distance in the histidine-tyrosine H bond is near 0·23 nm and the $Ca^{2+} \leftarrow O$ coordinate bond length about 0·17 nm.

It was concluded that the e.s.r. triplet resulted from decarboxylation of the acidic groups following electron-transfer processes. Electrons may be withdrawn from the glutamic-acid side chain as a consequence of coordination so that, when the structure donates an electron to replace one lost during irradiation, the γ-δ C—C bond will rupture to leave a free radical; this would explain the observed triplet. It was concluded that the superimposed singlet was situated on the O atom of the tyrosine residue owing to the loss of a hydroxyl proton.

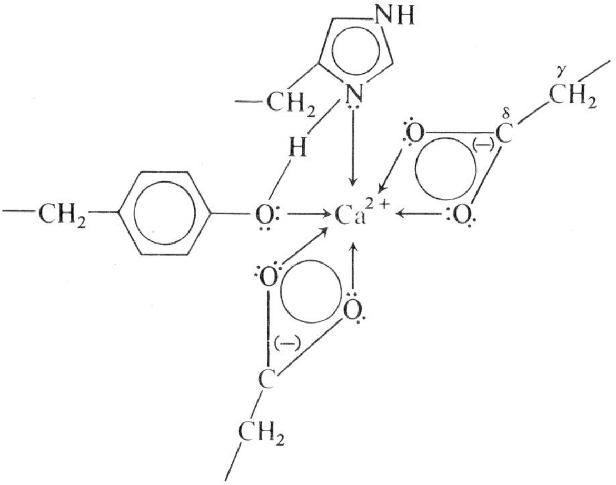

Fig. 8.48. Calcium co-ordination complex formed in clotted fibrin (Earland et al., 1972).

8.7. Conclusion

From the studies and results described in this chapter, it is clear that the conclusions drawn from e.s.r. can have far-reaching effects on our understanding of the structure and properties of biological materials. Further such examinations, together with the use of new methods such as spin-labelling techniques (Section 10.2.5), will ensure the fruitful application of electron spin resonance in this field in the future (Swartz et al., 1972).

References

Akasaka, K. (1964), *J. Chem. Phys.* **40**, 3100.
Allen, B. T. and Ingram, D. J. E. (1961), "Free Radicals in Biological System" (M. S. Blois, ed.), p. 215. Academic Press, London and New York.
Barkakaty, B. C. and Keighley, J. H. Unpublished.
Bogle, G. S. (1962). *Photochem. Photobiol.* **1**, 277.
Box, H. C. and Freund, H. G. (1964). *J. Chem. Phys.* **40**, 817.
Drew, R. C. and Gordy, W. (1963). *Rad. Res.* **18**, 552.
Earland, C., Keighley, J. H., Ramsden, D. B. and Turner, R. L. (1972). *Polymer* **13**, 579.
Eaton, W. C. and Keighley, J. H. (1969). *J. Text. Inst.* **60**, 1556.
Eaton, W. C. and Keighley, J. H. (1971). *Appl. Polym. Symp.* **18**, 263.
Edwards, M. L. (1961). *Nature* **190**, 1005.
Ghosh, D. K. and Whiffen, D. H. (1959). *Mol. Phys.* **2**, 285.
Gordy, W. (1959). *Rad. Res. Supp.* **1**, 491.
Henriksen, J. (1969). "Solid State Biophysics." McGraw Hill, Sevenoaks.

Ingram, D. J. E. (1958). "Free Radicals as Studied by Electron Spin Resonance." Butterworths, London.
Ingram, D. J. E. (1969). "Biological and Biochemical Applications of E.s.r." Hilger.
Keighley, J. H. (1968). *J. Text. Inst.* **59**, 470.
Keighley, J. H. (1971). *J. Text. Inst.* **62**, 511.
Keighley, J. H. and McKinley, M. I. Unpublished.
Kurita, Y. and Gordy, W. (1961). *J. Chem. Phys.* **34**, 282.
Mason, H. S. (1960). *Arch. Biochem. Biophys.* **86**, 225.
McCormick, G. and Gordy, W. (1958). *J. Phys. Chem.* **62**, 783.
Meybeck, A. and Windle, J. J. (1969a). *Rad. Res.* **40**, 263.
Meybeck, A. and Windle, J. J. (1969b). *Photochem. Photobiol.* **10**, 1.
Miyagawa, I. and Gordy W. (1960). *J. Chem. Phys.* **32**, 255.
Reisz, P. and White, F. W. (1966), *J. Amer. Chem. Soc.* **88**, 872.
Reisz, P. and White, F. W. (1967). *Nature* **216**, 1208.
Reisz, P. and White, F. W. (1967). *Rad. Res.* **32**, 744.
Reisz, P. and White, F. W. (1970). *Rad. Res.* **44**, 34.
Sinclair, J. W. and Hanna, M. W. (1967). *J. Chem. Phys.* **71**, 84.
Squires, T. L. (1963). "Introduction to Microwave Spectroscopy." Newness.
Swartz, H. M., Bolton, J. R. and Borg, D. C. (1972). "Biological Applications of Electron Spin Resonance." Wiley-Interscience, London.
Truby, F. (1962). *J. Chem. Phys.* **36**, 2227.
Weiner, R. F. and Koski, W. S. (1963). *J. Amer. Chem. Soc.* **85**, 873.

9. Mössbauer Spectroscopy
C. E. Johnson

9.1. Introduction

The Mössbauer Effect is the emission or resonant absorption of γ-rays by atomic nuclei in solids without energy loss or broadening due to the recoil of the nuclei. As a consequence, the radiation emitted by certain nuclei may be reabsorbed by identical nuclei. A new spectroscopic technique (sometimes called n.g.r.—nuclear gamma-ray spectroscopy) has therefore been opened up, where the spectrum results from the absorption of incident γ-radiation by nuclei in a specimen. The spectrum is observed by moving the source so that the γ-ray energy is shifted by the Doppler effect; individual lines result from transitions between different *nuclear* states in a way analogous to optical spectroscopy, in which the observed lines are due to transitions between different electronic states of an atom. The resulting spectrum (the number, energies and intensities of the lines) is affected by the electronic environment of the nucleus, i.e. by the electric and magnetic fields in the atom. Since it is sensitive to and characteristic of the binding of the atom in the solid, the spectrum provides a valuable tool for the study of the state of atoms in different chemical compounds. In recent years, n.g.r. has been applied extensively to measurements on biological molecules and these will form the subject of this chapter.

In general, the Mössbauer spectrum will show shifts and splittings of the lines due to three distinct and independent effects:

1. The *chemical shift* (also known by physicists as the isomer shift);
2. The *electric quadrupole splitting*;
3. The *magnetic hyperfine splitting*.

Each of these is useful in gaining an understanding of the environment of the Mössbauer nucleus, as will be described in detail in Section 9.5.

Like all measurements of its kind, Mössbauer spectroscopy is most useful in providing information on a biological molecule when it is combined with other spectroscopic measurements. It is closely related to the magnetic resonance techniques, n.m.r. (Chapter 7) and e.p.r. (Chapter 8), which also give essentially local information—the former about the environment of the resonant nuclear spins, the latter about electron spins. In principle, the Mössbauer technique is most closely related to n.m.r., since both are nuclear resonance phenomena, though the Mössbauer effect is not necessarily a magnetic resonance. The chemical shift in n.m.r. (Section 7.2.2.) is used in a similar way to the chemical shift in the Mössbauer spectrum, though their physical origins are different; in n.m.r., the shift is a magnetic effect, while in the Mössbauer effect it is electrostatic. N.m.r. in biochemistry is mainly used to determine the structures of protein molecules and may provide information on magnetic ions by measuring the field they produce at neighbouring protons. In practice, the Mössbauer effect gives the most complete information when it is observed for paramagnetic ions, and its use is then closely related to an e.p.r. measurement. In both these methods, direct measurements on the paramagnetic ion itself may be used (a) to study its chemical state and the type of bonding and (b) to obtain qualitative data on the local structure and symmetry in the neighbourhood of the ion.

In both n.g.r. and e.p.r. techniques, magnetic hyperfine interaction between the electronic and nuclear spins may be measured, and the two methods are complementary; in Mössbauer spectroscopy, the effective magnetic field at the nuclei due to the electrons is measured, while in e.s.r. the effective magnetic field at the electrons due to the nuclei is observed.

It should be emphasized that, for measuring chemical and solid-state effects, Mössbauer spectroscopy has neither the high sensitivity to a small number of atoms nor the great precision in measurement that n.m.r. and e.p.r. have. Its use lies mainly in the special kind of information it can give. (The absolute precision of the Mössbauer radiation, on the other hand, may be very high indeed—see Section 9.2.)

9.2. Resonant absorption of radiation

The problem of observing the resonant absorption of γ-rays by nuclei may be illustrated by considering first the analogous and better-known case of the resonant absorption of optical radiation by atoms. The well-known D lines of Na are emitted when an atom in an excited ^2P state decays to the ground ^2S state (Fig. 9.1a). (The splitting arises from spin-orbit coupling, which makes the energy of the $^2P_{\frac{3}{2}}$ state slightly different from that of the $^2P_{\frac{1}{2}}$ state.) If this radiation is passed through Na vapour, it can be resonantly absorbed by the Na atoms, which become excited resonantly from the ^2S to the ^2P state (Fig. 9.1b). The absorption process is just the reverse of emission and the energy required to excite the atoms is that of the incident radiation, since the same quantum levels are involved.

At first sight, it might appear that the same absorption process could occur for γ-radiation, with the energy levels and transitions in Fig. 9.1 (a) and (b) being those of nuclei rather than atoms. However, γ-rays are much more

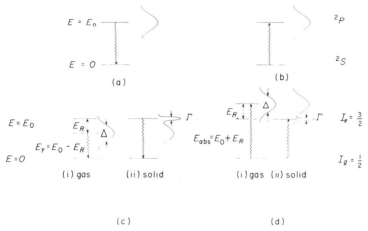

FIG. 9.1. Schematic diagram (not to scale) of energies and transitions involved in: (a) emission of light by excited Na atoms; (b) resonant absorption of Na light by Na vapour; (c) emission of γ-ray by excited Fe^{57} nuclei: (i) in a gas, (ii) in a solid; and (d) absorption of γ-ray by Fe^{57} nuclei: (i) in a gas, (ii) in a solid.

energetic (from a few kV to several million eV) and the effect of the recoil of the emitting and absorbing nucleus is not negligible, whereas the recoil for atoms is indeed negligible. Considering the nuclear case, the momentum p_R of the recoiling nucleus must be equal to the momentum $p = E_\gamma/c$ ($E_\gamma = \gamma$-ray energy, $c = $ velocity of light) of the γ-ray, by the conservation of momentum. The recoil energy of the nucleus is thus $E_R = p_R^2/2M = E_\gamma^2/2Mc^2$, where M is the mass of the nucleus. Since the recoil energy is small compared with the γ-ray energy, E_γ is approximately equal to E_0, the energy of the excited state above the ground state (i.e. the total energy available for the transition), so that $E_R \simeq E_0^2/2Mc^2$.

Therefore, the energy of the emitted radiation is less than the transition energy by E_R, since the recoil energy must be provided from the total energy E_0 available, i.e. $E = E_0 - E_R$ (Fig. 9.1c, i). Similarly, the energy required to excite the nucleus by the absorption of radiation would be $E_0 + E_R$ (Fig. 9.1d, i). Hence the energy of the emitted radiation is out of resonance with that required for absorption in an identical absorber by an amount of energy $2E_R$. The effect of the recoil depends upon whether the energy shift E_R is larger or smaller than the line-width Γ of the radiation.

The excited states have an inherent width or spread in energy $\Gamma = \hbar/\tau$ given by the Heisenberg uncertainty principle, where τ is lifetime of the excited state. The radiation is therefore broadened through the Doppler effect by the thermal motion of the atoms; the line is broadened by an amount $\Delta = 2\sqrt{(E_R \varepsilon)}$, where ε is the mean thermal energy of the atoms. For a gas at a temperature T, ε

$= \frac{1}{2}M\overline{v^2} \sim kT$, where $\overline{v^2}$ is the mean-square velocity and k is Boltzmann's constant.

Consider the effects of recoil for the atomic and the nuclear case. For atoms, e.g. Na, the radiation is of wavelength 600 nm (6000 Å), or energy E_0 of about 2 eV, so that the recoil energy E_R is about 10^{-10} eV. The thermal broadening is therefore 3×10^{-6} eV at room temperature. The lifetime of an excited atom is of the order 10^{-8} s, so that the inherent line-width Γ is about 10^{-7} eV. Hence the recoil energy and inherent width of the radiation are small compared to the thermal broadening, and the resonance absorption of atomic radiation is easily observed. However, for a 100 keV γ-ray emitted by a nucleus of mass 100, E_R is about 1 eV and $\Delta = 0.024$ eV. The lifetimes of low-energy nuclear states are comparable to those of atomic states, say 10^{-8} s or so, and thus the intrinsic width is very small ($\Gamma \sim 10^{-7}$ eV), and the ratio of the linewidth to the energy $\Gamma/E_0 \sim 10^{-11}$, i.e. the energy is defined extremely sharply. Thus the recoil energy shift E_R is much greater than the linewidth Γ, and resonant absorption of γ-rays by nuclei cannot normally be observed.

Two ways of observing nuclear resonance of γ-rays have been devised and successfully used by nuclear physicists. In one, the γ-ray source is mounted on a high-speed rotor and moved with velocities of 10^4 cm s^{-1} or more, so that the Doppler effect shifts the γ-ray energy $2E_R$ and compensates the energy loss due to the recoil of the nuclei. In the other, the source and absorber are kept stationary, and both are heated so that the line-width increases; in this way the overlap of the energies of the emitted and absorbed lines may become large enough for absorption to be observed.

It was while working on the second of these methods that R. L. Mössbauer (1958) discovered, that, under certain conditions, there is a finite probability that the recoil energy of the nuclei is *not* lost by the γ-rays. The energy profile of emitted γ-rays then looks like that shown in Fig. 9.1 (c) (ii); there is an unshifted and sharp line of width Γ superimposed on the shifted and broad distribution of width Δ. Similarly, if γ-rays are to be absorbed, their profile will look like that shown in Fig. 9.1 (d) (ii). Resonance absorption then occurs when a source of γ-rays and an absorber containing the ground state nuclei are used. Mössbauer discovered this for the 129 keV γ-ray in Ir^{191} and explained the result in terms of the binding of the nuclei in the solid state; for this work he was awarded the Nobel Prize in Physics in 1961.

9.3. The Mössbauer effect

Mössbauer discovered that, if the nuclei are tightly bound in a solid, the energy lost by the γ-rays when the nucleus recoils may be zero. In the previous description of the effect of recoil, it was assumed that the nuclei were free to recoil, as for example they would be in a gas where the recoil energy would appear as an additional kinetic energy of the emitting nucleus or atom. In a solid, however, it is not always possible for the nucleus to give up its recoil energy because of the quantized nature of the energy of the lattice vibrations (or phonons). If the recoil energy E_R is low compared with the binding energy of the solid, then the nucleus can only lose this energy by interaction with the crystal lattice, i.e. by exciting a phonon. Since energy can only be transferred to and from the lattice in discrete amounts, it will not always be possible to create a phonon if E_R is low; in this case, the γ-ray will not lose energy. In fact, since the whole solid will recoil, rather than just one nucleus, the energy loss will be $E^2/2M_{solid}c^2$ (where M_{solid} is the mass of the solid), which is completely negligible.

The probability f that the γ-ray will be emitted without energy loss may be expressed in terms of the amplitude of the lattice vibrations (assumed to be harmonic) by

$$f = \exp\left\{\frac{-4\pi^2 \langle x^2 \rangle}{\lambda^2}\right\}$$

where λ is the wavelength of the γ-rays and $\langle x^2 \rangle$ is the average value of the square of the amplitude of the atomic vibrations along the direction of emission. This probability is high when the energy of the γ-rays is low, for then λ is large, and for tightly bound solids at low temperatures, for then $\langle x^2 \rangle$ is small. At low temperatures, $\langle x^2 \rangle$ decreases and so f increases, but not indefinitely, since $\langle x^2 \rangle$ does not become zero at 0 K because of zero-point vibrations.

The conditions for the Mössbauer effect to be observable may be expressed conveniently by describing the solid in terms of the Debye model. In this model, the elastic properties of the solid are described in terms of the characteristic temperature θ_D, which is derived experimentally from low-temperature specific-heat measurements. The condition for observing the Mössbauer effect is that E_R should be of the same order of magnitude or less than both $k\theta_D$ and kT, which characterize the energy of the phonons.

When the condition $\lambda^2 \gg \langle x^2 \rangle$ or, which is the same thing, $E_R \ll k\theta_D$ and kT is satisfied, nuclear resonance of the γ-rays may be observed. For typical values of θ_D, γ-rays of energy of the order of 10 keV are required, though by making measurements of low temperatures this may be increased to 100 keV or more. So the Mössbauer effect is not a general property applicable to all nuclei, but is restricted to only those elements which have nuclei with low-energy γ-rays decaying to the ground state. The distribution of suitable nuclei is fairly random over the periodic table and is illustrated in Fig. 9.2. So far, over 40 nuclides have been discovered with low-energy γ-rays leading to the ground state which are suitable for Mössbauer-effect observations. Of these, Fe^{57} with a γ-ray energy of 14·4 keV is the easiest to observe and gives f-values close to one in suitable environments. E_R is only 2 meV, and $E_R/k = 20$ K, so that, for most solids, the Mössbauer effect is large even at room temperature. Since Fe is

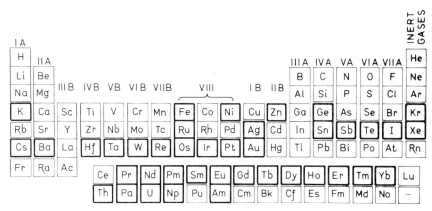

FIG. 9.2. The periodic table, with atoms in which the Mössbauer effect has been observed shown in heavy boxes.

a frequently occurring and important constituent of many biological molecules (haemoglobin, myoglobin, cytochromes, ferredoxins, etc.), this opens a useful field of study to supplement and complement other techniques; so far, most biological applications of the Mössbauer effect have utilized Fe^{57}. Usually, the source is kept at room temperature, but the absorber is in the form of a solution of the molecules which is made solid by cooling to low temperatures.

However, some other isotopes seem promising for biological work and the next few years will probably see more measurements using them, although it will usually be necessary to cool both source (to increase the f-factor) and absorber to low temperatures in order to obtain a spectrum. Perhaps the most promising of these isotopes are I^{127} and I^{129}.

The Mössbauer effect does not merely allow the observation of nuclear resonant absorption of γ-rays; it also involves the emission and absorption of γ-rays without interaction with the lattice. Therefore, for Mössbauer transitions, not only is there no recoil-energy loss E_R transferred to the lattice, but there is no broadening of the energy of the radiation by the phonons. The energy distribution of the γ-rays for a solid is shown in Fig. 9.1 (c) (ii) and Fig. 9.1 (d) (i) for emission and absorption, respectively. The natural or inherent linewidth Γ of the γ-ray is observed in the Mössbauer spectrum; this is of the order of 10^{-7} eV—much smaller than the thermal Doppler width $\Delta (\Delta \sim 10^{-2}$ eV) observed in the "conventional" γ-ray nuclear resonance kind of measurement. It is this narrow linewidth which makes the Mössbauer spectrum such a powerful source of information in solid-state physics, chemistry and biochemistry, since the small changes in energy which are associated with differences in chemical binding and the various electric and magnetic fields in solids can now be detected. These give rise to energy shifts and splittings of typically 10^{-6} eV; their effects on the spectrum and the information which may be deduced from them will be discussed in Section 9.5, following a description of the apparatus used to measure Mössbauer spectra.

9.4. Experimental apparatus

The nuclear resonant absorption is detected by varying the energy of the γ-rays using the Doppler effect and measuring the intensity of the transmitted radiation as a function of energy. Several commercial firms make Mössbauer spectrometers capable of the stability and precision necessary to carry out measurements on biological molecules. A typical experimental set-up is shown in Fig. 9.3. A mono-energetic source of radioactive Co^{57} (half-life 270 days) embedded in a palladium matrix is mounted on a transducer, which is driven with constant acceleration. If v is the velocity at any instant, the energy shift ΔE given to the γ-rays is

$$\Delta E = \frac{v}{c} E_0$$

If the source and absorber contain the Mössbauer isotope in the same chemical and physical state, nuclear absorption of the γ-rays will occur when the source and absorber are at rest relative to each other. When the source is moved, the resonance is destroyed and so the transmitted intensity of the γ-rays increases. If the counting rate is plotted against the velocity, the resonance line is traced out. For Fe^{57}, the half-life τ of the 14·4 keV state is $1\cdot 4 \times 10^{-7}$ s; the half-width of the resonance corresponds to a velocity of $0\cdot 2$ mm s^{-1}. To record the complete spectrum of a magnetic Fe compound requires velocities of up to 10 mm s^{-1}, i.e. much more modest velocities than the 10^4 cm s^{-1} required to compensate for the recoil shift in the "conventional" nuclear

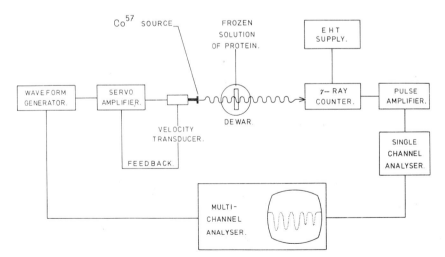

FIG. 9.3. Experimental arrangement of a Mössbauer spectrometer.

resonance measurements mentioned in Section 9.1. The waveform generator also provides pulses to advance the address of a multi-channel analyser operating in the multiscaler mode so that the spectrum may be recorded conveniently and automatically. A pick-up coil attached to the transducer provides feedback to keep the velocity linear. The absorber consists of a frozen aqueous solution of the biological molecules. In this form, a sample thickness of about 5 mm is suitable; if it is much thicker, the γ-ray will be too strongly absorbed by electronic processes. After transmission through the absorber, the γ-rays enter a counter and a single-channel analyser selects the 14·4 keV radiation and accumulates them in the multiscaler. The Mössbauer spectrum then appears on the oscilloscope display of the output from the analyser. Counting is continued until the spectrum gives the required detail. It is usually necessary to enrich the biological molecule in Fe^{57}, which is normally present in Fe to an abundance of 2·2%. If the depth of the Mössbauer absorption is 1%, it is necessary to accumulate 10^6 counts per channel to achieve a statistical accuracy of 0·1%; for strong sources (100 millicuries) and 256 channels, this requires measuring times of the order of a day.

9.5. The Mössbauer spectrum

The energy shifts and splittings in the Mössbauer spectrum give information about the electric and magnetic fields at the nuclei produced by their environment. Hence the Mössbauer spectrum may be used as a way of studying the electronic state of the atom containing the Mössbauer nucleus and also of acquiring some information about its immediate environment. The spectrum may also be used to identify a particular chemical species in a sample; through this, it may develop into a valuable tool for studying biochemical reactions.

As mentioned in Section 9.1., the Mössbauer spectrum may yield measurements of three independent properties (chemical shift, quadrupole splittings and magnetic splitting); the effects they produce on the Mössbauer spectrum, illustrated in Fig. 9.4, will now be described.

9.5.1. THE CHEMICAL (OR ISOMER) SHIFT

The energy of a γ-ray emitted from a nucleus varies with the chemical state (i.e. the oxidation state and the degree of covalency) of the atom. This gives rise to the *chemical shift*, which is the shift in the centroid of the Mössbauer spectrum of an absorber relative to that of a standard absorber measured under standard conditions (for Fe^{57}, it is usual to take the centre of the spectrum of metallic Fe at room temperature as the zero of the velocity scale). Since there are also shifts in energy due to changes in temperature and pressure (Josephson, 1960; Pound and Rebka 1960), it is important to be able to separate these from the experimentally determined shift in order to be able to derive meaningful chemical information from the data. The change in energy

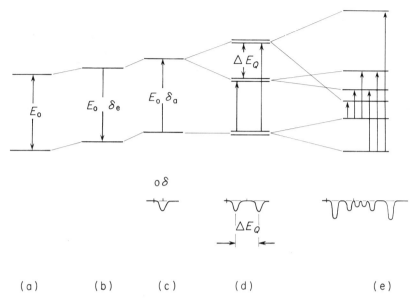

FIG. 9.4. Effect of chemical bonding on the nuclear energy levels of Fe^{57} and the resulting Mössbauer spectrum: (a) free nucleus; (b) emitting nucleus bound in a solid; (c) absorbing nucleus bound in a non-magnetic solid with cubic symmetry and lower electron density at nucleus than emitter; (d) non-magnetic absorber without cubic symmetry; and (e) magnetic absorber without cubic symmetry.

due to difference in chemical binding from compound to compound is small ($\sim 10^{-7}$ eV) but the high-energy resolution of the Mössbauer spectrum enables it to be detected. The first observation of the chemical shift was made by Kistner and Sunyar (1961) for Fe_2O_3.

The chemical shift is due to the electrostatic interaction between the electronic- and nuclear-charge densities in the atom. The electronic-charge density $\rho(r)$ may be written in terms of the electron wave function $\psi(r)$ as $\rho(r) = -e|\psi(r)|^2$, where $-e$ is the charge on the electron. When the atom is bound in a solid, the wave functions become modified and the density of electrons at the nucleus changes. Since the electrostatic potential V obeys Poisson's equation, $\nabla^2 V = -\rho(r)/\varepsilon_0$ (where ε_0 = permittivity of free space), the energy is increased compared to that of a bare nucleus. Only s-electrons have a finite density at the nucleus; if this is assumed to be a constant $|\psi(0)|^2$ over the radius R of the nucleus, the energy of the nucleus will be raised by $\dfrac{Ze^2}{6\varepsilon_0}|\psi(0)|^2 \langle R^2 \rangle$, where Z is the atomic number and $\langle R^2 \rangle = \int r^2 \rho_n \, dr$ is the mean-square radius of the nucleus (ρ_n is the nuclear-charge density).

The shift in the γ-ray energy arises because the nucleus has a different radius (R_e) in the excited (or isomeric—hence "isomer shift"—not to be confused with isomeric states of molecules) state from that (R_g) of the ground state. Accordingly, the overlap between the nuclear-charge density and the electron-charge density is different for the initial and final states of the γ-ray transition and the chemical shift is

$$\delta = \frac{Ze^2}{6\varepsilon_0} \{|\psi_a(0)|^2 - |\psi_e(0)|^2\} \{\langle R_e^2 \rangle - \langle R_g^2 \rangle\}$$

where, the subscripts a and e denote the wave functions for the absorber and emitter, respectively. If the source and absorber are in the same chemical state (i.e. $|\psi_a(0)|^2 = |\psi_e(0)|^2$), then the centre of the spectrum will occur at zero velocity. If, however, the absorber is in a different chemical state, it will have a chemical shift δ relative to the source, and the centre of the spectrum will occur at a velocity v different from zero. It is found from experiment that for Fe^{57}, $R_e < R_g$, so that an increase in electron density $|\psi(0)|^2$ produces a negative chemical shift.

For ionic Fe compounds, the chemical shift enables the oxidation state to be determined easily, since $\delta = 1.5$ mm s^{-1} for Fe^{2+} and $\delta = 0.5$ mm s^{-1} for Fe^{3+}, relative to metallic Fe. If there is strong covalent bonding (as in most biological molecules), δ decreases, and so its sensitivity as a method for measuring the oxidation state of Fe may be reduced. Table 9.1 shows some experimentally observed chemical shifts for some inorganic salts of Fe. The effect of covalency is seen from the decrease in the shift as H_2O molecules are replaced by Cl^- ions. The effect of co-ordination number is seen by comparing salts with tetrehedral co-ordination with those with octahedral co-ordination;

TABLE 9.1. Effect of ligands and co-ordination number on chemical shift

Ion	Co-ordination	Complex	Ligands	Chemical shift (mm/s)
Fe^{2+} (high spin)	octahedral	$FeSiF_6 6H_2O$	$6H_2O$	1.42
		$FeCl_2.4H_2O$	$4H_2O, 2Cl^-$	1.34
		$FeCl_2.2H_2O$	$2H_2O, 4Cl^-$	1.24
		$FeCl_2$	$6Cl^-$	1.20
	tetrahedral	$(NH_4)_2FeCl_4$	$4Cl^-$	1.05
Fe^{2+} (low spin)	octahedral	$K_4Fe(CN)_6$	$6CN^-$	0.06
Fe^{3+} high spin	octahedral	$FeCl_3$	$6Cl^-$	0.53
	tetrahedral	$(NH_4)FeCl_4$	$4Cl^-$	0.30
Fe^{3+} (low spin)	octahedral	$K_3Fe(CN)_6$	$6CN^-$	0.03

Shifts measured at 77 K, relative to α-iron at 290 K.

the latter have a smaller electron density at the nucleus. Table 9.2 shows the chemical shifts for Fe in some reduced HIPIP proteins; (high-potential iron protein) lies in the range of that for ferric salts. In these cases, the quadrupole splitting and, especially, the magnetic hyperfine splitting may often be used to decide between any ambiguities in the assignment of the chemical state. (Mössbauer chemical shifts are of course quite different from n.m.r. chemical shifts (Chapter 7).)

TABLE 9.2. Chemical shifts of Fe^{57} in reduced proteins

	Ligands	Chemical shift (mm s^{-1})
haemoglobin (de-oxygenated)	5N	0·90
rubredoxin	4S	0·60
ferredoxin	4S	0·56
HIPIP	4S	0·42
haemoglobin (oxygenated)	O_2, 5N	0·20

Shifts measured at 77 K relative to metallic Fe at 290 K.

9.5.2. THE QUADRUPOLE SPLITTING

The charge distribution in a nucleus with spin $I = 0$ or $1/2$ is spherically symmetrical but, in a nucleus with spin of 1 or greater, it will be non-spherical and the nucleus is said to have an electric quadrupole moment eQ. Q is given by $\int_{nucleus} (3z^2 - r^2) \rho_n dv$, and is positive if the nucleus is elongated along one axis like a rugby football, or negative if it is squashed like a pumpkin. The quadrupole is not aligned by (i.e. its energy is not changed by) an electric field directly, but it is aligned if the electric field has a gradient at the nucleus. If the environment of an atom has perfect cubic symmetry, there will be no gradient of the electric field at the nucleus. In most molecules, there will not be cubic symmetry and the resulting non-cubic ligand field at an atom will produce a gradient of the electric field given by

$$q = \frac{1}{e}\frac{\partial^2 V}{\partial z^2} = \frac{1}{4\pi\varepsilon_0} \int_{electrons} \frac{3z^2 - r^2}{r^5} \rho(r) dv$$

$$= \frac{1}{4\pi\varepsilon_0} \langle 3\cos^2\theta - 1 \rangle \langle r^{-3} \rangle$$

where z is the axis of the distortion; the nuclear levels will be split according to

$$\mathscr{H} = \tfrac{1}{4} e^2 Qq \left[I_z^2 - \tfrac{1}{3}I(I+1)\right]$$

and so the Mössbauer spectrum will be split.

For the Fe^{57} Mössbauer transition, the excited state has a spin of 3/2 and a positive quadrupole moment; hence the energy level will be split into two by the electric-field gradient. The ground state has a spin of 1/2 and so it has no quadrupole moment and the Mössbauer spectrum is split into two lines; the splitting $\Delta E_Q = \frac{1}{2}e^2 Qq$ is known as the quadrupole splitting.

The quadrupole splitting measures the departure of the environment of the Mössbauer nucleus from cubic symmetry. If the atom is surrounded by six ligands arranged in an octahedron with a small tetragonal distortion caused by pushing in the two ligand atoms along one axis (say the z axis) towards the central atom, then their negative charge will make the positive quadrupole moment align with the spin along the z axis, i.e. the nuclear state with $I_z = \pm 3/2$ will lie lowest, the sign of the electric-field gradient being negative. Thus, a determination of the sign of q enables the sense (elongation or compression) of the distortion to be established, the quadrupole splitting giving qualitative information about the local structure in the region of the Mössbauer nucleus.

As well as this kind of crystallographic data, the quadrupole splitting can be used to infer the oxidation state of Fe in biological compounds. For example, Fe^{3+} has a half-filled shell of $3d$ electrons and is in a ^6S state, i.e. the orbital angular momentum L is zero, the electrons are distributed in a sphere and the electric-field gradient is zero. While small admixtures of other states may be introduced by the chemical bonding in a molecule, in general the quadrupole splitting in Fe^{3+} ions is small, usually between 0 and 0·6 mm s^{-1}. The Fe^{2+} ion has six $3d$ electrons, and the extra electron outside the $(3d)^5$ core behaves like a single $3d$ electron; it is in a highly anisotropic orbital and gives rise to a large electric quadrupole splitting of 3 mm s^{-1} or more. A measurement of the quadrupole splitting ΔE_Q may, therefore, enable the oxidation state to be identified. Furthermore, a measurement of the sign of q enables the ground state orbital to be determined; if the quadrupole splitting is measured as a function of the temperature, the energies of the excited orbital states may also be estimated. Hence information on the splitting of the ionic levels by the ligand field may be deduced, which could be a valuable complement to magnetic hyperfine data.

Information about more complex structures represented by non-axial (orthorhombic, triclinic, monoclinic, etc.) distortions can in principal be deduced from measurements on single crystals of the electric-quadrupole interaction tensor.

9.5.3. MAGNETIC HYPERFINE STRUCTURE

Atoms or ions with an incomplete shell of electrons have a magnetic moment which can interact with the nuclear magnetic moment. In solids or molecules, most of the outer electrons of atoms are paired off in chemical bonding orbitals, so that magnetic properties are found only if the atom has an inner electron shell which is unfilled. This occurs in the $3d$ shell (e.g. for Fe) and for the $4f$ shell (rare-earth elements). The magnetic hyperfine interaction splits the

Mössbauer spectrum into several lines (from 6 to 23, or even more for Fe^{57}) spread over large energies (up to nearly 20 mm s^{-1} for Fe^{57}).

This splitting occurs only if there is a non-zero time average of the interaction. This can arise if the Fe is in a magnetic state (ferromagnetic, ferrimagnetic or antiferromagnetic), or is magnetized in an external magnetic field, or if the electron spin relaxation times are long. In general, the resulting spectra are complex, but simpler cases may occur if there is a well-defined axis of quantization in the atom, for example if there is magnetic order or if an external magnetic field is applied. The hyperfine interaction can then be represented as an effective magnetic field B_{eff} at the nucleus. This splits the excited state ($I = 3/2$) into four levels and the ground state ($I = 1/2$) into two; the resulting Mössbauer spectrum shows eight transitions which, in systems with a high enough degree of symmetry (e.g. metallic Fe), is reduced to six because of selection rules.

The effective field at the nucleus is made up of the sum of several contributions from interactions with different electrons in the atom; the three main contributions (Marshall and Johnson, 1962) will be discussed in turn.

9.5.3.1. *Core polarization*
This field, due to s-electrons which have a finite density at the nucleus and which are polarized by exchange interaction with the 3d electrons, is isotropic and is proportional to the spin (or magnetic moment) of the atom. It may be written

$$B_s = \frac{\mu_0}{4\pi} 2\beta \langle r^{-3} \rangle_{3d} \kappa \langle S \rangle$$

κ is a constant representing the core polarization, $\langle r^{-3} \rangle_{3d}$ is the mean value of the inverse cube of the radius of the 3d electron shell, β is the Bohr magneton ($=0.927 \times 10^{-23}$ JT^{-1}) and $\mu_0 = 4\pi \times 10^{-7}$ is the permeability of free space. It is not implied that the 3d electrons directly produce this field; the factor $\langle r^{-3} \rangle_{3d}$ comes into the other contributions which are from the d-electrons and it is convenient to express B_s in terms of it as well. When this is done, it is an experimental fact that the parameter κ is a constant of about 0·3 for the ions of the 3d group.

9.5.3.2. *Orbital field*
This is due to the orbital motion of the 3d electrons, which behaves like an electric current circulating round the nucleus. This produces a field

$$B_L = \frac{\mu_0}{4\pi} 2\beta \langle r^{-3} \rangle_{3d} \langle L \rangle$$

where L is the orbital angular momentum of the atom. In relating measurements to other magnetic data (e.g. e.p.r. or magnetic susceptibility), it is convenient to write this field in terms of the g-factor, i.e.

$$B_L = \frac{\mu_0}{4\pi} 2\beta \langle r^{-3} \rangle_{3d}(g-2) \langle S \rangle$$

9.5.3.3. *Dipolar field*

The $3d$ electrons have, in addition to charge, a spin magnetic moment, which interacts with the nuclei to give a dipolar field of the form

$$B_d = \frac{\mu_0}{4\pi} 2\beta \langle r^{-3} \rangle_{3d} \langle 3(S.\hat{r})\hat{r} - S \rangle_{3d}$$

(where $\hat{r}$ is the unitvector in the radial direction, r.

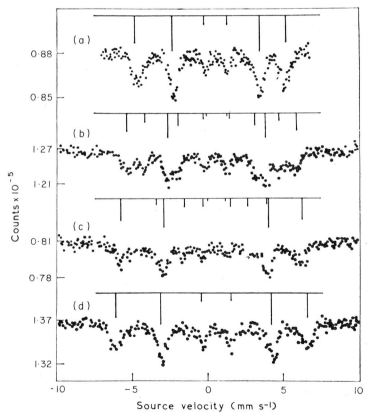

FIG. 9.5. Mössbauer spectra of haemoglobin fluoride (high-spin Fe^{3+}) at 4·2 K in applied magnetic fields of (a) 0·05 T; (b) 0·75 T; (c) 1·5 T; and (d) 3 T. Each spectrum is made up of almost symmetrical six-line patterns (Lang, 1967).

For a single electron or for atoms where all the electron spins are parallel (e.g. high spin Fe^{2+} or Fe^{3+}), this may be expressed as

$$B_{dz} = \frac{\mu_0}{4\pi} 2\beta \langle r^{-3} \rangle_{3d} \langle 3\cos^2\theta - 1 \rangle_{3d} \langle S_z \rangle$$

The orbital and dipolar fields are real magnetic fields (whereas the core polarization is a quantum-mechanical effect) and they are anisotropic. For the Fe^{3+} ion, there is no orbital moment (i.e. $L = 0$), so that $B_L = 0$, and, since the electron spin moments are spherically symmetrical, $\langle 3\cos^2\theta - 1 \rangle_{3d} = 0$ and $B_d = 0$, if covalent bonding effects are neglected. To this approximation, therefore, the hyperfine field arises only from core polarization. Experimentally, the hyperfine field in Fe^{3+} salts is found to decrease with increasing covalency of the ligands as Table 9.3 shows. From this table, B_s may be estimated for any molecule where Fe has any of these ligands, using the fact that the field is proportional to the spin S.

For the Fe^{2+} ion, $S = 2$ and B_s may be estimated to be 0·8 times the

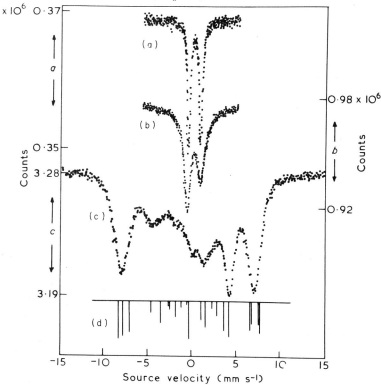

FIG. 9.6. Mössbauer spectra of haemoglobin cyanide (low-spin Fe^{3+}) at temperatures of (a) 195 K; (b) 77 K and (c) 4·2 K. Note the broadening of the spectrum as the temperature is lowered, and the low-temperature asymmetrical pattern, (d) gives the theoretical pattern.

TABLE 9.3. Effect of ligands on hyperfine fields (B_n in T) in Fe^{3+} ions

	Ligands	B_n
FeF_3	$6F^-$	62·2
$Fe_2(SO_4)_3(NH_4)_2 24H_2O$	$6H_2O$	58·4
Fe_2O_3	$6O^{2-}$	54·0
$FeCl_3$	$6Cl^-$	48·7
$(FeCl_4)^-$	$4Cl^-$	47·0
Fe^{3+} tris-dtc	$6S^{2-}$	46·0

corresponding Fe^{3+} field; B_L may be estimated from the g-value, and B_d from the quadrupole splitting, since $\langle 3\cos^2\theta - 1\rangle_{3d}$ also enters the expression for the electric-field gradient q and it follows that $B_d = \frac{1}{c^2}\beta q$, or $\pm \frac{4}{7}\frac{\beta}{c^2}\langle r^{-3}\rangle_{3d}$ for a pure orbital ground state. Given the g-values and the quadrupole splitting (including the sign of q), the hyperfine field in Fe^{2+} ions can be understood.

The oxidation state and the spin state of Fe may be most clearly distinguished from hyperfine field measurements. In high-spin Fe^{3+} ions, the hyperfine interaction is due to the core term only and is isotropic; a sharp six-line pattern is, therefore, observed in the Mössbauer spectrum measured in an external magnetic field, even in a polycrystalline specimen where the molecular

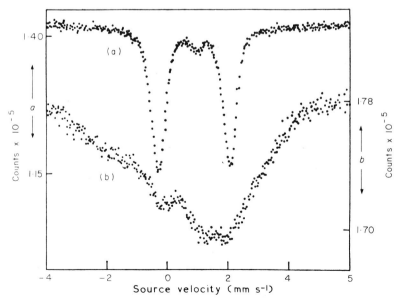

FIG. 9.7. Mössbauer spectra of reduced haemoglobin (high-spin Fe^{2+}) at 4·2 K in applied magnetic fields of (a) zero and (b) 3 T (Lang, 1967).

axes are randomly oriented relative to the magnetic field. Figure 9.5 shows some spectra of haemoglobin fluoride; the size of the hyperfine field varies from about 60 to 40 T according to the covalency of the ligands (see Table 9.1). Low-spin Fe^{3+} has a single $3d$ hole in a t_{2g} orbital state, so there will be orbital and dipolar fields. The hyperfine interaction will, therefore, be anisotropic and the spectrum of randomly oriented molecules in external fields will show broad lines with a markedly different pattern from those of high-spin Fe^{3+}; Fig. 9.6 shows spectra of haemoglobin cyanide.

High-spin Fe^{2+} ions show a large chemical shift and quadrupole splitting. In general, the hyperfine spectrum of randomly oriented molecules will have broad lines, but sometimes, because of the anisotropic susceptibility, sharp lines are observed. This combination of a large magnetic splitting with a large quadrupole splitting enables the Fe^{2+} ion to be recognized; Fig. 9.7 illustrates these features with the Mössbauer spectra of reduced haemoglobin. Low-spin

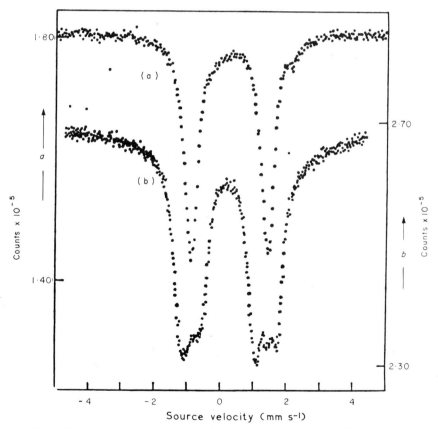

FIG. 9.8. Mössbauer spectra of oxygenated haemoglobin (low-spin Fe^{2+}) at 4·2 K (a) in zero applied field and (b) in an applied magnetic field of 3 T (Lang, 1967).

Fe^{2+} is easily diagnosed from its spectrum in a magnetic field; as $S = 0$, there is no magnetic hyperfine interaction, so that the effective field at the nucleus is just the external field applied and only a small splitting is observed. This is seen from Fig. 9.8, which shows the Mössbauer spectrum of oxygenated haemoglobin.

Thus the observation of magnetic hyperfine interaction in the Mössbauer spectra of Fe^{57} in biological molecules enables the oxidation and spin state of the Fe to be determined, and may further provide valuable data about its environment.

9.6. Applications to biology

Several kinds of applications of the Mössbauer effect to problems in biology have already been made. The most common and useful at present is the study of the electronic state of an atom in a biological molecule. The Mössbauer spectrum provides a powerful and detailed probe of the electronic state and local structure in the neighbourhood of the Mössbauer nucleus. The spectra can be observed and interpreted even when the complete structure of the molecule as a whole is unknown. Since the spectra are most easy to observe for the Fe isotope Fe^{57}, the technique is especially valuable for the study of Fe in biological molecules. Examples of the use for the study of haemoglobin compounds have already been mentioned (Figs 9.5–9.8). It frequently happens that the Fe atom is at the active centre of biological activity in the molecule, so that the information may help the understanding of the biochemical processes in which the Fe takes part. For instance, the change in the Mössbauer spectrum before and after a biochemical reaction may yield data on the electron transfer which takes place. Although the restriction of the Mössbauer effect to certain nuclei might seem to be a severe limitation of the method, many fruitful applications are possible (see from the following list of the most common Fe-containing biological systems):

Proteins which bond O_2	Enzymes	Proteins which store Fe
haemoglobin	haem enzymes	ferritin
myoglobin	Fe-S proteins	haemosiderin
haemerythrin	oxygenases	transferrins
		siderichromes

Since each of these types of system contains a large number of molecular species, it is clear that, even with Fe alone, there is no shortage of possible interesting measurements.

The Mössbauer spectrum may also be used as a "fingerprint" to identify the molecule in which the Mössbauer nucleus is bound. Observation of the spectrum is not limited to paramagnetic ions (as is e.p.r.). As more and more proteins are studied and their spectra become well characterized, n.g.r. spectroscopy could become a good method of analysing the products of a biochemical reaction and of following the course of the reaction in detail, providing some method is used (e.g. freezing) to arrest the progress of the reaction at various stages in a controlled way.

A further and ingenious use of the Mössbauer effect is as a detector of small velocities. This exploits the method of obtaining the Mössbauer spectrum by moving the source and absorber relative to each other with velocities of less than 1 mm s^{-1}. If the source or absorber has an additional small velocity superimposed on it, the Mössbauer line will broaden and the extra velocity may be deduced from the line-width. Applications to the measurement of the motion of the inner ear and to the breathing of ants have been successful. Of course, it is essential that the velocities which are to be measured are neither very small nor very large compared with the natural line-width of the Mössbauer line.

9.7. Measurements on biological molecules

Since the first measurements of the application of Mössbauer spectroscopy to biology on Fe in haemoglobin, reported in 1961 at the first international conference on the Mössbauer effect, a large volume of work has been done. It is not the object of this chapter to give a comprehensive review of the field; instead, it summarizes the main areas of work and gives some references where a fuller review may be found. These main areas are the haem proteins, the Fe-S proteins, haemerythrin and the Fe-storage systems.

9.7.1. HAEM PROTEINS

In these, the Fe has a fixed and characteristic environment known as the haem group (Sections 8.6.1 and 10.2.4); this is a relatively small molecule (Fig. 8.41) in which the Fe is co-ordinated to four N ligands and may also be joined to two further ligands lying above and below the plane. In haemoglobin and its derivatives, a protein molecule (a chain of amino acids) is attached via a histidine group to one of the non-planar ligands. The chemical state of the Fe atom is determined by the nature of the sixth ligand. In healthy blood, the Fe is ferrous; it has low-spin ($S = 0$) when oxygenated, and high-spin ($S = 2$) when reduced. Abnormal blood may contain ferric iron, which may have high-spin ($S = 5/2$), as in methaemoglobin (where the sixth ligand is a water molecule) and haemoglobin fluoride, or low-spin ($S = 1/2$) in haemoglobin cyanide. These spin states, originally determined from magnetic-susceptibility measurements, were later confirmed by e.p.r. measurements, which also enabled the orientation of the four haem planes in the molecule to be established. Haemoglobin and myoglobin were the first large biological polymers whose structure was completely determined by X-ray diffraction; this confirmed the orientation of the haem planes deduced from e.p.r.

measurement (p. 263). The Mössbauer spectra (Figs. 9.5–9.8) support the spin assignments and provide distinctive fingerprints so that measurements on the equilibrium between the spin states may be made. Since so much data was available on the haem proteins, it was natural that the first detailed studies using the Mössbauer effect should be carried out on them by Lang and Marshall (1966a, b). The initial measurements were carried out on haemoglobin and its derivatives obtained from rats which were fed on a diet containing enriched Fe^{57}. Lang (1967, 1970) has subsequently extended the work to include myoglobin (which stores and transports O_2 in muscles) and haem enzymes (cytochromes, cytochrome-c peroxidase). Haem proteins are also discussed in Section 10.2.4.1.

Because of the high degree of covalency in the bonds, the chemical shift and quadrupole splitting do not provide unambiguous information about the state of the Fe atoms, but the complicated magnetic interaction observed in the spectra enabled very detailed information to be deduced on the ligand field acting on the Fe. The experimental and theoretical techniques required to tackle and interpret the problem of finding out the electronic structure of the Fe in proteins from Mössbauer spectra has been developed mainly by Lang (1970).

9.7.2. Fe-S PROTEINS

These are proteins which generally contain an even number ($2n$) of Fe atoms which are believed to be co-ordinated to an equal number of labile (i.e. easily removed by chemical means) S atoms, and which may reversibly transfer n electrons (i.e. one per two Fe atoms) when oxidized or reduced (see also Section 10.2.4.2). The Fe-S proteins are of wide origin, occurring in plants, animals and bacteria; they have differing functions, e.g. photosynthesis, N_2 fixation, hydroxylation, digestion, etc. and they vary in molecular weight from 6000 to 300 000. They were originally discovered and recognized as belonging to a class of enzymes because they all showed an unusual e.p.r. signal with $g = 1.94$ when reduced, while no e.p.r. signal was observable in the oxidized state. The structure of one of these proteins, ferredoxin from the bacterium *Clostridium pasteruianum*, has been determined.

The Mössbauer effect has proved valuable in studying these proteins because spectra can be obtained in the oxidized (non-magnetic) state as well as in the reduced (magnetic) state. When oxidized, both Fe atoms were found to be practically in the same state but, after reduction, two spectra (Fe^{2+} and Fe^{3+}) were observed (Fig. 9.9), showing that only one of the Fe atoms had been reduced. As with haem proteins, more detailed information was obtained from magnetic hyperfine splitting; this was first observed for reduced samples of the flavoprotein xanthine oxidase, which has eight Fe atoms in a molecule of weight 300 000. Since then, work in several laboratories has been concentrated on the simpler 2-Fe 2-S proteins, for example the plant ferredoxins and

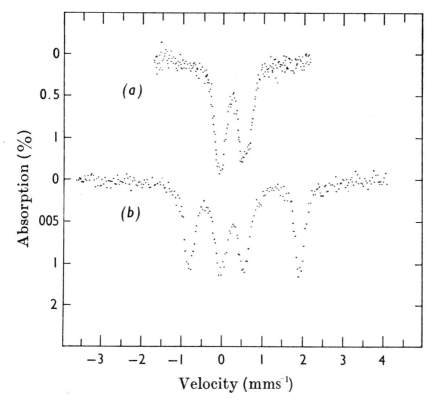

FIG. 9.9. Mössbauer spectra at 77 K of ferredoxin from *Scenedesmus*: (a) oxidized; and (b) reduced (Rao et al., 1971).

hydroxylases (putidaredoxin and adrenodoxin). All of these gave similar, though not identical, hyperfine spectra. The two Fe atoms were shown to be antiferromagnetically coupled together in the reduced state.

Other proteins containing Fe and S which have been studied included rubredoxin (one Fe and no labile S's per molecule), and the high-potential Fe protein (HIPIP) from *Chromatium*. General reviews of the Fe-S proteins (including their Mössbauer spectra) have been given by Tsibris and Woody (1970) and by Palmer (1973); see also p. 304.

9.7.3. HAEMERYTHRIN

This protein is the non-haem Fe protein responsible for O_2 transport in four different invertebrate phyla, including the sipuncilid worm, *Golfingia gouldii*. The molecule has a molecular weight of 108 000 and consists of eight sub-units, each containing two Fe atoms; when oxygenated, each sub-unit binds one molecule of O_2. Measurements have been made on deoxy, oxy and several oxidized (met) states. It was necessary to perform cyclic oxygenation and de-

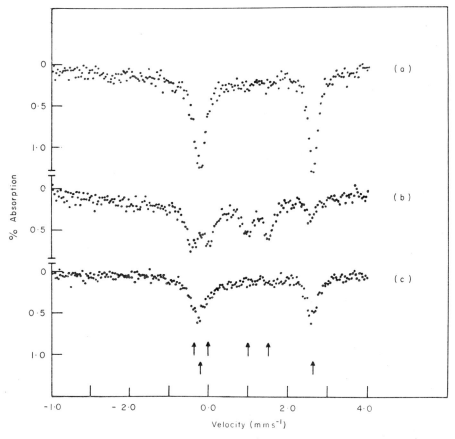

FIG. 9.10. Mössbauer spectra of haemerythrin (a) de-oxygenated (deoxyhemerythrin); (b) oxygenated (oxyhemerythrin), mixed with some de-oxygenated; and (c) after a further de-oxygenation (Garbett *et al.*, 1971).

oxygenation experiments to distinguish between oxidized and oxygenated forms (Garbett *et al.*, 1971); some spectra are shown in Fig. 9.10. De-oxyhaemerythrin gave the Mössbauer spectrum of a high-spin Fe^{2+} compound. Oxyhaemerythrin gave a Mössbauer spectrum consisting of two doublets of equal intensity, showing that the Fe atoms have different environments. Measurements in a magnetic field show that the Fe atoms are antiferromagnetically coupled together so that each sub unit is non-magnetic. The methaemerythrin complexes also were shown to be non-magnetic from the absence of hyperfine interaction in their Mössbauer spectra measured at low temperatures (4·2 K) and in large magnetic fields (3 T). These compounds appear to contain pairs of antiferromagnetically-coupled Fe^{3+} atoms in closely similar environments. By comparing the postulated structures of oxyhaemerythrin and de-oxyhaemerythrin, a possible mechanism for kinetic behaviour for the oxygenation equilibrium has been suggested.

9.7.4. IRON-STORAGE PROTEINS

Some work has been done on Fe-storage proteins, which generally contain ferric iron in an inorganic state. This includes ferritin (Boas and Window, 1966), haemosiderin (Johnson, 1971) and ferrichrome.

9.8. Conclusion—future prospect

The Mössbauer effect has proved a valuable addition to other well-established spectroscopic techniques for studying large biological molecules. For Fe-containing proteins, information on the state of the Fe atoms has been obtained, both when the Fe is in paramagnetic and diamagnetic states, and the interpretation of the spectra does not depend upon a knowledge of the molecular structure. Comprehensive reviews have been made for haem proteins by Lang (1970) and for the Fe-S proteins by Tsibris and Woody (1970). In the next few years a steady increase in this kind of work is to be expected, and an extension and development of the technique is likely to enable reactions of enzymes to be studied in detail.

References

Boas, J. F. and Window, B. (1966). *Aust. J. Phys.* **19,** 573.
Garbett, K., Johnson, C. E., Klotz, I. M., Okamura, M. Y. and Williams, R. J. P. (1971). *Arch. Biochem. Biophys.* **142,** 574.
Johnson, C. E. (1971). *J. Appl. Phys.* **42,** 1325.
Josephson, B. D. (1960). *Phys. Rev. Lett.* **4,** 341.
Kistner, O. C. and Sunyar, A. W. (1961). *Phys. Rev. Lett.* **4,** 212.
Lang, G. and Marshall, W. (1966a). *Proc. Phys. Soc.* **87,** 3.
Lang, G. and Marshall, W. (1966b). *J. Mol. Biol.* **18,** 385.
Lang, G. (1967). *J. Appl. Phys.* **38,** 915.
Lang, G. (1970). *Quart. Rev. Biophys.* **3,** 1.
Marshall, W. and Johnson, C. E. (1962). *J. Phys. Rad.* **23,** 733.
Mössbauer, R. L. (1958). *Z. Phys.* **151,** 124.
Palmer, G. (1973). *In* "Iron-Sulphur Proteins" (W. Lovenberg, ed.), Vol. 2, p. 285. Academic Press, New York and London.
Pound, R. V. and Rebka, G. A. (1960). *Phys. Rev. Lett.* **4,** 274.
Rao, K. K., Cammack, R., Hall, D. O. and Johnson, C. E. (1971). *Biochem. J.* **122,** 257.
Tsibris, J. C. M. and Woody, R. W. (1970). *Co-ordination Chemistry Reviews* **5,** 417.

10. Combined Applications and Other Techniques

D. W. Jones

10.1. Spectroscopic methods and biological polymers

Biological molecules are intrinsically more complicated than those of monomers or synthetic polymers, and their spectroscopic study generally involves greater difficulties in data collection and interpretation. The variety of interacting chemical groups present means that structural problems in physical biochemistry often require rather specific information at a number of sites within the biomolecule. For biological polymers, as with simpler molecules, structural information in more detail may be derived from diffraction measurements in the solid state. Moreover, the X-ray diffraction pattern is related rather directly (if not formally unambiguously) to the electron density (and hence to atomic positions) in the crystal; full analysis of, for example, an n.m.r. spectrum does not necessarily lead to a determination of the molecular structure. However, most biological activity occurs in solution, and it is important to be able to establish the similarity or otherwise of, for example, protein structure in the solid state and in aqueous solution; spectroscopy can help achieve this. By contrast, many non-spectroscopic biophysical methods give more gross information about the shape, size and other properties of the macromolecule.

There are two complementary approaches to the structural study of biological materials by diffraction or spectroscopy. On the one hand, interpretation of measurements on low mol. wt. material—e.g. amino acids and peptides, monosaccharides and oligosaccharides—can reveal detailed properties; on the other hand, interpretation of data from the biological material itself can provide a less complete picture of its molecular behaviour. As technique advances, the high information content of polymer spectra becomes more fully utilized. Even so, the confidence inspired by consistency of spectroscopic measurements with results from other methods on low-mol.-wt. models can provide a foundation for the interpretation of high-mol.-wt. spectra. This is not to underrate the significance of *differences* between spectra of the biopolymer and spectra of an assembly of its components.

The range of spectroscopic techniques described in the preceding chapters is illustrated by comparing applications of high-resolution n.m.r. with the traditional use of electronic and other spectroscopies for determining accurately a single parameter such as protein concentration. Electronic transitions sense changes in molecular orbitals of ground or excited states. A biomolecule usually has very few chromophores to sense and their bands are broad, although with a single chromophore, data can be very specific; even so, appreciable changes in other parts of the molecule may remain undetected.

The power of n.m.r. derives from an ability to sense several *kinds* of parameter. Thus shifts depend on electronic and molecular environments; couplings are indicative of interactions with neighbours; relaxation times are responsive to motions and other influences on magnetic dipolar interactions. N.m.r. can focus on specific, widely dispersed nuclei and, for one nuclear species, e.g. ^{1}H (but now similarly for ^{13}C), it can, in principle, sense protons in all the different amino acids in a polypeptide, with consequent complexity of the spectrum. N.m.r. probably has the greatest potential for approaching the degree of detail about the local structure of a biological polymer in solution revealed by diffraction of the crystalline state. While inferior to diffraction in wealth of stereochemical detail, n.m.r. has an ability, unlikely to be rivalled by diffraction even with synchroton X-radiation, of being able to follow conventional transitions in time, with consequent value for kinetic problems. Among the spectroscopic techniques, n.m.r., even by FT, is relatively insensitive (essentially, because of the small energy separation), however, and aqueous concentrations of biological materials are often low. Raman spectroscopy may offer considerable scope for the study of biological reactions (Freeman, 1974); it can also be complementary to n.m.r., e.g. in sensing disulphide bonds much more directly. Since both Raman and i.r. spectroscopy are concerned with molecular vibrations, they can be complementary in many chemical applications. With laser sources, photon counting and computing facilities for vibrational analysis, Raman spectroscopy is now available, as has been seen in Chapter 3, for biopolymers. By comparison with i.r. spectroscopy, it has the additional advantage of being able to follow directly changes in conformation between the intact solid state and that in solution in H_2O or D_2O. Conformational studies of biopolymers accessible by spectroscopic techniques include not only the spatial distribution of atoms and groups and their intramolecular and intermolecular interactions. They can also include variations of thermodynamic properties with solvent and temperature, for example determined from following the formation of —CO...HN— H bonds in polypeptides by i.r. amide absorption (Section 2.2.3.2). Infrared is at its best at the primary and secondary levels of, for example, protein structure; u.v. spectroscopy can contribute more at higher levels of association.

Any physical method concentrates attention on one aspect of the structure; this restriction is perhaps less obvious for X-ray diffraction (which senses *all* the electrons in the crystal structure) than for spectroscopic methods (e.g. ^{1}H n.m.r. probes the regions near H nuclei only). Chapters 2–9 have considered each spectroscopic method in turn and described how it can enable structural information to be extracted about biological molecules. For a full structural picture, it is generally desirable to combine several spectroscopic techniques and especially so for molecules of the complexity of biological polymers. Some examples of converging attacks on biological systems are mentioned in the following sections, although in most cases opportunity is taken to introduce or highlight the application of one particular technique. Attention is drawn to relevant review articles and it is hoped that the reader may thereby gain some feeling for the most likely spectroscopic techniques to apply to the solution of a particular problem.

10.2. Combined spectroscopic applications to biological systems

10.2.1. CONFORMATION OF POLYPEPTIDES AND PROTEINS:
HELIX/RANDOM-COIL TRANSITIONS

The conformation of homogeneous biopolymers in solution and their variation with pH, for example, can be studied by several spectroscopic techniques, especially ORD/CD, Raman and n.m.r.; usually the conformational assignment from spectroscopy is less explicit than it is from a diffraction analysis in the solid state. While Raman spectra may suffer from fluorescence of the sample, near-i.r. spectroscopy of the overtone N—H bands in polyaminoacids, for example, can facilitate study of protonation or solvation by solvent of intramolecular H bonds; suspected reduction in helicity should preferably be tested by CD/ORD. Infra-red absorption spectroscopy can help to establish conformation by following the spectrum of the chromophore group through chemical changes, for example to give an H-bonded variant in a dimer; or, less dramatically, as neighbouring groups in the polymer are varied chemically, for example, the H-bonded peptide group in different polypeptides (Hanlon, 1970). The higher i.r. frequencies (wavenumbers $3000\,cm^{-1}$ or more) sense functional groups through stretching frequencies and overtones, while the far i.r. frequencies (wavenumbers below $500\,cm^{-1}$) depend much more on the skeletal structure; in between, i.r. senses both chemical and conformational effects.

While it has the advantage of being non-destructive, sequence analysis by n.m.r. in polypeptides is of limited value in that the side-chain resonances which characterize the amino acids are insensitive to the primary structure. N.m.r. has been suggested as a rapid means of examining large numbers of closely related protein hydrolysate samples, when available in moderate quantity, in search of small differences in amino-acid composition. In the absence of secondary or tertiary structure, and with due allowance for the formation of the peptide link, the spectrum of a polypeptide or protein (possibly sharpened by "deconvolution") might be expected to be closely approximated by a superposition of n.m.r. spectra of constituent amino acids. Such syntheses have been made of high-field n.m.r. spectra of denatured proteins such as lysozyme, ribonuclease, insulin, gelatin, even for the relatively heterogeneous wool proteins (Dale and Jones, 1974; Jones, 1976), and of the Raman spectrum of lysozyme (Section 3.2).

With very short peptides, as with amino acids, the relative lifetimes of conformations about the C_α—C_β bond, termed rotamers, may be determined from the coupling constants (Bartle et al., 1972; Jones, 1976). For a number of small cyclic peptides, and some linear ones, the combination of n.m.r., ORD and i.r. spectroscopy has led to the unambiguous determination of solution conformations.

The linear pentadecapeptide antibiotic gramicidin A, which has the primary structure HCO-Val-Gly-Ala-Leu-Ala-Val-Val-Val-Trp-Leu-Trp-Leu-Trp-Leu-Trp-NHCH$_2$CH$_2$OH, is a relatively small biomolecule with a shape closely associated with function. Its selective effect on ion permeation of lipid bilayers arises from a 4-Å-diameter ion-conducting channel formed by two or more gramicidin A molecules. Thus understanding of its action depends on a rather detailed three-dimensional knowledge of the conformation of this polypeptide in solution. Urry (1973) has described how this conformation has been defined spectroscopically in three stages; optical spectroscopy and ORD/CD can often indicate the presence and extent of helicity, whereas n.m.r. can establish which residues are present in a helix and their conformation. In the case of gramicidin A, helix conformation was suggested by hypochromism, or decreased intensity by mutual polarization, at 190 nm in the electronic absorption spectrum in trifluoroethanol with increasing order. The CD pattern is characteristic of a left-handed rather than a right-handed helix, while the n.m.r. C_αH—NH coupling constants of 8 Hz, rather than 2 or 3 Hz, identify it as a β- rather than an α-helix. The new β-helix proposed for the transmembrane channel contains 6·3 residues/turn, with peptide C—O groups alternately pointing parallel and anti-parallel to the helix axis. Recently, Raman spectroscopy has also been directed at the A' polypeptide as a probe of H-bonding and hence of conformation.

Transitions between conformations of polypeptides and proteins in solution, for example in the denaturation process, have received much study by ORD (Section 6.4) and n.m.r., especially 220 MHz ^{1}H. Both techniques involve appreciable difficulties of interpretation in this field and they have on occasion seemed to conflict, for example, about the temperature at which a transition occurs (Liu and Anderson, 1970). Whereas ORD is sensitive only to structure, p.m.r. is also sensitive via relaxation times to motional rates. Distinct chemical shifts have sometimes been assigned to helical and random-coil forms; alkyl peaks in native lysozyme occur at very high field because of ring-current shifts from nearby aromatic rings in trytophan, tyrosine, etc. Combination of the two techniques can be fruitful in helping to overcome the ambiguities arising from the several causes of n.m.r. line broadening. Use of ^{13}C n.m.r. offers hope of detecting more chemical-shift changes at the transitions than with ^{1}H resonance.

It may be mentioned here that several spectroscopic techniques have been applied to the study of helix-coil transitions in polynucleotides and nucleic-acid systems (such as the ribonucleic acids RNA, and deoxyribonucleic acid DNA), of which the monomers consist of three components: an organic base, phosphoric acid and a sugar. The denaturation by dimethylsulphoxide of t-RNA, involving disruption of base stacking, can be followed (Schweizer, et al., 1973) by 220 MHz ^{1}H n.m.r. Since, as has been seen with proteins, i.r. spectroscopy can detect H bonds (Section 2.4.5), it can differentiate well between pairings of unsaturated bases, such as adenine-uracil and guanine-cytosine. CD/ORD measurements (Section 6.5) can sense base stacking as well, while spin-labelling (see Section 10.2.5) allows the change in mobility at the particular point of the biomolecule to which the label is attached to be monitored by e.s.r. during a conformational transition.

For transfer ribonucleic acid (t-RNA) and related polynucleotides, for which the appreciable Watson-Crick base-pairing helps to reduce the number of structural alternatives (Kearns and Shulman, 1974), 220 and 300 MHz n.m.r. has enabled the H bonds responsible for the secondary structure in solutions to be established.

10.2.2. STRUCTURE AND ACTION OF PROTEINS AND ENZYMES: PARAMAGNETIC PROBES

Among the globular proteins, which can range in molecular weight from 12 000 to 250 000, the extremely effective metabolic catalysts known as enzymes form a vitally important class. The very few of these (the first was lysozyme) which have had their three-dimensional structures established represent a minute fraction of the 10^6 or 10^7 which may exist in a mammal.

The unique properties of enzymatic (e.g. ribonuclease) and electron-transport (e.g. cytochrome c) proteins appear to derive from the way the polypeptide chains can be folded so as to bring particular functional groups into specific mutual geometrical arrangements. Thus a significant problem is to monitor the changes in this conformation that are induced by environmental changes (McDonald and Phillips, 1970). A short general review is available (Knowles, 1972) on the application of magnetic resonance methods (n.m.r. and e.s.r.) to the study of enzyme structure and action. For the future, there is a prospect of n.m.r. providing considerable dynamic information about biochemical processes in enzyme systems, a distinct advantage over diffraction methods. As Dwek (1973) points out, there are several distinct aspects of biological molecules and in particular of enzyme systems, to which the n.m.r. approach may be applied:

1. ^{1}H, ^{13}C and ^{19}F resonance of the macromolecule itself, e.g. lysozyme with the help, where known, of the crystal structure, chemical-shift differences may be interpreted between material untreated and modified, either chemically or by the presence of paramagnetic-probe ions.

2. Relaxation rates of nuclei of small ligand molecules (at high concentration) interact with macromolecules (at smaller concentrations); with fast exchange and paramagnetic-ion probes, distances between ligand nuclei and the paramagnetic centre can be measured.

3. Resonance of solvent water; again, with paramagnetic ions, relaxation-rate measurements can lead to information about distances from metal ions to water molecules in the hydration sphere.

4. Study of bound ions, exchange rates and binding constants, via resonance of less-sensitive $I > \frac{1}{2}$ nuclei, including ^{23}Na, ^{19}K and ^{35}Cl.

5. Effect of *covalent* paramagnetic spin labels on nuclear relaxation rates (in addition to the e.s.r. spectrum from the spin label, discussed in Section 10.2.4) permits distance determinations, for example.

Because of their involvement in biochemical processes, metal ions play a vital role in the structure of many biological polymers; in others, metal ions may be deliberately introduced as so-called extrinsic probes. Paramagnetic metals allow e.s.r. to be brought to bear (see Sections 8.6.1 and 10.2.4.1 for discussion on haemoglobin). They also reduce *nuclear* relaxation times (Mildvan and Cohn, 1970) and cause distinctive contact shifts (Section 7.4 and Phillips, 1973) in the n.m.r. spectra of biological molecules, effectively increasing chemical-shift resolution. Cohn and Reuben (1971) have described the combination of e.s.r. and n.m.r. by the use of paramagnetic probes (especially Mn^{2+}) as activators (substituting, for example, for the natural activator Mg^{2+}) in relaxation-time studies of the action of phosphoryl transfer enzymes (Section 7.4.2). Among the advantages of this approach to the structure, configuration, surroundings, dynamics and metal-role at the active site are that the procedure is non-destructive, equilibrium is undisturbed by the small energies involved, and assignment is much more specific than in optical spectroscopy.

N.m.r. measurements with lanthanide pseudo-contact-shift probes have enabled R. J. P. Williams and colleagues (Campbell *et al.*, 1975; Levine and Williams, 1975) to eliminate many conformational possibilities of a mononucleotide or an enzyme in solution and even to deduce a preferred conformation. The essence of the method is that the shifts (in all the protons) induced by the Eu^{3+} or Nd^{3+} ion (with short electron-relaxation times) depend on the distance, r (actually r^{-3}), between atomic nuclei and ion and on one or two (according to symmetry) angles involving r; the broadening (which may be induced by another lanthanide, such as Gd^{3+}, with a longer electron-relaxation time) is independent of angle and depends only on distance (actually r^{-6}). For pairs of nuclei in the molecule or complex, the ratio of shifts (or of line-widths) observed is compared with ratios calculated for postulated conformations (and for various positions of the metal ion if its location is not certain).

An example of the value of lanthanide-probe ions (binding here to phosphate groups) in establishing molecular conformation in solution is provided (Barry et al., 1974) by the mononucleotides 9-β-D-

AMP

TMP

ribofuranosyladenine-5′-monophosphate (AMP) and 1-β-D-deoxyribofuranosylthymine-5′-monophosphate (TMP). With the aid of interactive displays, measured ratios of dipolar shifts, and of broadenings, for different pairs of nuclei within each complex, enabled the most probable conformations to be determined from a vast array of possible molecular shapes. The conformations derived in aqueous solution turned out to be closely similar to those derived by diffraction for the crystalline structures.

10.2.3. BINDING OF DRUGS AND OTHER SMALL MOLECULES TO BIOMOLECULES

The action of many drugs appears to be initiated by relatively weak binding to certain sites in macromolecules or cellular structures (Fischer, 1971, 1973); the mechanism has similarities with other reversible biological interactions, such as between enzymes and substrates. At the *receptor* (macromolecule) end of the interaction, it would evidently be desirable to know which are the sites and in what ways they are (perhaps rather subtly) changed in structure and conformation when binding occurs. Despite appreciable difficulties, n.m.r. can investigate this to a limited extent, for example by focusing attention on a small region of the high-resolution spectrum that may contain well-resolved peaks. *Changes*, induced chemically or physically, may also be followed by n.m.r., and deuteriation is now being exploited so that, for example, in a large synthetic polypeptide very few undeuteriated H's are left for unequivocal identification by n.m.r. However, valuable contributions to knowledge of binding to the biomolecule can be obtained more readily from changes, whether in chemical shifts, coupling constants, or relaxation rates (Section 7.5), in the n.m.r. of the *small molecule* (usually present in much higher concentration than the macromolecule). Hence binding may cause the n.m.r. chemical shift of the small molecule to change by as much as 1 ppm, or 100 Hz in 100 MHz. If, then, the small molecule is exchanging at an appreciably faster rate than $10^2 s^{-1}$, a weighted average spectrum of the two forms, bound and free, will be seen. This rapid-exchange situation, in which the exchange rate, $1/T_{ex} > 1/T_{2\,bound} > 1/T_{2\,free}$ gives rise to a single n.m.r. line, is an important one for biological interactions, with typically $10^2 < 1/T_{2\,bound} < 10^4 s^{-1}$; it may be possible to observe the effect of 10^{-5}M binding sites.

Slow (or non-existent) exchange will result in the simultaneous presence of spectra of both bound and free forms relaxing independently, while intermediate exchange rates can yield line broadening. It should be remembered that if, as is likely, only 1% of the small molecules are in the bound form, chemical-shift changes may escape detection. When bound to the homogeneous protein lysozyme, favourable for study partly because of its solubility, a number of inhibitor sugars show broadened spectra. Many inhibitors display binding shifts (comparatively large because binding is close to an aromatic ring); differentiation between several contributions to the broadening has enabled the lifetimes of the inhibitor/lysozyme complexes to be determined.

The change in relaxation rate of the small molecule (Fischer, 1973) occurs because its rotational motion is reduced by the interaction with the biomolecule and its correlation time (the time τ_c, for a perceptible degree of motion) increased; relaxation times, T_1, of free and bound forms are inversely proportional to the corresponding correlation times. In a system described by several very different correlation times, the shortest will have most effect on the relaxation time, so that increases in T_1 can be associated rather specifically with those parts of the small molecule most affected by binding. In attributing selective broadening effects to complexing of the small molecule to the protein or other biomolecule, care must be taken to eliminate several other, mostly less specific, possible sources of broadening (Hollis, 1972).

Fluorescence spectroscopy (Section 5.3.2.3) is another technique for studying the interaction of drug and dye molecules with nucleic acids and proteins (Chignell, 1972), provided the drug or macromolecule is fluorescent or can be made so by complexing with a fluorescent label. As discussed in Chapter 6, the optical activity of a system involving a ligand binding to a protein or nucleic acid can also change, because of a change either in structure or optical activity of protein (Section 6.6) or in structure or activity of ligand (Section 6.7 and Chignell and Chignell, 1972).

10.2.4. COMBINED STUDIES OF PARAMAGNETIC BIOLOGICAL MACROMOLECULES

10.2.4.1. *Haemoproteins*

The structure and function of haemoproteins and particularly of haemoglobin (mol. wt. 64 500), the respiratory protein of red blood cells, have been extensively studied by diffraction, magnetic susceptibility, and several spectroscopic techniques, so that haemoproteins probably are the class of biological molecules best established by physical methods. The haem chromophore has accessible π^* electronic transitions (e.g. Section 6.6) with absorption bands in the visible region around 560, 530 and 420 nm, while the presence of ferric ion can lead to e.s.r. (Section 8.6.1) and Mössbauer spectra (Section 9.7.1). In oxygenated solutions, i.r. spectroscopy suggested (with the support of Mössbauer spectroscopy) that an O bridge between the ions linked the haems in the haem dimer. Resonance Raman spectra show the displacement of the ions from the haem plane; recently, the resonance Raman spectra of haemoglobins (Freeman, 1974) have been studied in water. It has

even been suggested that a variable-frequency dye laser could enable blood abnormalities to be diagnosed according to whether a resonance Raman spectrum is detected when the laser is tuned to a specific absorption frequency.

With a background of X-ray crystallographic and magnetic susceptibility measurements, correlations between e.s.r. single-crystal spectra and Mössbauer spectra have been very valuable complements in the study of haemoproteins (Holtzman, 1972). Magnetic resonance studies of haemoglobin have been summarized by Maričić (1975). E.s.r. spectra cannot always be observed in haemoproteins, either because of reduction of high-spin $(S = 5/2)$ ferrihaemoprotein to low spin $(S = 0)$ ferrohaemoprotein, or because spin-lattice relaxation broadens the peaks excessively. Where it *is* applicable, e.s.r. can be more precise than Mössbauer spectroscopy (Lang, 1971), but the latter (particularly with variable-temperature measurements) can yield a greater variety of information, mainly about paired electrons (through sensitivity to field gradient). For weakly paramagnetic materials such as the haemoproteins, high-resolution n.m.r. provides a valuable means of determining the magnetic susceptibility from rather limited amounts of material (Bartle et al., 1975).

In that its detailed use has been confined to a limited number of nuclear species, Mössbauer spectroscopy (Chapter 9) has had a restricted field of application among biological polymers. In addition to the applications of ^{57}Fe resonance to Fe-containing proteins mentioned here and discussed in Section 9.6 and 9.8, Holtzmann (1972) has suggested activity studies of proteins by ^{67}Zn, ^{133}Cs or ^{40}K spectra, study of Xe interaction with proteins by ^{129}Xe or ^{131}Xe spectra (from ^{129}I or ^{131}I) and ^{127}I or ^{129}I (from ^{127}Te or ^{129}Te) spectra to study tyrosine in enzyme/substrate interactions.

10.2.4.2. Fe-S proteins

The term Fe-S protein (Beinert, 1973) applies to Fe proteins in which S is a ligand of the Fe and is not held simultaneously by a more powerful ligand such as porphyrin. Despite their wide occurrence and involvement in many life-sustaining functions such as photosynthesis, N fixation and electron transport, Fe-S proteins as a class were discovered only relatively recently. Their absorptions in the visible/u.v. were much weaker than for haem proteins and their structures have proved more amenable to e.p.r., n.m.r. and Mössbauer spectroscopies (Section 9.7.2). The observation of an e.p.r. signal with a g-value of 1·94 in non-haem protein from mammalian, and later bacterial, sources was an important factor in establishing the existence of this class of catalysts for electron transfer at low potential levels. This distinctive signal was not at first observed with the ferredoxins, despite their similarities in other ways. With higher concentrations of ferredoxins and lower temperatures, the eventual observation of this signal helped to unify and stimulate the field of Fe-S proteins.

Resonance techniques have made a considerable contribution towards understanding of the structure of the Fe centre in Fe-S proteins. In the relatively simple e.p.r. spectrum of the plant ferredoxin protein P. putida (often known as putidaredoxin), nuclear hyperfine splitting indicates two Fe nuclei with a total spin of $\frac{1}{2}$. Changes in the h.f.s. induced by isotopic substitutions (for example, ^{32}S has nuclear spin $I = 3/2$) demonstrated that the unpaired spin also interacts with the two labile S atoms as well as with some (probably four) stable S atoms from the protein backbone.

10.2.4.3 Non-haem Fe-S proteins: rubredoxin

Among the biological activities of proteins containing Fe and S are photosynthesis, N fixation and probably biochemical electron transport via multiple redox states of the Fe. As some compensation for the disadvantage of poor stability, such proteins, with moderate mol. wts, may be susceptible to study by e.s.r., n.m.r., Mössbauer and optical spectroscopy (absorptions at 450 and 555 nm), and C.D.

In contrast to the so-called Fe-S proteins (Sections 10.2.4.2) which yield "inorganic" S as H_2S with acid, rubredoxin, the simplest non-haem Fe protein, contains "non-haem" Fe coordinated, roughly tetrahedrally, via S atoms to four cysteine residues (out of a total of about 50 amino acids). Anisotropy of the e.s.r. absorption points to some distortion of the four S atoms from tetrahedral symmetry. According to Mössbauer spectroscopy (Section 9.7.2) and CD, oxidized rubredoxin is high-spin ferric ($S=5/2$) and one-electron reduced rubredoxin is high-spin ferrous ($S=2$).

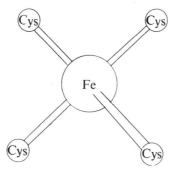

220 MHz p.m.r. spectra (Phillips et al., 1970) of oxidized and reduced rubredoxin in the region -2.0 to $+10.0$ ppm differ markedly from each other and from that expected for a random-coil structure without paramagnetic Fe-effects. The high-spin Fe appears to broaden contact interactions so much that they cannot be positively identified in either redox state of rubredoxin.

10.2.5. STUDY OF MEMBRANE STRUCTURE BY MAGNETIC RESONANCE: SPIN LABELS

In the solid state, the unpaired electron in an N p orbital of a stable nitroxide free radical, $=N-O$, can exhibit a "through-space" dipolar anisotropic hyperfine coupling to the nucleus, so that the simple three-line e.s.r. spectrum is angular-dependent. If a derivative of a biological molecule is synthesized with such a nitroxide free radical in a specific orientation, small quantities can be introduced with (it is hoped) comparatively little disturbance to the otherwise diamagnetic surroundings. The radical can then act as a spin label or probe of the motion of that part of the molecule to which the radical is attached (although establishment of precisely where the label is located can be difficult). Thus, for example, labelled phospholipids (fatty-acid esters containing a phosphate group and generally a nitrogenous base), which are introduced in an excess of unlabelled molecules, can enable the translational and rotational motion to be studied in cell walls, membranes of lipid vesicles; this includes, for example, following the influence of adding chemicals of physiological interest.

The use of spin labels (mentioned in Section 7.4.2) is thus an extension of the more general principle of investigating structure and function of biological macromolecules by attaching "reporter" or "probe" molecules. These can be either (a) passive detectors or markers which do not disturb the system but are themselves affected by changes in the surroundings; or (b) active or perturbing probes which are designed to influence their surroundings, e.g. by causing large chemical shifts as in the n.m.r. paramagnetic probes (Section 10.2.2). (There is, of course, a danger that a label thought to be passive does in fact change the biological action; bulky spin labels must alter the system somewhat.) Organic free radicals (detected by e.s.r.—Chapter 8) have advantages over optically absorbing labels (detected by fluorescence and optical absorption—Chapter 5) in greater sensitivity to surroundings and applicability to optically opaque systems such as membrane suspensions. In general, if a biomolecular system is varied, then an attached spin label can sense, near the attachment point, changes in polarity (which affects the hyperfine coupling constant a_0, for example), conformation or motion. With membranes, sensitivity limitations on detection of spin labels in monolayers and bilayers favour the use, where possible, of relatively large multilayers in which the (perturbing) probe concentration is kept acceptably low.

When the local anaesthetic benzyl alcohol is added to enythrocyte membranes spin-labelled with 10^{-4}M of a steroid nitroxide (I), the e.s.r.

I

spectrum first sharpens, indicating a fluidizing effect of the perturbing agent and then, at higher (about 200 mM) concentrations of alcohol, exhibits a new component. By taking into account the occurrence of many highly immobilized binding sites on separated membrane-protein, Metcalfe (1970) deduced that this new spectroscopic component corresponded to highly immobilized spin-label binding sites freshly exposed as a result of irreversible disruption of structure of the membrane. These conclusions were consistent with complementary n.m.r. and partition-coefficient studies.

In the above example, it is inferred that motion (detected via the e.s.r. spectrum of the spin label) of the probe molecule benzyl alcohol follows motional changes in the membrane. Thus one can detect whether or not the membrane stays "intact", i.e. retains its biochemical or physiological functions. P.m.r., perhaps at its most informative when applied to small molecules in solution, can also sense membrane structure indirectly through the motions of a small probe such as benzyl alcohol, partitioned into the membrane. With both aromatic and methyl ^{1}H signals initially broadened by the membrane (as compared with aqueous solution), the probe relaxation rates, $(1/T_2)_{bound}$, deduced from the line-widths, $\Delta v_{\frac{1}{2}}$, are found to decrease with increasing alcohol concentrations and to pass through a minimum; this corresponds to maximum mobility of the alcohol molecules, presumably because the membrane structure is disordered. At higher alcohol concentrations, the relaxation rate increases owing, Metcalfe found, to the increasing domination of the contribution from the membrane protein (with broader lines for the benzyl alcohol) over the lipid component of the membrane. Evidently, when the relaxation-rate curve starts to increase, the membrane has incurred some irreversible structural change from its initial intact state, perhaps the disruption of crucial lipid/protein interactions.

Phospholipid model membranes have also been examined by high-resolution n.m.r., with and without paramagnetic shift probes. The glycerophospholipid lecithin from egg yolk can be dispersed in D_2O by

$$\begin{array}{l} CH_2OCOR \\ | \\ CHOCOR \\ | \qquad\quad O^- \\ CH_2O-P{\overset{\displaystyle\diagup}{\underset{\diagdown}{}}}OCH_2CH_2\overset{+}{N}Me_3 \\ \qquad\qquad O \end{array}$$

(R is a fatty acid)

sonication to give 200–300 Å semi-permeable membrane bilayers (vesicles) of oriented molecules that enclose an aqueous phase. Introduction (by preparation in $MnSO_4$ solution) of paramagnetic Mn(II) ions within the vesicles substantially broadens the choline $-N^+Me_3$ signal from its normal 6 Hz (motionally narrowed from the 20 Hz of CH_2); addition of $MnSO_4$ outside the vesicles does not broaden the $-N^+Me_3$ ^1H signal, indicating impermeability of the bilayer membranes to Mn(II). They are also impermeable to Eu(III); additions of the ion-shift probe $Eu(NO_3)_3$ into the external phase leaves one $-N^+Me_3$ peak at the same shift while another $-N^+Me_3$ peak is shifted to high field and broadened. In this way, n.m.r. is able to differentiate non-destructively between molecules in the internal and external surfaces of the lecithin-containing membrane.

Optical activity techniques have also been applied to the study of membranes and suspensions of polypeptides. Presumably because of their particular nature, such aggregate macromolecular systems possess optical-activity features distinct from, but related to, those of polypeptides and proteins in solution. Light-scattering distortions may be responsible for apparent inconsistencies. While it is difficult to extract the inherent CD of the chromophores from the observed CD of suspensions, measured membrane optical activity does not indicate that protein structure in membranes differs from that of globular proteins in solution.

Another kind of label, that of small molecules of known chemical structure which exhibit fluorescence, can be attached to macromolecules, including membranes. When bound to erythrocyte membranes, the fluorescent label 1-anilinonaphthalene-8-sulphonic acid (ANS) (Section 5.5) is influenced by cations and by local anaesthetics. Other fluorescent labels which can be incorporated into specific regions of the membranes (e.g. hydrocarbon region or bilayer interface) should facilitate study of drug/membrane interactions (Chignell, 1972). Levine (1972) has reviewed diffraction and spectroscopic studies of membrane structure.

10.2.6. POLYSACCHARIDES

Cellulose, the most abundant natural polymer, and its derivatives have been subject to extensive spectroscopic study, reviewed in a number of articles in the volumes (Bikales and Segal, 1971) recently added to Ott's "Cellulose and Cellulose Derivatives". Poor solubility in n.m.r. solvents and other difficulties with hydroxyl groups have restricted the use of high-resolution n.m.r. (Barker and Pitman, 1971) but selective deuteriation has been applied to bacterial cellulose (Gagnaire and Taravel, 1975). Broad-line (Section 10.3.2.1) and pulsed n.m.r. have been concerned especially with water sorption (e.g. Child and Jones, 1973). Much more extensive use has been made of i.r. spectroscopy (Section 2.5.2), especially to complement diffraction information on structure and anomeric configuration (Blackwell and Marchessault, 1971), but also for measurements of the state of order in the celluloses, often tackled via deuteriation, and the effects of chemical treatments. E.s.r. (Arthur, 1971), applied to observe the influence of high-energy radiation and charring on cellulose, might also be useful for pyrolysis and flame-proofing studies.

Muzzarelli (1973) has summarized spectroscopic measurements on alginic acid, chitin and chitosan, regarded as natural chelating polymers. In acid hydrolysis of chitan, elemental analysis $(C_8H_{13}O_5N)_n$ and p.m.r. indications of N-acetylglucosamine, glucosamine and acetic acid as the only products were consistent with a fully acetylated form of poly-N-acetylglucosamine; with optical-rotation data, chitan was established as β-linked 2-acetamido-2-deoxy-D-glucan. Chitan has narrower OH stretching bands than most biopolymers; the much sharper i.r. bands in chitan than in the less pure and less fully acetylated chitin are attributed to a less ordered structure in chitin than in chitan. With chitan, as with cellulose, lack of absorption about 3500 cm^{-1} shows that all hydroxyl groups are H-bonded. Although few polysaccharides have accessible chromophores for ORD to provide useful conformational indications, optical rotation at a single wavelength can be related to the mutual inclination of adjacent glycosidic bonds.

10.3. Emergence of new or modified spectroscopic techniques

In this section, very brief mention is made of some recent variants of established methods and of some spectroscopic techniques not covered in detail in earlier chapters.

10.3.1. NEW SPECTROSCOPIC DEVELOPMENTS

10.3.1.1. *Nuclear magnetic resonance spectroscopy*

The advantages for biological systems of carrying out n.m.r., at *high fields*, for example, ^1H resonance at 300 MHz made possible by means of superconducting coils, have already been pointed out in Chapter 7. Synthesis of substantially deuteriated polypeptides is laborious and expensive but can be valuable for unequivocal spectral assignment. However, the major relevant development in contemporary n.m.r. is the arrival of Fourier-transform pulsed, as distinct from continuous wave, n.m.r. spectrometers (Section 1.5.3). The increased sensitivity allows ^1H to be studied at lower concentrations (10^{-4} M) and enables less sensitive nuclei such as ^{13}C to be examined. Furthermore, relaxation times of individual nuclei can be measured with ease (Shaw, 1974).

^{13}C n.m.r. spectra (mentioned in Section 7.6.2), generally recorded by Fourier-transform spectroscopy (Section 7.3.3), have a wider range (around 350 p.p.m) of shifts than for ^1H, with an appreciable sensitivity to changes in the chemical environment. For proton-bearing carbons in large and medium-sized molecules, the T_1 values are dipolar in origin and do not have intermolecular contributions; they are, therefore, excellent for deriving correlation times. If concentration problems can be overcome, ^{13}C n.m.r. should be useful for studying strongly bound ligands which are exchanging relatively slowly. Lee et al. (1973) and (Feeney (1975) have summarized the advantages of Fourier-transform and especially ^{13}C n.m.r. in the study of biological membranes. T_1 relaxation times, usually determined by intramolecular dipolar interactions, can be used to indicate their relative degrees of motion.

Brief mention may be made of two very different kinds of n.m.r. experiment, well outside conventional spectroscopy. First, Mansfield and Grannell (1973) have proposed an n.m.r. "diffraction" process in solids whereby each lattice point is uniquely identified (by an extremely large, linear Zeeman-field gradient) and gives a spike in the absorption curve. Modulation of the n.m.r. transient response could, in principle, enable the lattice of the resonating nuclei to be reconstructed; potential application to fairly regular biophysical structures (cell membranes, fibrous structures) is proposed. Lauterbur (1973) has taken up an earlier and related idea whereby narrow liquid-water n.m.r. lines in biological systems are utilized to derive the spatial distribution (or n.m.r. "zeugmatogram") of the sample from the n.m.r. absorption line-shape in a linear magnetic-field gradient. Second, it has been found with animals and, more recently, with man, that n.m.r. spin-lattice (T_1) relaxation times of water tend to be higher in malignant than normal tissue. N.m.r. may provide a means of diagnosing and following the growth of cancerous tumours (Damadian et al., 1973); it has also been used for monitoring blood flow in humans (see Jones and Child, 1976). Section 10.3.2.1 introduces a further application of n.m.r.

10.3.1.2. E.s.r. in flowing systems

E.s.r. (Chapter 8) can be applied to the study of rapid reactions involving free radicals with half-lives, $\tau_{\frac{1}{2}} > 5$ ms, if flow systems are used which maintain a high enough steady-state concentration of short-lived intermediates. The stopped-flow e.s.r. kinetic technique, first applied to enzyme-substrate redox reactions, involves two fluid lines feeding a four-jet liquid-flow mixing chamber and a quartz flat cell inserted in a conventional e.s.r. spectrometer. As a step towards an investigation in solution of free radicals in pathological and physiological processes (though here at pH 1–2 and concentrations $10^{-2} - 10^{-3}$M), the technique has been applied to S-containing radicals (see Section 8.5.3) produced by the chemical (Ce^{4+} ion) oxidation of cysteine and related thiols (Wolf et al., 1969). An improved e.s.r. stopped-flow system, incorporating rapid-action solenoid valves with flow rates up to 25 cm^{3}s^{-1}, has been described for determinations of accurate kinetic data for dynamically-generated free radicals involved in radiation damage and/or protection to cells (Kertesz and Wolf, 1973).

10.3.2. ADDITIONAL TECHNIQUES

10.3.2.1. *N.m.r. of solids and of water in biological materials*

Chapter 7 deals largely with the application of *high-resolution* n.m.r. to solutions of biological polymers. Early applications of n.m.r. to solid polymers were made with continuous-wave *broad-line* spectrometers, since direct magnetic dipolar interactions, present in the solid state but averaged out in solution, causes substantial line broadening. Typically, as the temperature increases, the line widths in solid samples tend to diminish from the value appropriate to a rigid lattice, and nuclear resonance transitions with fairly sharp decreases in width, corresponding to increased molecular motions, may be observed well below the melting point. For dry celluloses at a given temperature, attempts have been made to relate the second moment of the absorption curve to crystallinity as measured by X-ray diffraction. With moisture present, the ^{1}H n.m.r. absorption line (usually presented as the first differential of the absorption in broad-line spectra) may be composite, with a narrow water line emerging above the broader one associated with protons in the polymer phase. This can be utilized to give an indication of moisture content in, for example, wool, cellulose and corn products (see Jones, 1976).

Relaxation times, measured by pulsed n.m.r. provide a better means of tackling the difficult problems associated with the state of water in biological systems. Brief summaries of n.m.r. studies on moist biological polymers have been given by Lynch and Marsden (1971) for wool, by Barker and Pittman (1971) and by Forslind (1971) for cellulose, and by Walter and Hope (1971) for living biological cells and model systems.

10.3.2.2. *Mass spectrometry*

Mass spectrometry is one of the oldest and most precise instrumental techniques of analysis; it is not covered in detail since it does not constitute a spectroscopic method in the same sense as those to which separate chapters have been devoted. Mass spectroscopy (m.s.) involves ionization of the sample, for example by electron bombardment, separation (by trajectory in electric and magnetic fields) of the ions formed according to their mass/charge (m/e) ratios, and measurement of their abundances. Although destructive, the m.s. can be measured on as little as 10^{-6}g. The requirement that a sample should have a vapour pressure of at least 10^{-6} torr in practice limits molecular weights to about 2000. Thus, although m/e ratios as high as 1700 have been measured, application to large biological polymers has been slight. For primary-sequence determinations of amino acids in peptides and proteins, free peptides have been converted to more volatile derivatives. The application to biomolecules (rather than biological polymers) has been reviewed by Arsenault (1973), and to amino acid sequencing by Biemann (1972) in this field, computerized data-handling is essential.

10.3.2.3. *Neutron inelastic scattering spectroscopy*

In the last few years, the coherent *elastic* scattering of beams of low-energy (thermal) neutrons, from a liner accelerator or nuclear reactor of fairly high flux, has begun to be applied to the crystal structures of biological molecules (Schoenborn, 1972). This is neutron diffraction, which is much more sensitive than X-ray diffraction to the scattering (and hence positions in the crystals) of atoms of low atomic number, especially D and, to some extent, H. The elastic-scattering process involves the atomic nuclei but it leaves unchanged the energies of neutrons in the beam and nuclei in the sample.

In the *inelastic* scattering of neutrons by atomic nuclei, there is a change of energy and, as with Raman inelastic scattering of photons, the vibrations of the nuclei in the scattering sample may either gain (up-scattering or anti-Stokes) or lose (down scattering) energy (and the neutrons, of course, lose or gain energy, respectively). Neutrons with energies comparable with those of lattice vibrations evidently provide an especially sensitive probe of such motions. In contrast to Raman spectra, neutron inelastic incoherent scattering (n.i.i.s) spectra are not subject to the requirement of an oscillating polarizability; nor are vibrations accessible by n.i.i.s. limited by the selection rule of a change in electric dipole moment appropriate to i.r. absorption. This independence of optical selection rules is a consequence of interaction of the neutron with the nucleus rather than with the electron distribution. It makes neutron spectroscopy particularly valuable for the study of low-frequency (10–1000 cm^{-1}) vibrations in polymers (Safford and Naumann, 1967) and, potentially, of the weak "hydrophobic" forces within and between chains in biological polymer structures (White, 1970).

In any scattering experiment, both energy and momentum are conserved (taking into account the initial and final states of both the radiation field and the molecular system). Now a thermal neutron, with its relatively large mass and with a wavelength, λ, of about 1 Å, has a much greater momentum than that of an optical photon. A second characteristic of neutron spectroscopy, then, is the (controllable) extent of momentum transfer, in addition to energy transfer. Consequently, by following the dependence of scattered intensity on energy and momentum transfers, with n.i.i.s. both the frequency and wavelength of vibrational waves in the whole structure of a polymer can be measured.

The scattering cross-section characterizing the interactions of a given nuclear isotope with a neutron beam has two components, corresponding to two kinds of inelastic scatterings. Coherent n.i.i.s. has some analogy with electromagnetic scattering but allows derivation of the dispersion curve of frequencies versus wave vector ($k = 2\pi/\lambda$); longer wavelength optical spectroscopy provides such information only for k near zero. Incoherent scattering arises when the scatterer is an isotopic mixture or has non-zero nuclear spin; it yields the full density spectrum (i.e. versus frequency) of the vibrational states. The coherency of the scattering depends markedly on the species of nucleus. For H, which scatters very much more strongly than other nuclei, most is incoherent; for other nuclei, coherent scattering is stronger than incoherent. Thus, for H-containing polymers, incoherent scattering from H will be dominant unless either it is eliminated by deuteriation or the processes can be separated by polarization analysis (Allen, 1973; Allen and Wright, 1975).

In addition to experimental limitations, this new branch of molecular spectroscopy involves complex complementary theoretical calculations. However, in its combination of frequencies (or energies) and momenta, n.i.s. is a unique branch of spectroscopy for the study of molecular phenomena on a time scale 10^{-11}–10^{-12} s. Of the solid and liquid polymers studied so far, most have been synthetic, for which, for example in polypropylene, the n.i.s. frequency distribution has agreed well with those predicted for helical conformations. With the availability of higher neutron fluxes and improved discrimination between coherent and incoherent scattering, there is a prospect of greater application of n.i.s. spectroscopy to the study of binding forces in biological polymers (White, 1972; Egelstaff, 1974).

10.3.2.4. Miscellaneous techniques

When a photon of high energy, $h\nu_0$, causes a molecule to ionize, the kinetic energy, $\frac{1}{2}mv^2$, of the ejected electron can be measured. The ionization energy, I_{mol} of the molecular orbital from which the electron was ejected can thus be deduced from the equation

$$h\nu_0 = I_{mol} + E_{vib}^{ion} + \tfrac{1}{2}mv^2$$

where E_{vib}^{ion} represents the vibrational energy of the ion formed. This is the principle behind the branch of electron spectroscopy known as *photon-electron spectroscopy* (p.e.s.) (Baker and Betteridge, 1972). If the source, $h\nu_0$, is of relatively *low energy* (l.e. p.e.s.), for example the He 584Å (21·21 eV) line in the vacuum-u.v. region, then ionization potentials smaller than this, from the outer molecular orbitals (whether bonding or not), can be measured in what is therefore called *molecular* photo-electron spectroscopy. If $h\nu_0$ is of *high energy*, in the soft X-ray region, for example K_a X-radiation from Mg or Al, then the protons can also eject inner-shell electrons. Photo-electron spectroscopy from X-ray sources (X.p.e. or H.e.p.e.s.) is often referred to as electron spectroscopy for chemical applications or analysis (e.s.c.a.), since the binding energies of the core electrons both characterize the atoms of any element (except H) and also (through chemical shifts) are sensitive to electronic effects; lacking facility for atomic analysis, however, u.v. p.e.s. has only limited analytical potential.

Photo-electron spectroscopy is one of the major spectroscopic developments of the last decade. Complexity severely limits the usefulness of molecular p.e.s. spectra from moeities (apart from metals) of high molecular weight and the great majority of p.e.s. measurements have been made on relatively small molecules.

X.p.e.s. is particularly sensitive to isolated heavy atoms in a matrix of light atoms. For metal-containing proteins, X.p.e.s. can help establish the oxidation state(s) of the metal ions, a facility which could be particularly valuable for a diamagnetic complex with low Mössbauer activity (Section 10.2.4). X.p.e. spectra of three bacterial ferredoxins show $2p_{3/2}$ ionization energies corresponding to a single kind of Fe (III), while the width of the line indicates low rather than high spin (Liebfritz, 1972; Andrews et al., 1975). In the bimetallic enzyme erythrocuprein (Jung et al., 1973), or superoxide dismutase, greater intensity of the $2p_{3/2}$ levels of Cu than Zn have been interpreted as arising from closer proximity of the Cu binding site to the cleft or surface of the enzyme. While it may be early for this kind of biological application of X.p.e. (Millard, 1974) both molecular and X-ray p.e.s. have considerable potential for the study of bonding at surfaces, mainly because of the small depth from which ejected electrons can escape.

γ-radiation from nuclear energy levels can be utilized in a radioactive-tracer labelling technique (less likely to perturb the system than most labels—Section 10.2.5) to study rotational correlation times, internal motions, and conformational changes in biological polymers *in vivo* (Meares and Westmoreland, 1972). In a "cascade" sequence of γ-ray emission, a metastable ^{111}Cd nucleus in the $I = 7/2$ state (itself the result of ^{111}In electron-captive decay) decays first to an $I = 5/2$ state and then to a stable ($I = \frac{1}{2}$) ground state of ^{111}Cd. The coincidence counting rate between the two γ-rays may depend on the angles between their directions of propagation, and this angular correlation can be perturbed by external electric fields, for example by binding of the nucleus to a macromolecule. Since nuclear relaxation times may be measured, preliminary experiments suggest that *perturbed angular correlations of γ-radiation* provide, in principle, a sensitive means (since it utilizes radioactive traces) of labelling and studying motional properties of biomolecules *in vivo* (Graf et al., 1974).

If molecules in solution are orientated by hydrodynamic flow, it is possible to detect anisotropic molecules properties from the dichroism between light polarized parallel and perpendicular to the flow. *Flow dichroism spectra*, reviewed by Wada (1973) for biopolymers, can resolve overlapping absorption spectra of chromophores with different orientations (e.g. in viruses) and can provide another means of investigating binding of small molecules to biological polymers (e.g. intercalation of polycyclic hydrocarbons to DNA).

Whereas with normal vibrational Raman spectroscopy very many vibrational frequencies are excited in a biomolecule, with *resonance Raman spectroscopy* specific enhancement can be obtained for those vibrational modes associated with a particular group of atoms of the chromophore. This is achieved (Spiro, 1974) by having the incident radiation tuned to an electronic absorption frequency of the sample; coupling with the electronic transitions associated with this band enhances the vibrational transitions which can, therefore, be used to follow structural changes at sites of biological function. Resonance Raman scattering indicates that diamagnetic oxyhaemoglobin is low-spin Fe (III) (rather than Fe (II)), with similar frequencies to those of cyanohaemoglobin (Spiro, 1975).

References

Allen, G. and Wright, C. J. (1975). *In* "Physical Chemistry: International Review of Science, Series 2," Vol. 8, Macromolecular Science, p. 223, Butterworth, London.

Andrews, P. T., Johnson, C. E., Wallbank, B., Cummack, R., Hall, D. O. and Krishna, Rao, R. (1975). *Biochem. J.* **149**, 471.

Arsenault, G. P. (1973). *In* "Experimental Methods in Biophysical Chemistry" (C. Nicolau, ed.). p. 3, Wiley, London.

Arthur, J. C. (1971). *In* "Cellulose and Cellulose Derivatives" (N. M. Birkales and L. Segal, eds), 5, Part 5, p. 937. Wiley-Interscience, New York.

Allen, G. (1973). *In* "Chemical Applications of Thermal Neutron Scattering" (B. T. M. Willis, ed.), p. 97, University Press, Oxford.

Baker, A, D., and Betteridge, D. (1972). "Photo electron spectroscopy: Chemical and Analytical Aspects". Pergamon, Oxford.

Barker, R. H. and Pittman, R. A. (1971), *In* "Cellulose and Cellulose Derivatives" (N. M. Birkales and L. Segal, eds), Vol. 5, Part 4, p. 181. Wiley-Interscience, New York.

Barry, C. D., Glasel, J. A., Williams, R. J. P. and Xavier, A. V. (1974). *J. Molec. Biol.* **84**, 471.

Bartle, K. D., Jones, D. W. and L'Amie, R. (1972). *J. Chem. Soc. Parkin* **II**, 650.

Bartle, K. D., Jones, D. W. and Maričić, S. (1975). *Biochem. Educ.* **3**, 48.

Biemann, K. (1972). *In* "Biochemical Applications of Mass Spectrometry" (G. R. Waller, ed.), p. 405. Wiley-Interscience, New York

Beinert, H. (1973). *In* "Iron-Sulphur Proteins" (W. Lovenberg, ed.), Vol. 1, p. 1. Academic Press, London and New York.

Bikales, N. M. and Segal L. (1971). "Cellulose and Cellulose Derivatives", Vol. 5, Parts 4 and 5. Wiley-Interscience, New York.

Blackwell, J. and Marchessault, R. H. (1971). *In* "Cellulose and Cellulose Derivatives" (N. M. Bikales and L. Segal, eds), Vol. 5, Part 4, p. 39. Wiley-Interscience, New York.

Brundle, C. R. (1971). *Applied Spec.* **25**, 8.

Campbell, I. D., Dobson, C. M. and Williams, R. J. P. (1975). *Proc. Roy. Soc. A.* **345**, 41.

Chignell, C. F. (1972). *In* "Methods in Pharmacology, 2, Physical Methods" (C. F. Chignell, ed.), p. 33. Appleton-Century-Crofts, New York.

Chignell, C. F. and Chignell, D. A. (1972). *In* "Methods in Pharmacology, 2, Physical Methods" (C. F. Chignell, ed.), p. 111. Appleton-Century-Crofts, New York.
Child, T. F. and Jones, D. W. (1973). *Cellulose Chem. Tech.* **7,** 525.
Cohn, M. and Reuben, J. (1971). *Accounts Chem. Res.* **4,** 214.
Dale, B. J. and Jones, D. W. (1974). *Text. Res. J.* **44,** 778.
Dwek, R. A. (1973). "Nuclear Magnetic Resonance in Biochemistry". Clarendon Press, Oxford.
Damadian, R., Zaner, K., Hor, D., DiMaio, T., Minkoff, L. and Goldsmith, M. (1973). *Ann. New York Acad. Sci.* **222,** 1048. Many other recent applications of magnetic resonance to biology and medicine are covered in pp. 1–1124 of this Journal.
Egelstaff, P. A. (1974). *In* "Spectroscopy in Biology and Chemistry" (S.-H. Chen and S. Vip, eds), p. 269. Academic Press, New York and London.
Feeney, J. (1975). *In* "New Techniques in Biophysics and Cell Biology" (R. H. Pain and B. J. Smith, eds) **2,** 287. Wiley, London.
Fischer, J. J. (1971). *In* "Methods in Pharmacology" (A. Schwartz, ed.) Vol. 1, p. 431. Appleton-Century-Crofts, New York.
Fischer, J. J. (1973). *In* "A Guide to Molecular Pharmacology-Toxology". (R. M. Featherstone, ed.), Vol. 1, Part 2, p. 583. Dekker, New York.
Forslind, E. (1971). *In* "NMR: Basic Principles and Progress", Vol. 4 (P. Diehl, E. Fluck and R. Kosfeld, eds), p. 145, Springer-Verlag, Berlin.
Freeman, S. K. (1974). "Applications of Laser Raman Spectroscopy", Chapter 11. Wiley-Interscience, New York.
Gagnaire, D. and Taravel, F. R. (1975), *FEBS Lett.* **60,** 317.
Graf, G., Glass, J. C. and Richer, L. L. (1974). *In* "Protein-Metal Interactions" (M. Friedman, ed.), p. 639.
Hanlon, S. (1970). *In* "Spectroscopic Approaches to Biomolecular Conformation" (D. W. Urry, ed.), p. 161. Amer. Dental Assn.
Hollis, D. P. (1972). *In* "Methods in Pharmacology: 2, Physical Methods" (C. F. Chignell, ed.), p. 191. Appleton-Century-Crofts, New York.
Holtzman, J. L. (1972). *In* "Methods in Pharmacology: 2, Physical Methods" (C. F. Chignell, ed.), p. 157. Appleton-Century-Crofts, New York.
Jones, D. W. (1976). *In* "Recent Advances in Fibre Science" (F. Happey, ed.). Academic Press, London and New York (in press).
Jones, D. W. and Child, T. F. (1976). *Adv. Mag. Res.* **8,** 123.
Jung, G., Ottnad, M., Bohenkamp, W., Bremser, W. and Weser, U. (1973). *Biochim. Biophys. Acta* **295,** 77.
Kearns, D. R. and Shulman, R. G. (1974). *Accounts of Chem. Res.* **7,** 33.
Kertesz, J. Ç. and Wolf, W. (1973). *J. Phys. E* **6,** 1009.
Knowles, P. F. (1972). *Essays in Biochem.* **8,** 79.
Lang, G. (1971). *In* "Magnetic Resonance in Biological Systems" (C. Franconi, ed.), p. 163.
Lauterber, P. C. (1973). *Nature* **242,** 190.
Lee, A. G., Birdsall, N. J. M. and Metcalfe, J. C. (1973). *Chem. Brit.* **9,** 116.
Levine, B. A. and Williams, R. J. P. (1975). *Proc. Roy. Soc. A.* **345,** 5.
Levine, Y. K. (1972). *Prog. Biophys. Molec. Biol.* (J. A. V. Butler and D. Noble, eds), Vol. 24, Chapter 1. Pergamon, Oxford.
Liebfritz, D. (1972). *Angew. Chem. internal. Edit.* **11,** 232.

Liu, K. J. and Anderson, J. E. (1970). *J. Macromol. Sci.-Rev. Macromol. Chem.* **C5,** 1.
Lynch, L. J. and Marsden, K. H. (1971). *Search* **2,** 95.
Mansfield, P. and Grannell, P. K. (1973). *J. Phys. C. Solid State Phys.* **6,** L422.
Maričić, S. L. (1975). *In* "Magnetic Resonance in Chemistry and Biology" (J. N. Herak and K. J. Adamić, eds). Marcel Dekker, New York.
Meares, C. F. and Westmoreland, D. G. (1972). *In* "Cold Spring Harbor Symposia on Quantitative Biology", Vol. 36, p. 511. Cold Spring Harbor, New York.
McDonald, C. C. and Phillips, W. D. (1970). *In* "Fine Structure of Proteins and Nuclei Acids". (G. D. Fasman and S. N. Timasheff, eds), p. 1. Marcel Dekker, New York.
Metcalfe, J. C. (1970). *In* "Permeability and Function of Biological Membranes" (L. Bolis, A. Katchalsky, R. D. Keynes, W. R. Lowenstein and B. A. Pethica, eds), p. 222. North-Holland, Amsterdam.
Mildvan, A. S. and Cohn, M. (1970). *Advan. Enzymol.* **33,** 1.
Millard, M. M. (1974). *In* "Protein-Metal Interactions". (M. Friedman, ed.), p. 589. Plenum Press, New York.
Muzzarelli, R. A. A. (1973). "Natural Chelating Polymers". Pergamon Press, Oxford.
Phillips, W. D. (1973). *In* "N.M.R. of Paramagnetic Molecules: Principles and Applications", Chapter 11. Academic Press, New York and London.
Phillips, W. D., Poe, M. and Weiher, J. F. (1970). *Nature* **227,** 574.
Safford, G. J. and Naumann, A. W. (1967). *Adv. Polymer Sci.* **5,** 1.
Schoenborn, B. P. (1972). *In* "Cold Spring Harbor Symposium on Molecular Biology", Vol. 36, p. 569. Cold Spring Harbor, New York.
Schweizer, M. P., Chan, S. I. and Crawford, J. E. (1973). *In* "Physico-Chemical Properties of Nucleic Acids" (J. Duchesne, ed.), Vol. 2, p. 187. Academic Press, London and New York.
Shaw, D. (1974). *J. Phys. E. Sci. Instrum.* **7,** 689.
Spiro, T. G. (1974). *Accounts Chem. Res.* **7,** 339.
Spiro, T. G. (1975). *Proc. Roy. Soc. A.* **345,** 89.
Urry, D. W. (1973). *Research and Development* **24,** [3] 30.
Walter, J. A. and Hope, A. B. (1971). *Prog. Biophys. Mol. Biol.* **23,** 3.
White, J. W. (1970). *J. Macromol. Sci.-Chem.* **A4,** 1275.
White, J. W. (1972). *In* "Polymer Science" (A. D. Jenkins, ed.), Vol. 2, p. 1744. North-Holland, Amsterdam.
Wada, A. (1973). *Appl. Spec. Rev.* **6,** 1.
Wolf, W., Kertesz, J. C. and Landgraf, W. C. (1969). *J. Mag. Res.* **1,** 618.

Subject Index

A

Absorbance, 13, 35, 156
Absorption process, 2, 148
AB spin system, in n.m.r., 200
Adrenodoxin, Mössbauer spectra, 291
Alanine, crystal e.s.r. spectra, 243
Alginic acid, 309
α–helix,
    far i.r. spectra, 137
    fibrous proteins, i.r. spectra, 55
    Raman scattering, 89
Amino acids, *see also* names of specific amino acids
    CD spectra, 182
    composition determination by n.m.r., 297
    electron spin resonance spectroscopy, 241ff
    γ–irradiation and e.s.r. spectra, 242
    Raman spectroscopy, 86
    relative lifetimes of conformers, 298
    study by n.m.r., 195
AMP (9–β–D–ribofuranosyladenine–5′–monophosphate), n.m.r. study by lanthanide probes, 301
AMX spin system in n.m.r., 200
Anharmonic oscillator, 21
Anilinonaphthalene–8–sulphonic acid (ANS), fluorescence properties, 163
Anisotropic electronic splitting, in electron spin resonance, 227, 231ff
Anti-Stokes line, 81
Apodization (far infra-red spectroscopy), 129–130

Arginine, electron spin resonance spectra, 257
A.T.R. (attenuated total reflection), 49–50
AX spin system in n.m.r., 200

B

Balanced-bridge e.s.r. spectrometer, 240
Beer–Lambert law, 12, 35
Binding of drugs to biomolecules, n.m.r. and fluorescence spectroscopy, 301
Binding sites,
    e.s.r. spin labels, 306
    study by n.m.r., 301
Bloch equations, in n.m.r., 212
Blood, e.s.r. study of action, 262
Blood clotting, e.s.r. study, 264
Blood flow, monitoring by n.m.r., 310
Bohr magneton, 217
    in e.s.r., 222
Boltzmann distribution of energy states, 5, 192, 224
Bovine serum albumin,
    study by n.m.r., 197

C

^{13}C n.m.r. spectra, 215, 309
Calcium, effect on fibrin clots, e.s.r. study, 268
Cancerous tumours, n.m.r. diagnosis, 310

CAT (computer of average transients), 15, 204
Cavities, sample, for e.s.r. measurements, 236
Cellulose, 308
 infra-red spectroscopy, 75
 moisture determination by n.m.r., 311
Chemical shift,
 in n.m.r. spectroscopy, 1, 193, 194
 in Mössbauer spectroscopy, 271, 278–281
Chitan, 309
Chitin, 309
 infra-red spectroscopy, 77
Chitosan, 309
Chlorophyll, light absorption, 153–155, 160
Chemical exchange, study by n.m.r., 212–214
ĊH radical, e.s.r. spectra, 254
Chromophore interaction mechanisms,
 μ–m, 169
 coupled-oscillator, 169
 one-electron, 169
Chromophoric groups, 9, 165ff, 169, see also electronic absorption spectroscopy
Chymotrypsinogen, CD spectra, 171, 179
Circular dichroism (CD), 9, 165–188
 measurement, 167
 polypeptide and protein structure, 170ff
Combination bands, 31
Complex high-resolution n.m.r. spectra, 201
Computer averaging, in n.m.r., 204; see also CAT
Conjugated molecules, electronic spectra, 147
Contact shifts, n.m.r., 211, 218
Core polarization, in Mössbauer spectroscopy, 283
Corn products, moisture determination, by n.m.r., 311
Correlation time, τ_c, 12, 201, 202, 217, 302
Cotton effects, positive and negative, 165

Creatine kinase, study by n.m.r., 208
Crossed-coil probe, in n.m.r., 204
Crystalline field, electronic splitting in, 226
Crystallinity,
 of cellulose, n.m.r. measurement, 311
 by electron spin resonance, 229
 far infra-red spectroscopy, 139
 infra-red measurement, 52, 67
Cyanohaemoglobin, 315
Cysteine, e.s.r. spectra, 260
Cysteine, irradiated, e.s.r. spectra, 243–244
Cystine, irradiated, e.s.r. spectra, 243–248
Cytochrome c,
 Mössbauer spectra, 290
 paramagnetic n.m.r. shifts, 211
 study by magnetic resonance, 299

D

Denaturation of proteins,
 study by CD spectra, 181
 study by ORD, 172
Derivative curve,
 in e.s.r. spectra, 239
 in infra-red spectra, 63
1-β-D-deoxy-ribofuranosylthymine-5'-monophosphate, see TMP
Depolarization ratio (Raman spectroscopy), 83
Detectability of a spectrum, 14
Deuteriation,
 amino acids, e.s.r. spectra, 242
 effect on e.s.r. spectrum, 255
 effect on infra-red spectrum, 50
 far infra-red spectra, 135
 value in n.m.r. of polypeptides, 309
Dichroic ratio, 38; see also polarized light
Dichroism,
 circular (CD), 165–188
 far infra-red spectroscopy, 139 ff.
 infra-red, 52
Diphenylpicrylhydrazyl (DPPH), 240
Dipolar field, in Mössbauer spectroscopy, 284

INDEX

Dipolar interaction,
  in electron spin resonance, 229
Dipolar relaxation,
  in nuclear magnetic resonance, 201, 216
Disulphide bridge, in proteins, e.s.r. spectra, 260
Disulphide cross links, e.s.r. spectra, 243
DNA,
  CD spectra, 176
  interaction of aminoacridines, study by Cotton effects, 185
  nuclear magnetic resonance, 199, 299
  intercalation, flow dichroism spectra, 314
  Raman spectra, 114
Doppler effect, in Mössbauer spectroscopy, 271, 273
Double-beam spectrometers, 15, 34
Double resonance, in n.m.r., 214
  saturation transfer, 216
Drude equation, 168, 171
Drug/membrane interactions, fluorescent labels, 308
Drugs, binding to biomolecules, 301–302
  study by n.m.r., 301
Dye-protein binding, fluorescence studies, 160

E

Electric quadrupole moment, 281
Electric quadrupole splitting, in Mössbauer spectroscopy, 271
Electromagnetic spectrum, 3, 191
Electron magnetic moment, 11, 221
Electron spectroscopy,
  e.s.c.a., 313
  high energy (h.e.p.e.s.), 313
  low energy (l.e.p.e.s.), 313
Electron spin angular momentum, 222
Electron spin magnetic moment, 10, 222
Electron spin quantum number, 7, 9, 222
Electron spin resonance (e.s.r.), 10, 221–270
  spectral characteristics, 224–234
  spectrometers, 234–240

Electronic absorption spectroscopy, 145–164
Electronic-charge density, in Mössbauer spectroscopy, 279
Electronic spectroscopy, 9, 145–163
  environmental effects, 160
Electronic splitting,
  in electron spin resonance, 225
Emission spectroscopy, 145–164
Energy levels, 2, 145
  electronic, 146
  in n.m.r., 191
  nuclear, in Mössbauer spectroscopy, 279
Erythrocyte membrances, e.s.r. spin labels, 306
Enzymes,
  creatine kinase (CK), study by n.m.r., 208
  Mössbauer spectra, 288
  optical rotatory properties, 175
Enzymes, structure and action,
  combined spectroscopic study, 299
  study by magnetic resonance, 299
Enzymes, haem,
  Mössbauer spectra, 290
Enzyme systems, n.m.r. study, 300
Erythrocuprein, X-ray photo-electron spectroscopy, 314
Erythrocyte membranes, fluorescent labels, 308
E.S.R., see electron spin resonance
Exchange narrowing, in e.s.r., 229
Exchange rates, drug binding studies by n.m.r., 302
Excitation coefficients (in electronic spectroscopy), 155
Excited state, 6
  energy transfer, 154
Extinction coefficient, 13, 35, 156

F

Far infra-red spectroscopy, 119–144
  early history, 120
  filters, 123

Fourier-transform, 125, 128, 143
grating spectrometers, 124
interferometers, 125
microwave techniques, 125
polarized spectra, 137ff
transparent materials, 120–123
Ferredoxin,
bacterial, X-ray photo-electron spectroscopy, 314
magnetic resonance study, 304
Mössbauer spectroscopy, 281, 290
Ferrihaemoprotein, 303
Ferrohaemoprotein, 303
Fe-S proteins, 290
combined spectroscopic study, 304
Mössbauer spectra, 288, 293
Fibrinogen, e.s.r. study, 264
Flow dichroism spectra, 314
Flowing systems, e.s.r. in, 310
Fluorescence, 6, 152
assay, 157
excitation spectra, 157
polarization, 158
sensitized, 154
Fluorescence spectroscopy,
study of drug and dye interaction, 302
Fourier transform spectroscopy, 16,
far infra-red, 125, 128, 143
in n.m.r., 204
Franck-Condon principle, 150
Free electron in e.s.r., 224
Free-induction decay (f.i.d.) in n.m.r., 204
Free radicals, 221ff
Free-spin g-value, 226

G

γ-irradiated amino acids, e.s.r. spectra, 244
γ-rays, 8
angular correlation, 314
in Mössbauer spectroscopy, 272
g-factor, 13, 217, 223ff
Gaussian functions, 41, 229
Globular proteins, combined spectroscopic study, 299
Glycine, crystalline e.s.r. spectra, 241–243
Gramacidin A, combined spectroscopy study, 298
Group theory, 26
Gyromagnetic ratio, 192, 217

H

^{1}H n.m.r. 189–218, 299–302, 311
Haem proteins, Mössbauer spectroscopy, 289
Haemerythrin, Mössbauer spectra, 288, 291
Haemoglobin,
e.s.r. crystal-field splitting, 264
e.s.r. studies, 262–264
Mössbauer spectroscopy, 281
Mössbauer spectra, 284–288
paramagnetic n.m.r. shifts, 211
combined spectroscopic studies, 303
Haemoglobin cyanide, Mössbauer spectrum, 285
Haemoglobin fluoride, Mössbauer spectrum, 284
Haemoproteins,
combined spectroscopic studies, 303
conformational changes by CD spectra, 182
determination of magnetic susceptibility, 304
extrinsic Cotton effects, 181
Hair, see also wool,
infra-red spectra, 57
Hamiltonian equation, for e.s.r., 231
Heisenberg uncertainty principle,
in e.s.r., 228
in Mössbauer spectroscopy, 273
Handedness of helices,
study by CD, 298
study by ORD or CD, 175
Helix content, from ORD and CD spectra, 170
Helix/random-coil transitions, combined study by spectroscopic techniques, 297

INDEX

High-field (superconducting) magnets, 204, 309
Hydrogen bonds,
  infra-red spectroscopy, 51
  polypeptide infra-red spectra, 60
Hyperfine coupling, in n.m.r., 208, 218
Hyperfine splitting, e.s.r., 229–234
Hypochromism in gramicidin A, 298

I

INDOR (internuclear double resonance), 216
Induced emission, 5
Infra-red spectroscopy, 9, 17–79
  attenuated total reflection (A.T.R.), 49
  derivative spectroscopy, 41–48
  derivative spectroscopy (in proteins), 63
  deuterium substitution, 51
  differential, 40
  polarized, 37
  specimen preparation, 48
  spectrometers, 22, 33–35
  theory, 17–32
Inhibitors, drug binding studies by n.m.r., 302
Intensity of spectral line, 12,
  in electron spin resonance, 225
  in nuclear magnetic resonance, 193
Interferogram, computation of, 128
Interferogram, far i.r., production of, 126
Internal conversion, 6, 151
Intersystem crossing, 151
Ionizing radiation, radicals induced by, e.s.r. spectra, 253
Iron, co-ordination number, 280ff
Iron, oxidation state, 280ff
Iron, spin state, 285ff
Iron-sulphur proteins, *see* Fe-S proteins,
Irradiated materials, *see* electron spin resonance,
Isomer shift, in Mössbauer spectroscopy, 278–281

J

Jablonski diagram, 148

K

Keratin,
  derivative infra-red spectrum, 64
  feather and pig, e.s.r. study, 256
  ethylene diamine complexes, 69–74
Kinetics, by e.s.r., 310
Klystron oscillator, in e.s.r., 236
K.M.E.F. (keratin, myosin, epidermin, fibrinogen) group, e.s.r. spectra, 250ff
Kronig-Kramers transforms, 168

L

Lanthanide probe ions, *see also* shift reagents,
  n.m.r. study of nucleotides and enzymes, 300
  n.m.r. study of AMP, 301
  n.m.r. study of TMP, 301
Lanthanide shifts in n.m.r., 210, 212
Lasers (in Raman spectroscopy), 85
Lecithin, 307
Lifetimes of states, 6, 155, 274
Linewidth, 13,
  in e.s.r. spectroscopy, 227
  in Mössbauer spectroscopy, 276
  in n.m.r. spectroscopy, 201
Lipid-protein interactions, study by magnetic resonance, 307
Lorentzian curves in n.m.r., 193
Lorentzian functions, 41–43
Lysine, electron spin resonance spectra, 257
Lysozyme,
  CD spectra, 179
  drug binding, n.m.r. study, 302
  Raman spectrum, 88, 101
  study by n.m.r., 197

324

INDEX

M

Magnetic hyperfine splitting, in Mössbauer spectroscopy, 271, 282
Magnetic susceptibility, weakly paramagnetic materials, 304
Magnets,
  for e.s.r., 240
  for n.m.r., 190, 203
Mass spectrometry, 311
Matrix rank analysis (in CD spectra of proteins), 174
Maxwell-Boltzmann equation, in e.s.r., 224; see also Boltzmann distribution
Melanin in wool fibres, e.s.r. spectra, 251
Membranes,
  optical activity, 308
  structural study by magnetic resonance, 305
Methionine, electron spin resonance spectra, 257
Michelson interferometer, 125
Modulation, in e.s.r., 238
Moffitt-Yang relation, 168, 172
Moisture contents, n.m.r. determination, 311
Molecular photo-electron spectroscopy, 313
Mössbauer spectroscopy, 8, 271–293
  comparison with n.m.r. and e.p.r., 272
  history, 274
  Mössbauer effect, 274–277
  spectrometers, 277–278
Motion, detection by Mössbauer effect, 289
Multiple resonance in n.m.r.,
  double resonance, 214
  nuclear Overhauser effect, 215
  spin decoupling, 214
Multi-scan averaging, 204
  see also CAT
Myoglobin, Mössbauer spectra, 288

N

Neutron diffraction, 312

Neutron inelastic scattering (n.i.s.), 312
N.I.S. see neutron inelastic scattering, 312
Nitroxide free radical, $=N{-}0$, 305
Noise decoupling, in n.m.r., 215
Nuclear-charge density, in Mössbauer spectroscopy, 279
Nuclear gamma-ray spectroscopy, see Mössbauer spectroscopy,
Nuclear hyperfine splitting, e.s.r., 229–234
Nuclear induction, 204
Nuclear magnetic moment, 11, 192
Nuclear magnetic resonance (n.m.r.), 11, 189–219
  broad-line, 311
  chemical shift, 193, 194
  "diffraction", 310
  Fourier-transform (FT) spectroscopy, 204, 309
  high-field, 309
  high-resolution, 193–219
  history, 189
  multiple resonance experiments, 214
  novel experiments, 310
  paramagnetic shifts, 209
  spectrometers, 190, 202–207, 309
  spin-spin splitting, 193, 200
  symbols and definitions, 217
Nuclear Overhauser effect (NOE), 215
Nuclear relaxation, 201, 216
Nuclear shielding in n.m.r., 194
Nuclear spin, 11, 192, 229
Nucleic acids, see also DNA, RNA
  extrinsic Cotton effects, 185
  helix-coil transitions, combined spectroscopic study, 299
  nuclear magnetic resonance, 199
  optical rotatory properties, 176
  ORD and CD, 170, 175
  Raman spectra, 104
Nucleotides,
  nuclear magnetic resonance, 199
  optical activity, 176
  Raman spectra, 107

O

Optical activity, 165–188
  asymmetry, 169
Optical density, 35, 156
Optical rotatory dispersion (ORD), 9, 165–188
  measurements, 166
Orbital field, in Mössbauer spectroscopy, 283
Overtone bands, 31
  fibrous proteins, 63
Oxidation state, from Mössbauer spectrum, 282
Oxyhaemoglobin, resonance Raman scattering, 315

P

Paramagnetic interactions, in n.m.r., 208–214
Paramagnetic biological macromolecules, combined spectroscopic studies, 303
Paramagnetic probes, 208
  magnetic resonance study of proteins and enzymes, 299
Paramagnetic relaxation effects, in n.m.r., 208
Paramagnetic shifts, in n.m.r., 209
Paramagnetism, in electron spin resonance, 221–270
Penicillamine, electron spin resonance, 259
Peptides,
  α-helical content from CD spectra, 176
  electron spin resonance spectroscopy, 241
  far infra-red spectroscopy, 133
Perturbed angular correlations of γ-radiation, 314
Phase-sensitive detection, 15,
  in e.s.r., 238
Phosphate binding, 301

Phospholipids, study by spin labels, 305
Phosphorescence, 7, 152
Phosphoryl transfer enzymes, use of paramagnetic probes, 300
Photochemical reactions, 153
Photo–electron spectroscopy, 313
Photosynthesis, Fe-S proteins, 304
Pigmented wool, e.s.r. spectrum, 252
Pleated-sheets in fibrous proteins, e.s.r. study, 256
Point group, 26
Polarization of fluorescence, 158
Polarized light, 9, 37ff, 165ff
Polarizers for infra-red spectroscopy, 39
Poly(adenylic acid) (poly A), Raman spectroscopy, 110–115
Polyglycine,
  e.s.r. spectra, 243, 256
  Raman spectra, 97
Poly-L-alanine (PLA),
  crystal e.s.r. spectra, 243
  far infra-red spectroscopy, 136
  infra-red spectra, 56
  Raman spectra, 90
Poly-L-cystine, e.s.r. spectra, 248
Poly-L-glutamic acid, e.s.r. spectra, 243
Poly-L-leucine, crystal e.s.r. spectra, 243
Poly-L-lysine, CD spectra, 173
Poly-L-proline, e.s.r. spectra, 243
Polynucleotides,
  helix-coil transitions, combined spectroscopic study, 299
  Raman spectra, 104
Polypeptides, see also names of specific polypeptides
  amide infra-red bands, 54
  β-conformation (by Raman spectroscopy), 96
  ĊH radical, e.s.r. study, 233–235
  conformation, combined spectroscopic study, 297, 298
  electron spin resonance spectroscopy, 241
  infra-red dichroism, 56
  infra-red vibrational interaction, 59
  ORD and CD spectra, 170
  Raman spectra, 93

Raman spectra in aqueous solution, 99
random-coil, vibrational spectroscopy, 99
sequence analysis by n.m.r., 297
Polysaccharides, 308–309,
 infra-red spectra, 74–77
 optical rotatory properties, 186
Position of spectral line, 13, n.m.r., 193
Potential–energy curves, 149, 151
Probability coefficient, 5
Proline, electron spin resonance spectra, 257
Proteins, *see also* Fe-S proteins; *see also* names of specific proteins
 α/β structure, study by e.s.r., 256
 amide infra-red bands, 54
 chain conformation by infra-red spectroscopy, 67
 chain conformation by Raman spectra, 88
 conformation, combined spectroscopic study, 297, 298
 extrinsic Cotton effects, 179
 extrinsic Cotton effects of ligands, 183
 iron-storage, Mössbauer spectroscopy, 293
 Mössbauer spectroscopy, 281, 288
 n.m.r. chemical shifts, 196
 Raman spectra, 101–104
 resolution of n.m.r. spectra, 199
 secondary structures and ORD, 170–176
 study of binding by extrinsic Cotton effects, 183
 water absorption (infra-red spectroscopy), 65
 X-ray photo-electron spectroscopy, 314
Proton relaxation rates, 208
Proton (^{1}H) n.m.r., 11, 189–218, 296, 299, 300, 309, 311
Pseudo-contact shifts, n.m.r., 211, 218
Pulse methods in n.m.r., 204, 309
Putidaredoxin, Mössbauer spectra, 291

Q

Quadrupole splitting, in Mössbauer spectroscopy, 281
Quantization in a magnetic field, 192, 222
Quantum number,
 electron spin, 222
 nuclear spin, 11, 192
 rotational, 19
 vibrational, 21
Quantum yield, 156
Quenching in electronic spectroscopy, 153

R

Radiative emission, in e.s.r., 224
Radiofrequency spectroscopy, 11, 189–219, 221–270 *see also* nuclear magnetic resonance (n.m.r.)
Raman spectroscopy, 10, 81–118
 advantages for biological applications, 84, 296
 comparison with neutron scattering, 312
 dispersion curves (for poly-L-alanine), 91
 polarization, 82
 resonance, 303, 314
 spectrometers, 85
Rayleigh scattering, 10, 81, 86
Receptor interactions, 301
Recoil energy, in Mössbauer spectroscopy, 273
Red blood cells, *see* haemoproteins,
Relativistic correction, in e.s.r., 224
Relaxation processes, 5, 201, 227
Relaxation rates,
 drug binding studies by n.m.r., 302
 enhancement, in n.m.r., 208
Relaxation times,
 in electron spin resonance, 227
 in nuclear magnetic resonance, 201
Reporter molecules, 305–308
 Cotton effects in DNA or RNA, 184
Resolution of spectral lines, 14
 derivative infra-red spectroscopy, 41ff

INDEX

Resolution of spectrometers, 14
Resonance cavity, in e.s.r., 236
Resonance Raman spectroscopy, 303, 314
Resonant absorption of γ-rays, *see* Mössbauer effect,
Respiratory protein, *see* haemoproteins,
9-β-D-ribofuranosyladenine-5′-monophosphate, *see* AMP,
Ribonuclease,
    circular dichroism (CD) spectra, 173, 179
    paramagnetic n.m.r. shifts, 211
    Raman spectra, 102
    study by n.m.r., 190, 197
    study by magnetic resonance, 299
RNA,
    CD spectra, 177
    nuclear magnetic resonance, 199, 299
    Raman spectra, 115
Rotational spectra, 18
Rotational strength, 170, 171
Rubredoxin,
    combined spectroscopic study, 304
    Mössbauer spectroscopy, 281
    p.m.r. spectra, 305
Russell–Saunders $(L + S)$ coupling, in e.s.r., 223

S

Sample cavities, e.s.r., 236
Sample spinning (in n.m.r. spectrometers), 202
Saturation, 5,
    broadening, 229
    and nuclear Overhauser effect (NOE), 215
    transfer in n.m.r., 216
Schrödinger equation, 19
Selection rules, 6
    gross, 17
    rotational transitions, 20
    vibrational spectra, 20
Sequence analysis of polypeptides by n.m.r., 297

Sequence determination in peptides, by mass spectrometry, 311
Shape of spectral line,
    in electron spin resonance, 229
    in n.m.r., 193
Shift reagents, in n.m.r., 212
Signal-to-noise, 14,
    in e.s.r., 238
Signal-to-noise enhancement, 14
    double-beam technique, 15, 22, 34
    Fourier-transform spectroscopy, 16, 125ff, 204, 309
    lock-in detection, 15
    signal averaging (CAT), 15
    in n.m.r., 204
Silk, e.s.r. spectra, 250, 255
Silk fibroin,
    far infra-red spectroscopy, 137
Singlet state 7, 9, 147
    in e.s.r., 226
Skeletal vibrations, far infra-red spectroscopy, 131
Solvent effects, electronic spectroscopy, 161
Spectrum accumulation, in n.m.r., 206
Spin decoupling in n.m.r., 214
Spin density, in e.s.r., 230
    in proteins, 258
Spin labels, 305–307
Spin-lattice relaxation time, T_1, 6,
    in electron spin resonance, 227
    in nuclear magnetic resonance, 201
Spin-orbit coupling, 272
    in electron spin resonance, 229
Spin-spin coupling, n.m.r., 200
Spin-spin relaxation time, T_2, 6,
    in electron spin resonance, 229
    in nuclear magnetic resonance, 201
Spinning sidebands, 195
Spontaneous emission, 5
Stokes line, 81
Stopped-flow e.s.r., 310
Sulphur-containing amino acids, e.s.r. spectra, 243
Superconducting magnets for n.m.r., 190, 203

Superoxide dismutase, X-ray photo-electron spectroscopy, 314
Symmetry, element, 23
operation, 23

T

T_1 relaxation time, 6
  in e.s.r., 227
  in n.m.r., 201
T_2 relaxation time, 6
  in e.s.r., 229
  in n.m.r., 201
Tetramethylsilane (TMS) reference in n.m.r., 195
Thermal equilibrium, in n.m.r., 192
Thermal populations, 5
  see also Boltzmann distribution of energy states
TMP, n.m.r. study by lanthanide probes, 301
Transfer ribonucleic acid, see (t-RNA),
Transient n.m.r. signals, 204
Transition metal, in e.s.r., 226
Transition probability, 5
  in electronic spectroscopy, 156
  in n.m.r., 192
Transitions spectroscopic, 2
  n-π*, 147
  π-π*, 147
Transition moment, 5
Transmission e.s.r. spectrometer, 236
Triplet states, 7, 9, 147
  e.s.r., 225
t RNA,
  denaturation, combined spectroscopic study, 299
  n.m.r. spectra, 198

U

Ultraviolet spectroscopy, 8, 145–164, 165–188
Unpaired electrons, see electron spin resonance,
  in n.m.r., 208

V

Visible spectroscopy, 8, 145–164, 165–188

W

Width of spectral line, 13,
  in electron spin resonance, 227
  in n.m.r., 193
Wool,
  electron spin resonance, 251
  electron spin resonance, g-values, 261
  moisture determination by n.m.r., 311
Wool proteins,
  study by high-field n.m.r., 298

X

Xanthine oxidase, Mössbauer spectra, 290
X-ray diffraction, comparison with spectroscopic methods, 2, 295–297
X-ray photo-electron spectroscopy, 313

Z

Zeugmatography, 310